# The Pinyon Jay

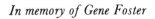

*In memory of Gene Foster*

# The Pinyon Jay

Behavioral Ecology of a Colonial
and
Cooperative Corvid

*by*
John M. Marzluff and Russell P. Balda

*Illustrated by*
Tony Angell
*and*
*Caroline Bauder*

T & A D POYSER
London

First published 1992 by T & AD Poyser Ltd
Print-on-demand and digital editions published 2010 by T & AD Poyser, an imprint of
A&C Black Publishers Ltd, 36 Soho Square, London W1D 3QY

www.acblack.com

Copyright © 1992 T & AD Poyser

ISBN (print) 978-1-4081-3693-5
ISBN (epub) 978-1-4081-3692-8
ISBN (e-pdf) 978-1-4081-3691-1

A CIP catalogue record for this book is available from the British Library

This is a print-on-demand edition produced from an original copy.

It is produced using paper that is made from wood grown in managed sustainable
forests. It is natural, renewable and recyclable. The logging and manufacturing processes
conform to the environmental regulations of the country of origin.

Printed and bound in Great Britain

MIX
Paper from
responsible sources
FSC® C013604

# Contents

# List of Photographs

# List of Figures

# List of Tables

# Preface

As I gaze out of my window I see a group of over 40 corvids fighting persistently for a mouthful of deer. Many wear individual color tags and I notice that the membership of this rowdy group is in constant turmoil. The Common Ravens, 30 meters from my window, provide a stark contrast to their taxonomic relatives, Pinyon Jays, that are the subject of this book. One rarely sees fighting among foraging Pinyon Jays and this corvid remains in stable groups for life.

We have attempted to satisfy a diverse group of readers with this book. Our goal is to provide a solid foundation of data, analyses, and insights for professional readers that is written in a style light enough to keep the attention of the interested amateur. To this end, we present many of our findings in standard scientific tables and figures, and we rely on the use of accepted statistical procedures to substantiate the robustness of our results. Lay readers need not dwell on these mathematical constructs, but we must present them so that other professionals can make their own judgments about our claims. Although we attempt to shield uninterested readers from heavy analyses by placing many statistics in tables and figure legends, we have tried not to simplify the presentation of results and theories. Instead, we have attempted to generalize these discussions so that amateurs might better grasp the points we have attempted to make. Whenever possible, we have related our findings to human analogues, not to suggest that Pinyon Jays behave like humans (or *vice versa*), but to make our results more enjoyable and, we hope, less forgettable.

Our book is organized into two broad sections. The first 4 chapters aim to acquaint readers with the natural history of Pinyon Jays and their relatives. The major theme of these chapters is that reliance on seeds of the pinyon pine has produced the unique behavioral and morphological adaptations that characterize Pinyon Jays. For the remainder of the book we present the results of our, now 20-year-long, study of a flock of jays living in and around Flagstaff, Arizona – the Town Flock. Individuals in this flock have been uniquely color-banded since 1972 and intensively studied from that time through 1987. Color-banding allowed us to eavesdrop on the lives of thousands of jays. Without color bands most of the information in our book would have been unobtainable. If one theme stands out from our observations of the Town Flock it is that variety is the spice of Pinyon Jay life. This flock inhabits an extremely variable montane environment which precludes many stereotyped responses usually attributed to "bird brains". Accordingly, we stress that learning and memory are important adaptations. The age composition, and especially the sex composition, of this flock is extremely variable and we suggest that the latter has major consequences for the options pursued by young jays.

Our discussions of the Flagstaff Town Flock include new summaries of previously published results and unpublished results. We statistically analyze this new material and summarize significant results from previously published material. We have avoided repeatedly citing our published work. We invite the interested reader to consult the following references which served as primary sources for the chapters that follow. Chapters 1 and 5 summarize Balda and Bateman 1971, Balda *et al.* 1972, and Marzluff and Balda 1988a. Chapter 4 relies upon Marzluff 1987. Chapter 6 extends the information in Balda and Balda 1978. Chapter 7 summarizes Marzluff and Balda 1988b, 1988c. Chapter 8 is based on Balda and Bateman 1972, Marzluff and Balda 1988a, and Marzluff 1988a, 1988b. Chapter 9 extends information in Bateman and Balda 1973, Marzluff 1983, 1985, and Marzluff and Balda 1990. Chapter 11 summarizes Marzluff and Balda 1989. Chapter 12 extends Marzluff and Balda 1988a, 1990. Where appropriate, we have introduced chapters with quotes from early naturalists who studied Pinyon Jays so that the reader might gain a historical background for our observations.

A long-term study such as this requires an incredible support team. It is my pleasure to acknowledge those who gave so freely of their time, ideas, data, and moral as well as monetary support as I learned about Pinyon Jays. The most crucial ingredient for a successful long-term study is consistent effort throughout the study by a few persons. Russ Balda, Gene Foster, and Jane Balda provided the consistent monitoring and administrative support needed to insure a cohesive study. Many students, myself included, joined the study for shorter periods as we attacked our own special interests. Much of our knowledge of Pinyon Jay biology is a direct result of the graduate studies done by Dan Cannon, Larry Clark, Diana Gabaldon, and Pat McArthur. I thank them for granting me free access to their unpublished works. My research would not have been possible without the friendship and help of Katharine Bartlett, Gene Foster, Bill and Judy Burding, Elaine Morrall, and the other "feeder watchers" in Flagstaff. I thank them for allowing me to use their homes as blinds for observing, trapping, and banding jays. Katharine Bartlett continues to aid our research by providing access to Gene Foster's records. Mike Munoz, Mike Horn, and John Luginbuhl volunteered many hours to help me chase jays. I thank them for their help and company in the field and lab. Peter Price and Graydon Bell have read more than their share of Pinyon Jay manuscripts and through their comments they have broadened my insight into Pinyon Jay biology. Graydon's teachings about statistical inference have been especially helpful. Without the tools he taught me the analyses I present would not have been possible.

I am especially indebted to my parents, Joseph and Elizabeth Marzluff. They always were interested in my pursuits and continually supported my endeavors with encouragement and finances. I would never have developed the love I have for biology if they had not allowed me to catch porcupines in western Kansas, chase deer and bighorn sheep in Montana, and go unpunished when the occasional snake, frog, or hamster escaped in the house.

My interest in biology was nurtured and directed by 3 early mentors. I thank Ken Highfill for showing me the mountains, Stan Roth for immersing me in field ecology, and Dick Hutto for introducing me to formal ornithology and suggesting that I pursue graduate school in Flagstaff. More recently, my wife, Colleen Marzluff, has acted as a well-informed sounding board for many of my ideas. She has financially supported much of the research I present, but more importantly, she has provided the all-important moral support necessary for the completion of this project. She has given me the space I needed to write and rewrite this book. Her parents, Al and Alice Sanford, have been great friends and loyal supporters. Financial support for my research was also provided by Henry Hooper, graduate dean of Northern Arizona University, Sigma Xi, The Chapman Memorial Fund, The Wilson Ornithological Society, and Achievement Rewards for College Scientists.

Many colleagues commented on the prose presented on the following pages. I thank Craig Benkman, Randy Breitwisch, Jerry Brown, Sylvia Hope, Kris Johnson, David McDonald, and Marvin Wilkening for their advice, encouragement, and insights. The lifetable analysis would not have been possible without David McDonald. I thank him for his generous donation of time to analyze our data. Emilee Mead and Robyn O'Reilly drafted the technical figures. Terry Vaughan, Caroline Bauder and Tony Angell contributed lively artwork that kept me motivated when no end to this project was in sight. Tom and Zetta Wojcik provided computer and office facilities that kept me warm and productive in Maine. Lastly, I thank Andy Richford for believing in the Pinyon Jay story and allowing me the freedom to write in a comfortable style.

I moved to Flagstaff in 1980 and first encountered the Town Flock on a casual walk from my apartment. I was looking for lizards and cacti, the real essence of Arizona, when a flock of blue birds quickly passed in front of me. I must have been preoccupied because I dismissed them as Western Bluebirds and kept searching for lizards! Little did I know then, that I would soon be on a first name basis with each and every member of that Pinyon Jay flock. I hope you enjoy learning about this fascinating bird as much as I have.

John Marzluff
Raven Haven, Maine
Fall, 1991

# Introduction

Although this book is about an incredible bird, it is also a story about many people who shared in the excitement of the search for answers about this fascinating creature. This story could not be written without their help.

It all started in early spring of 1968 when Dr Gary Bateman came to my office to report that a large group of blue birds had been feeding in his yard the day before. At that time Gary lived in a rural area near Flagstaff, called Doney Park. I told him that if they were Pinyon Jays he would not see them again as they are notorious wanderers and always on the go. The next day he again reported Pinyon Jays feeding in his yard with his chickens. I found that somewhat strange and even wondered if these birds were in fact Pinyon Jays. I decided I must see for myself and made a trip out to Doney Park the next day. I found the birds just across the road from Gary's ranch, and to my surprise they were in the midst of nest-building. Within a two-hour period I located eight nests in a rather small area. As a young, eager ornithologist I was beside myself with excitement and curiosity. The next day I went back to the same area to mark the nest I had previously found and in the process located an additional 6 nests. In less than four hours I had located 14 nests, all under construction. The birds were very conspicuous and industrious, paying us no heed as we watched them construct their nests. The situation was perfect for Gary and me to study the breeding biology of this colonial nester. We were both aware that very little was known about this strange bird. But first I had to make an important phone call to Pocatello, Idaho. Some years before, when Dr J. David Ligon and I were working on our doctoral research projects in the Chiricauhua Mountains of south-eastern Arizona, we often discussed research problems we would like to tackle if we ever were lucky enough to return to the southwest. He brought up the Pinyon Jay as being a relatively unknown bird that he hoped to work on someday. Thus, I felt he should know of my good fortune, and I must ask him if he minded my working on the biology of the Pinyon Jay. He was both happy for me and a little envious (he told me later) of my discovery. Some years later Dr Ligon made major contributions to our knowledge about Pinyon Jays. The next day I asked Gary Bateman if he would like to participate in the study and he welcomed the opportunity. Thus, the longest study of my career was launched. Little did we know just how long it would be, or how it would end, or how many people would participate in it, or what we would discover.

For the next three years we studied and observed everything we could about these plain looking (but not plain acting!) blue birds in Doney Park. We located nests, marked eggs, recorded hatching dates, recorded egg temperatures (surely under the influence of the late Dr S. Charles Kendeigh, my PhD advisor and friend), weighed young, followed the flock around as best we could, mapped the nest locations, mapped the home range, and tried to color-band the birds. Because the birds nested in the cold winter we asked questions about adaptations for winter nesting. We were very excited to discover that early in the nesting season birds placed their nests on the south side of trees to take advantage of the morning sun. Once we discoverd the bird's attraction to pinyon pine seeds we had a whole new set of questions to challenge us.

Students in my ecology and ornithology courses were all introduced to the Pinyon Jay and spent many hours (some on cold, windy, snowy, days) mapping the colonies, watching nests, and searching on hands and knees for the winter food stores the birds seem to find readily in the litter and debris. In this latter task we would have starved if we had to live off what we were able to find. Students affectionately (I hope) referred to my General Ecology course as Pinyon Jay Biology I and Ornithology as Pinyon Jay Biology II. In any event they discovered new things about the ecology and natural history of the bird, together with Gary and me. We all shared in the excitement of uncovering a new fact.

At this time a number of undergraduate students volunteered to help with the study and we readily agreed as we had our hands full. Most wrote research papers and received a credit or two for their efforts. At least one of these saw the light, caught fire, and went on to do a doctoral thesis on another social corvid in Mexico. Another student worker turned out to be a hobbyist who collected bird eggs for an avid collector in California. I was unaware of this fact at the time. I once returned to a nest tree to find that the branch the nest was situated on had been neatly sawn off. Branch, nest and eggs were nowhere in sight. I later saw that clutch of eggs in a collection in San Diego. We tried to follow the flock with car, bicycles, motorcycles, skis, and snowshoes. One year on Easter weekend Steve Vander Wall agreed to record nest and egg temperatures throughout the entire night. Steve showed up the next morning, chilled to the bone, weary, and wet, as we had an unexpected snow storm that night. Steve was the only person who could find more nests than I could and I continually lost our beer wagers. Steve was also the best tree climber in the Pinyon Jay business. You may have seen Steve's recent comprehensive book on food caching. Mike Morrison spent many snowy hours and days watching Pinyon Jays go to roost and measuring the temperature of water in vials attached to branches. I've never stopped to count how many ornithologists got their early start with Pinyon Jay I and II.

Shortly thereafter I learned about a flock of Pinyon Jays that were regular visitors to a feeding station run by Gene Foster and Katharine Bartlett within the city limits of Flagstaff. This flock had been coming to their elaborate feeding station since the early 1960s. They apparently arrived in Flagstaff during the era that Federal Agencies were bent on destroying the Pinyon-Juniper Woodland in northern Arizona. Thousands of Pinyon Jay flocks must have been displaced – most perished but the Town Flock found a safe haven in Flagstaff. Their feeding station was not just a platform where they scattered seeds for the birds, but an entire backyard where at least 15 different types of food were put out to satisfy the individual tastes of the different species, and

even different individuals. I will never forget the extensive mealworm sorting procedures they established to ensure that the Pygmy Nuthatches had their supply of young, soft, pink, juicy mealworms! It is not surprising that the flock of Pinyon Jays was a regular visitor to the station. Early on we did a set of displacement experiments with these birds, and to our surprise almost every displaced adult bird returned to Gene and Katharine's backyard even though we had displaced them miles away. The same was not true, as Nick Udvardy and I found out the next summer, for younger birds.

In the late 1960s a number of other important long-term studies were underway in the US with social corvids: John Crook's important paper on the ecological correlates of social behavior in birds was popular and stimulating, and W. D. Hamilton's 1964 seminal paper had just appeared. During this time, E. O. Wilson almost single-handedly created a new sub-science of biology popularly known as Sociobiology. The ornithological world anxiously awaited to see if bird studies would provide support for Hamilton's ideas. This was a very exciting time for many ornithologists as intense discussion and debate was the order of the day. During the early years of our study the most common questions asked by my ornithological colleagues were "do Pinyon Jays have helpers, and if so who are they, and do they really help?" If I answered those questions once I answered them a hundred times! My answers were not always simple, and not always direct. It appeared as though Pinyon Jays simply had not read about all those exciting and interesting sociobiological theories. As we were only later to discover, the social system of the Pinyon Jay is not simple, and helping behavior is only one of many interesting and complex behaviors imbedded in the social fabric of a Pinyon Jay flock. Nature is never simple once we begin to probe, and often clear pictures do not emerge nor are all findings and answers neat and clean.

It is our hope that this book will entertain, enhance and enlighten the reader about this unique and complex creature. We hope the reader has as much fun reading this book as we had with this bird! It would be very interesting to know how the second edition of this book would read after another 20 years of study on this fascinating creature.

Russell P. Balda
Flagstaff, Arizona
Fall, 1991

CHAPTER 1

# A bird addicted to seeds

*The Pinyon Jay "is an eminently sociable species at all times, even during the breeding season, and is usually seen in large, compact flocks, moving about from place to place in search of feeding grounds, being on the whole rather restless and erratic in its movements; you may meet with thousands in a place to-day and perhaps to-morrow you will fail to see a single one."*

(Bendire 1895)

Sounds from a distance punctuate the woodland stillness and drift in before the birds arrive: *Aaah, Kraw, Kraw, Aah, Aaaah, Aah, Aah, Krawk-Kraw-Krawk!* Flashes of blue begin to appear through the uppermost branches of deep green conifers. Two, five, ten, fifty, nearly two hundred noisy Pinyon Jays emerge to engulf the trees and ground in a sea of blue. The flock moves steadily in one direction, birds at the back leap-frogging over those at the front. Several birds remain perched, scattered among the trees, attentively watching for potential predators lured to this conspicuous blue buffet. A human onlooker's senses are overwhelmed with color, noise, and motion. Moments later all is again quiet, the blue army seemingly vanishing into thin air leaving the observer wondering: What were those blue birds? Where did they come from? Where did they go? Did I really see so many birds for just an instant?

Most who have spent time in the southwestern United States have experienced such an introduction to one of North America's most unusual birds. The Pinyon Jay (*Gymnorhinus cyanocephalus*) is a drab blue bird the size of a Blackbird or an American Robin (90–125 g). The only breaks in an adult jay's blue plumage are a subtle white throat patch, black outer webs of the primary wing feathers, and black legs, feet, and bill. The top of an adult's uncrested head and its facial (malar) regions are especially deep, cyanine blue. The other body, wing, and tail feathers are a paler, flax blue. Young of the year (juveniles) have mouse-gray body feathers, few or no glossy blue feathers in the malar region, and no white throat patch. One-year-old birds (yearlings) have grayish blue body feathers, glossy azure-blue facial feathers, clove-brown primary and secondary wing feathers, and white throat patches.

Pinyon Jays are solidly built birds with relatively large heads, long (28–37 mm) and pointed, chisel-like bills, and short tails. Birds attain adult proportions within 6 months of leaving the nest. Males slightly outweigh, and have slightly longer bills than, females (average males weigh 112.1 g and have 35 mm long bills compared to females averaging 98.9 g in weight with 31.9 mm

*Three Pinyon Jays pause alertly in a ponderosa pine tree. (J. Marzluff).*

bills). There is, however, considerable overlap in size between the sexes with many larger females outsizing smaller males.

The Hopi Indians refer to the Pinyon Jay as a "bird of war" because of its bold behavior in the face of dangerous predators (Northern Goshawks, Cooper's Hawks, Prairie Falcons, Great-horned Owls, Coyotes and Grey Foxes). Threats by these predators are rebuffed by the sheer number of mobbing birds. In reality, however, Pinyon Jays are anything but war birds; they are committed pacifists. They live in large co-operative flocks and rarely fight among themselves. They do not defend property or food. They rarely prey upon the eggs or nestlings of other birds. Instead, they are specialized vegetarians and seasonal insectivores that occasionally take an unsuspecting young rodent, lizard, or snake.

PINYON JAY COUNTRY

Close association with pinyon pine (*Pinus edulis*) woodlands and the nutritious seeds these pines bear led to the name "pinyon" being given to these jays. The distributional range of Pinyon Jays closely parallels the range of pinyon pine-juniper woodlands of the southwest and inter-mountain regions of the United States (Fig. 1). However, Pinyon Jays range farther north than their namesake tree and pinyon pines occur much farther south than the bird. Even within pinyon-juniper country, flocks occasionally spill over into higher elevation ponderosa pine (*Pinus ponderosa*) forests. Ponderosa pines reach heights greater than 30 m and grow on dry sites in open, park-like stands at elevations of 2200–2800 m. The pinyon-juniper woodland is a "pygmy forest" of widely scattered, short (usually less than 6 m tall), bushy trees. This woodland is found at elevations of 1700–2200 m where moisture is barely sufficient to sustain tree growth.

Variable climate characterizes mountain forests inhabited by Pinyon Jays. Quiescent volcanoes and cinder cones, flanked by pine forests, rise up to 2400 m above dry desert grasslands and chaparral 1200 m above sea level. Dry desert air cools and warms more rapidly than does moist air. The result is a striking temperature gradient paralleling change in altitude and dramatic daily swings in temperature at any one elevation. On dry mountains average temperatures drop 10°C for every kilometer gain in elevation (MacArthur 1972). More impressive is the common 25–30°C daily fluctuation in temperature at a given location. Summer temperatures rarely top 38°C and winter temperatures rarely dip below −5°C. Moist Pacific and Gulf of Mexico air masses waft into this temperature regime, rise in altitude with strong convection systems, and drop sudden torrential summer rains and winter snows. Afternoon rain storms, some quite violent, are almost daily occurrences during the summer (July to early September) "monsoon" season. Winter snows are heaviest in February, March, and April, often dumping 0.5–1 m in single day-long storms. Precipitation is striking in this desert environment, but it is the exception. Clear, brilliant blue skies are the rule. The ponderosa pine forests around Flagstaff, Arizona, for example, receive nearly 90% of possible solar radiation each year.

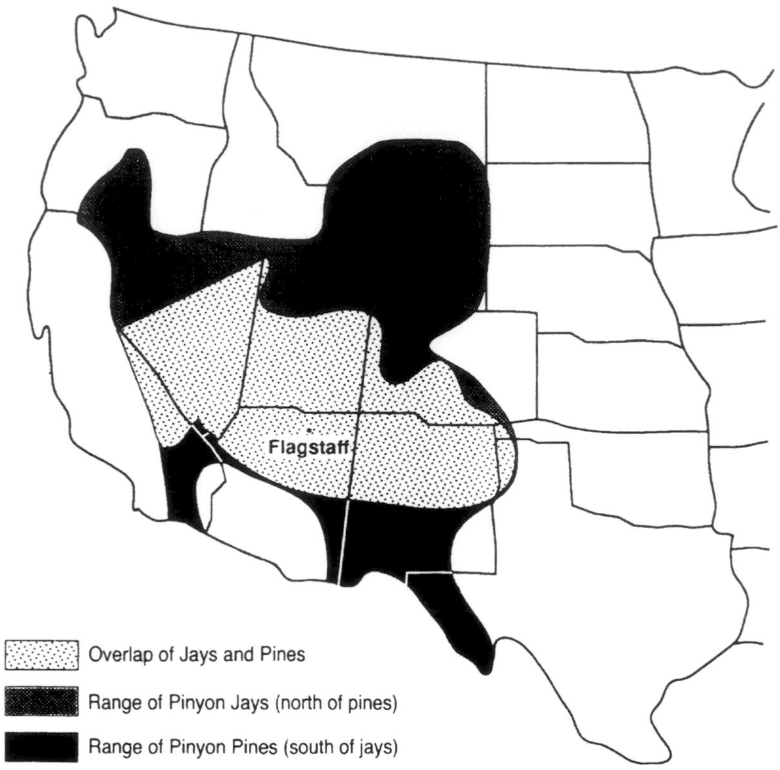

Figure 1    *Distributional range of Pinyon Jays and pinyon-juniper woodland in the USA.*

Annual variation in temperature and precipitation are pronounced in Pinyon Jay country. We illustrate two important variations in Figures 2 and 3. Cold winter temperature is presumably stressful to jays who must maintain warm, constant body temperatures while expending additional energy flying and searching for food. Ponderosa pine forest winter temperatures average well below freezing some years, but well above freezing other years (Fig. 2). Pinyon-juniper woodlands are warmer than ponderosa forests, but equally variable (Fig. 2). Spring snowfall is another important component of the environment: jays nest at this time and heavy snows destroy nests and heighten the energy demands of breeding birds. Ponderosa forests vary dramatically in this respect receiving nearly 2.4 m of snow in some springs and less than 0.2 m in others (Fig. 3). Pinyon-juniper woodlands rarely have substantial spring snowfall (Fig. 3).

Superimposed upon this variable climate is a variable abundance of pine seeds each year (Fig. 4). Ponderosa and pinyon pines are masting species; they produce enormous "bumper" seed crops at irregular intervals. In these bumper years almost every tree in an area is laden with cones bearing edible seeds. In other years the same trees produce virtually no seeds, but other areas may

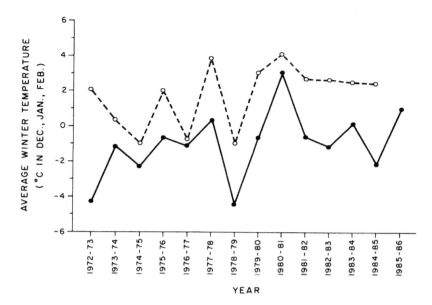

Figure 2    *Annual variation in average winter temperature in ponderosa pine forests (solid line) and pinyon-juniper woodlands (dotted line) in northern Arizona. Data are from Flagstaff, AZ (ponderosa pine forest) and Winslow, AZ (pinyon-juniper woodland). From Marzluff and Balda (1988a).*

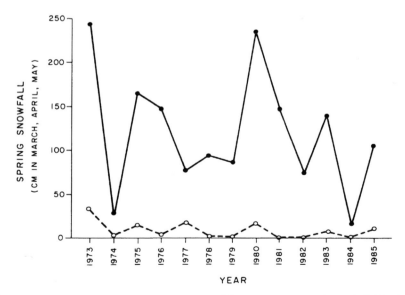

Figure 3    *Annual variation in total spring snowfall in ponderosa pine forests (solid line) and pinyon-juniper woodland (dotted line) in northern Arizona. These data are from the same sites as temperature data in Figure 2. From Marzluff and Balda (1988a).*

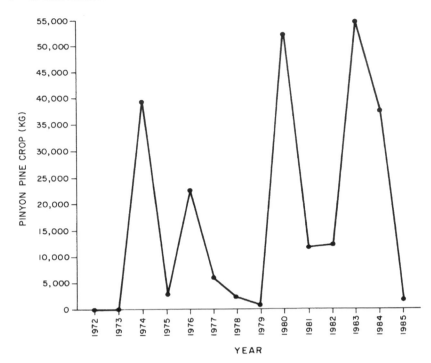

Figure 4    *Annual variation in the amount of pinyon pine seeds sold by Native American Indians to a Flagstaff, AZ purchasing agent. From Marzluff and Balda (1988a).*

harbor productive trees. In truly exceptional years huge cone crops occur over hundreds of square miles. For example, C. Anderson, a seed merchant in Flagstaff, Arizona who purchases seeds collected by native American Indians, reports that 1980 yielded one of the largest crops of seeds ever harvested in north central Arizona (52,000 kg). Yet, in 1979 production in this same area was very low (500 kg; Fig. 4).

This variable production of seeds poses several problems for Pinyon Jays. They must be able to sort among the many unproductive stands to locate currently productive areas. They must collect a specialized crop produced only in the fall and then utilize it throughout the unproductive portion of the year. They must develop specialized behavioral and morphological tools to harvest, transport, cache, and recover seeds, but must not overspecialize, so that when crops fail over wide areas other foods can be efficiently utilized. As we will argue throughout this book, reliance upon this unpredictable food source and life in a variable environment have been major environmental forces shaping the demography and natural history of this unique jay. We begin this argument by showing how seed dependency may have led to the evolution of a unique annual cycle and a unique social system.

## THE ANNUAL CYCLE OF A SEED ADDICT

Here we shall briefly introduce the Pinyon Jay by describing the major activities of a flock throughout the year (Fig. 5). We shall extend many of these topics in the chapters that follow.

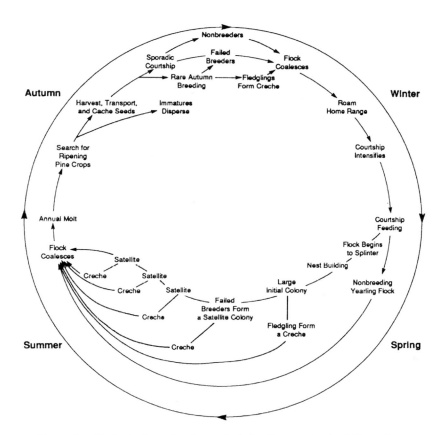

Figure 5   *Flock cohesion and major events during the annual cycle of Pinyon Jays.*

*Fall*

Pinyon pine cones that have been slowly developing for three years begin to swell and ripen in mid-July to early August. The sight of these golfball-sized, green, and tightly closed cones literally drives Pinyon Jays crazy. Their daily routine for the next 2–3 months is entirely dedicated to harvesting seeds and transporting them to familiar grounds where they are poked (usually one per site) 1.5 cm into the thick layer of pine needles covering the ground. These hidden seeds are not visible from above. This latter behavior, known as

*A Pinyon Jay in a pinyon pine tree heavily laden with open and tightly closed cones. (J. Marzluff).*

caching, has evolved to allow seeds produced only in the fall to fuel the flock's economy throughout the winter and early spring.

Flocks search as a group for ripening cones and loose seeds within the confines of their large (2300 ha, the equivalent of 23 sq km) home ranges. In some years, when cone crops are small and irregularly distributed on the home range, the flock wanders widely in search of abundant crops elsewhere. It may often join with other flocks at concentrations of ripening pinyon pine cones to form aggregations numbering thousands of birds (Bent 1946). Flock home ranges are often contiguously distributed throughout suitable habitat but, because borders are undefended, wandering flocks are free to search each other's ranges. Once seeds are located, flock members first eat, and then fill their elastic throats and mouths with seeds. Suddenly a rhythmic *Kaw-Kaw-Kaw* is given, members of a flock assemble, necks swollen as if with goiters, and fly as a group back to familiar terrain on their home range to individually cache the spoils of their hunt. A complete eating, harvesting, and caching cycle takes about an hour when nearby cones on a flock's home range are exploited and may take most of the day when food sources up to 10 km distant are tapped.

Ground rhino horn is a supposed aphrodisiac for some humans, but green cones of the pinyon pine are proven aphrodisiacs for some Pinyon Jays! A Pinyon Jay's sexual desires fade as its sexual tissues (gonads) regress after spring and summer breeding. However, in a warm pinyon-juniper woodland of central New Mexico when green pinyon cones are abundant, a most unique and unusual event occurs. In June and July jays' gonads halt their normal downward spiral and again enlarge. As the very successful harvest continues through the fall, some adults with enlarged gonads reduce their association with the flock and breed in an inconspicuous colony. The ability of abundant green cones to sexually arouse Pinyon Jays was investigated by J. David Ligon (1971, 1974a, 1978). He determined that the sight of green cones elicited gonadal growth even in the absence of the usual stimuli to breed such as increasing day length and warm weather. Green cones are even "sexy" enough to arouse captive jays! Arousal, of course, need not entail conscious thought. It is simply adaptive to breed when food is plentiful. Therefore, the ability to physiologically prepare for breeding by rapidly enlarging the gonads as abundant food is being prepared, would be favored by natural selection.

Jays born the previous summer may not fully participate in stocking the flock's larder. On several occasions we have observed gray juveniles watch adults cache seeds and then proceed to the exact spot and apparently recover the cache. We have also observed juveniles harvesting seeds and caching them alongside adults. However, early fall is a time for more than harvest; it is a time of juvenile independence. Juveniles establish their social positions within the flock by frequent aggressive encounters. This can be a major turning point in a juvenile's life as it wanders between neighboring flocks, eventually remaining with its natal flock or emigrating from it to become a permanent member of another flock.

## Winter

After the fall seed crop is depleted or snow prevents further caching, and juveniles have dispersed, the flock roams about its home range. A home range is defined as the area in which an animal (or social group) lives, except while dispersing or migrating (Wittenberger 1981). In winter, Pinyon Jays may visit some areas of their home range two or three times a day and visit others only once or twice a week. All the flock members roost clustered midway up in the foliage of three or four large, clumped trees. Most often they roost on the south side of the tree to take advantage of the last warming rays of sun before the cold night. Each tree may hold up to sixty sleeping birds. Just prior to sunrise awakening jays can be heard calling from their communal roost. Calling increases as the sun rises and flock members begin swirling flights above the tree tops. The conspicuous calling at this time may act to rally the flock whereas the rapid circling flights may function to increase the birds' metabolic rates and body temperatures (Balda *et al.* 1977). After nearly half an hour of noisy, swirling flights the flock leaves together and flies a considerable distance (at least 0.5 km) to begin foraging. Soon thereafter, jays are commonly seen along roadsides gathering gravel for their gizzards and foraging along the ground in typical leap-frog style. There is a great deal of movement from ground to tree and back during the initial feeding period each morning. At any time, about 40% of the flock forages on the ground. Jays search for insects and pine seeds by slowly walking along, stopping frequently to cock their heads, and thrust their bills to turn over rocks and pieces of wood, probe crevices and pine litter, or flake off loose bark from fallen trees and stumps. When insects are captured they are held at the tip of the bill, beaten senseless across rocks,

*A crowd of jays mob a Great-horned Owl.*

tree limbs, or bare soil and then eaten. Seeds are either swallowed whole and held in the throat or pounded open on a nearby rock or branch and eaten.

Northern Flickers, European Starlings, and Steller's Jays often join the foraging flock, presumably to take advantage of its well developed sentry system (Balda *et al.* 1972). One of the earliest observations of Pinyon Jay behavior was that feeding flocks are commonly surrounded by four to 12 sentries (Cary 1901). Sentries are positioned in exposed and concealed locations in high vantage points. They remain motionless and silent until the approach of an intruder; then the sentinel(s) nearest the potential predator give loud rhythmic *Krawk-Krawk-Krawk* calls, repeating the series two or three times. This warning call (termed a Multiple Rack, see Chapter 4) usually causes the entire flock to cease feeding immediately and fly up into the trees where the birds attempt to locate the predator visually. Feeding resumes in a few minutes if the warning is not repeated, but if the sentinel, now flying frantically within meters of the intruder, continues calling, 15 or more flock members converge to co-operate in mobbing the predator. Accipiter hawks, Great-horned Owls, gray foxes, coyotes, and domestic cats are vigorously attacked by diving, loudly squawking birds. This mob scene usually causes these predators to flee, with many jays pursuing them for long distances, scolding continuously for up to 45 minutes. The remaining flock members fly several kilometers away to resume foraging in a new site. We do not know the age, sex, or dominance status of all sentries, nor do we know how long each is on duty between foraging bouts of its own. We assume most jays share this duty, as dominants, subordinates, adults and yearlings have all been observed "on guard duty".

Winter days end as they begin with noisy flights around the roost trees. Birds seldom, if ever, roost where they last fed in the evenings. Instead, long noisy flights of 2–6 km often precede roosting. Birds arrive on the roosting grounds one to two hours before roosting and begin settling in trees. Roost sites may be selected to minimize exposure to cold weather as most birds roost on southerly aspects of trees. As birds settle in to roost they tuck their bills over their shoulders into their back feathers (Fig. 6). If warning calls signal an intruder's approach as roosting is beginning, the entire flock flies off to select another roost site. When most birds settle in the roost, however, warning calls rarely cause movement of the flock. On one occasion a flock departed its roost well after sunset for an unknown reason.

Subtle courtship behaviors may occur throughout the fall and winter. Courtship feeding, usually of females by males, increases within the foraging flock from mid-November through December. Early feedings involve adults and juveniles and consist of relatively quiet exchanges of food between almost motionless partners. By mid-December courtship becomes more obvious and noisy. The flock now consists of monogamous pairs from the previous breeding season and birds with new mates to court. Unmated birds include yearlings preparing for their initial reproductive effort as well as adults whose previous mates have died. Mated pairs often forage side by side or perch quietly in trees on the periphery of the foraging flock. Courtship feeding now involves females chasing males and begging for food from them by crouching, fluttering their wings, and uttering loud, raspy *Chirrs*. Wild chase flights involving three to 12 birds clucking as they dodge and dive in and out of trees become common

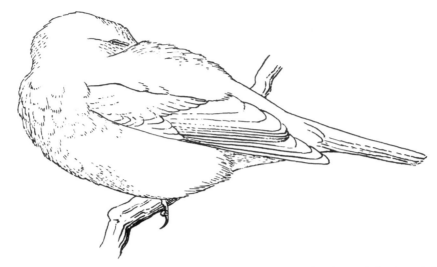

Figure 6    *Posture of sleeping Pinyon Jays. This posture may reduce heat loss as the head is tucked across the back to minimize exposed surface area and feathers are erected to increase insulation. From Balda* et al. *(1977).*

through January and February. All age classes participate in chases which may take birds up to 500 m from the foraging flock and may last up to 15 minutes. These chases may represent the initial stages of courtship for unpaired jays, as many gray birds participate, but identifying rapidly flying individuals makes interpretation tenuous at best.

### Spring

As the end of February approaches, noisy courtship feeding intensifies and pairs of birds isolate themselves hundreds of meters from the main foraging flock. Perennial and newly established pairs begin carrying pine needles and small twigs in their bills. The flock's activities now dramatically change to include hours of nest site selection and courtship on the "traditional" nesting grounds within the home range. Early morning and evening foraging by the flock and roosting remain unchanged, but now up to 40% of each day may be spent on the nesting grounds. In contrast, during late fall and early winter only about 20% of each day is confined to these 20–70 ha portions of the home range. Pairs visit many potential nest sites, often placing a needle or twig on a branch, cooing softly, and pointing their bills at the site. Unmated birds forage around the potential colony site and may continue to initiate courtship and participate in chase flights. Courting birds continue to leave and re-enter the flock for most of the day.

Nest and colony site selection may initially last for weeks until a suitably undisturbed location is settled upon and nest building begins in earnest. Nest-building by the majority of birds over 2 years old commences during late February to mid-March. During nest site selection one member of the pair, usually the male, carries a stick from site to site. Often four or five sites are "blessed" with the same stick before it is left at one or dropped. These sites may or may not be visited by the same pair again. Often some sites end up with a pile of sticks that an untrained observer may mistake as the beginnings of a nest. In contrast, once a site is selected both members of the pair immediately begin carrying sticks to a single site. Under ideal conditions they can build a complete nest in five to seven days. Most pairs in the colony are actively engaged in nest-building at the same time. They constantly advertise their location with calls that sound like "near". Nest-building, like early courtship, occurs only in spurts, usually in mid-morning, after a few hours of foraging. Building ceases for the day if a predator, such as a Northern Goshawk or a Prairie Falcon, disturbs the colony. Heavy snowfall can interrupt nest construction for up to five days. During these hiatuses members of the flock continue to roost together and feed as a unit for an hour or two each morning, sporadically during the day, and again each evening.

The onset of courtship, nest construction, and egg laying is dependent upon the previous fall's pine crop and the current spring's snowfall. As mentioned, following exceptional pine crops some pairs in warm pinyon-juniper woodlands of New Mexico may nest in the fall. More commonly, nesting is several weeks to a month earlier in springs following bumper fall crops than following poor crops. Flocks living in colder ponderosa pine forests often have early nesting attempts delayed by heavy snows.

Complete nests are about 30 cm in diameter and composed of a loosely packed outer layer of 6–40 cm long twigs with a dense, 2 cm thick, inner cup of finely shredded herbs, grasses, and rootlets. Unlike Steller's Jays, whose territories often lie within Pinyon Jay colonies, Pinyon Jay nests rarely include pine needles in the cup lining and never include a mud layer between the coarse stick layer and the fine cup layer. If human rubbish is scatterd throughout the traditional nesting grounds some of it may be incorporated into nests. We have seen wads of mattress lining, bits of decomposed rags, pieces of diapers, and paper towels in jay nests. Initially it looked as though some of our refuse might have been of use to the jays. However, such is not the case as these materials are quick to absorb moisture. After a snow storm, natural materials in the nest lining dry out quickly, but moisture clings to man-made materials. Moist nest lining freezes and thaws every day, thus cooling the eggs and the bellies of incubating females. Most females with these artificial materials in their nests eventually abandon them and start over.

Copulation and egg-laying follow a few days after the nests are complete. A usual clutch consists of four aqua-green, brown-flecked eggs. Reproductive effort is closely adjusted to the size of the previous fall's pine crop. Large clutches of five eggs are common after abundant crops and small clutches of three eggs are common after poor crops. During the incubation period, females usually leave their eggs only for two or three brief defecation, stretching, and feeding bouts per day. Their mates provide virtually all of their food during hourly visits to the colony. These males forage as a group away from the nesting colony and synchronize their return visits to the nests. As a result, a nesting

*A typical clutch of four eggs. The nest's dense layer of insulating grass and rootlets also is noticeable inside the looser outer layer of sticks. (J. Marzluff).*

colony is an inconspicuous patch of forest or woodland except for three to five minutes in each hour, when feeding occurs.

Naked, pink young hatch after 17 days of incubation. Females brood their young almost continuously until they are ten days old and capable of regulating their own body temperatures. Periodically, brooding females will stand on the nest cup and probe deeply into the nest lining to remove blood-sucking fly larvae that live in the nest lining and take daily blood meals from the bare-breasted nestlings. The males supply all the food for the young and over 80% of the food for the female during this brooding period.

As the nestlings age they become darkly pigmented and feathered. Both parents share feeding and sanitization duties, but before feminists rejoice, we must point out that males provide the bulk of the food and females provide most of the cleaning. Females often remain at the nest for 10–40 minutes after feeding to probe the coarse stick layer in an effort to remove the parasitic fly larvae. The young fledge from a now flattened nest 21 days after they hatch.

Communal feeding of some nestlings typically begins when brooding by the females ends. Occasionally, nests are attended by 3 birds: the parents plus a non-breeding, yearling son. These sons, referred to as "helpers", fully participate in rearing the nestling, their actions closely paralleling their father's.

Let's digress a little to consider why some jays help to rear their siblings rather than breed themselves. We will discuss many reasons later in Chapter 9, but first we wish to introduce an idea that may help explain the occurrence of many co-operative behaviors. Individuals having the greatest representation of their genes in future generations can be thought of as winners in the evolutionary game; their traits multiply through the ages, just as savings in a bank grow by compound interest. Individuals replicate their genes by producing offspring. In sexually reproducing species, descendents carry half of their mother's genes and half of their father's, therefore parents can be thought of as passing on 50% of their genes each time they produce one offspring. More distant relatives also share genes because they have common ancestors in previous generations. The more closely related two individuals are, the more recent is their common ancestor, and hence the more genes they share (for example, grandchildren hold on average 25% of their genes in common with their grandparents, but only 12.5% in common with their great-grandparents). W. D. Hamilton (1964) realized that relatives may help each other because in so doing benefactors indirectly pass on their genes. Helpers at the nest, for example, aid in the rearing of their full siblings which carry, on average, half of their genes (each sibling has a 50:50 chance of inheriting any gene from either parent). On average, an individual increases its genetic representation in future generations just as rapidly by helping to produce siblings as it does by producing offspring. Therefore, if helping increases the number of young that parents are able to produce, the extra siblings represent a genetic payback to the helper. If this payback is greater than the helper could attain by breeding, then helping is the best way to pass on genes; this method is known as "indirect selection". (Often this is referred to as "kin selection", but we follow Brown (1987) and use the term "indirect selection" in reference to helping behavior because the siblings helped are not direct descendents of the helpers. "Kin selection" refers to helping direct descendents such as offspring as well as helping collateral descendents such as siblings.)

Back to a flock's breeding activities. The first spring nesting attempt of the year includes all pairs grouped into a loose colony, sometimes called a neighborhood. Usually only one nest is placed in a tree and neighboring nests are typically 5–30m apart. In three cases where two nests occurred in the same tree, the trees were very tall and one nest was in the lowest branches while the other was near the top. As nest predators and spring storms destroy nests, unsuccessful pairs leave the original colony and re-nest in smaller satellite colonies. Satellites form within a week of failure, often several kilometers from the original site. They usually include all pairs failing within a span of a week or two. We have documented the formation of 12 satellites by members of one flock through their six-month-long breeding season. Many pairs may make four or five nesting attempts in a single season.

*Summer*

Progression of the breeding season throughout the summer leads to a gradual splintering of the flock into numerous, spatially independent, feeding aggregations. Aggregations occasionally encounter each other and briefly mix, but quickly separate to pursue their independent goals. Successful breeders from the same colony feed together. Their actions are synchronized and revolve around the hourly visits to the colony to feed incubating females or growing nestlings. Males that forage together while their mates incubate also roost together along the colony periphery. Then, as nestlings attain thermal independence, parents from a colony roost together.

Breeders successfully fledging young typically do not re-nest, but continue to associate as a subgroup while caring for their young for up to a month after fledging. All the young fledged from a colony form another aggregation by gathering into a creche. The loud, raucous begging calls given by the dependent young serve as an attractant for these young to gather together. Once we played a tape of these calls to a group of young as we walked by them. They followed us as long as we played the tape! A creche is the Pinyon Jay equivalent of a nursery school. It is a group of mobile young, fledged from all the nests in a colony, that remains together while still dependent upon their parents. The creche is guarded by a few adults while the remaining adults forage and then return to feed the crechlings. Parents and crechlings readily recognize each other and only a few mistaken feedings occur. Creching is an important initial indoctrination into social life.

Many yearlings form yet another aggregation of non-breeders that forage as a group throughout the day and roost together at night. They occasionally associate with other flock members, notably foraging adult male groups, as they roam through nesting colonies and creches. These teenage gangs may contain mated pairs as well as single birds.

The annual molt of breeders abruptly signals the end of the breeding season and the close of the annual cycle. Few nests are initiated beyond early July and most jays are molting by late July. Molting and breeding are energetically expensive and must be separated during the year (Ligon and White 1974). Non-breeding yearlings have already molted into a more adult-like plumage in April and May. Breeders, non-breeders, and crechlings begin to coalesce into a single flock with coordinated daily activities in late July. The last subgroups to rejoin the flock are the few successful late breeders and their now free-flying young. The late summer molt equips jays with both a new set of airfoils for the extended flights which characterize the upcoming pine seed harvest and renewed insulation to retard heat loss during the upcoming winter.

## Variable foods pose an evolutionary dilemma and produce unique strategies

Organisms utilizing seasonal and/or spatially variable foods must eventually encounter a dilemma: either evolve generalized diets and feed on whatever foods happen to be available, or evolve specialized diets and search widely for areas where favorite foods are abundant, store them for use in lean times, and use them to enhance reproduction and survivorship. The various jays inhabiting southwestern pine forests have evolved to both ends of this specialist-generalist continuum. Pinyon Jays have specialized and evolved several attributes enabling efficient and year-long exploitation of pine seeds while two of their relatives, Scrub Jays and Steller's Jays, have remained generalists and utilize seeds opportunistically. Another common response to this dilemma is simply to migrate to areas where food is available, but no western jays are true migrants.

We will never know why Pinyon Jays specialized. A likely prerequisite, however, is a strong flight capacity. Perhaps an ancestral population of proto-Pinyon Jays acquired, through the luck of random mutation, longer wings and stronger flight muscles. These jays could fly farther and expend less energy on flight than other jays having the usual short, rounded wings. These accomplished aerialists were probably already capable of opening thick-hulled seeds like acorns and pine seeds. Specializations in the articulations of Pinyon Jay jaws are found in many New World jays, suggesting that these adaptations for opening tough seeds were present in ancestral jays (Zusi 1987). Strong flight capacity would free ancestral Pinyon Jays to wander farther and farther in search of abundant pine crops. This would have been an important breakthrough in the evolution of specialization because it would enable jays often to encounter widely dispersed, but locally abundant crops. Jays with poor flight capacity would rarely encounter abundant crops, so natural selection would favor generalized traits which allowed such jays to utilize a wide variety of foods on their smaller home ranges.

Foraging widely in search of seeds is probably responsible for one of the Pinyon Jays' most unusual attributes, lack of territoriality. Many birds defend all-purpose territories that provide all their nutritional, nesting, and roosting needs. Scrub Jays and Steller's Jays may defend such territories. Birds that do not defend all-purpose territories typically at least defend feeding or nesting territories. Territoriality allows the defender or group of defenders exclusive use of the resources found therein. However, some resources are not economically worth defending. Jerram Brown (1969) formalized this idea by arguing that the costs of territorial defense must be exceeded by the benefits of exclusive use before territoriality will evolve. Large stands of pine trees would require defense in order to provide reliable seed crops to a jay flock. Defending the borders of such vast areas would be likely to occupy much of an individual's or a group's foraging time and insects and small nocturnal rodents living within the territory would meanwhile damage or consume many seeds. Territoriality is therefore unlikely because defense costs would be so extreme that the benefits, even assuming that all avian competitors could be excluded, would be negated.

Rather than physically exclude seed competitors through territoriality, Pinyon Jays evolved numerous morphological and behavioral adaptations for rapid exploitation of seeds. Sharp bills allow tightly closed cones to be opened before they ripen and open naturally, exposing their valuable seeds to other jays, titmice and rodents. The Pinyon Jay esophagus is much like a balloon: the muscles relax allowing the throat to stretch as more and more seeds are stuffed in. This enables 40–50 seeds to be collected at a time and then transported to a safe storage area elsewhere. In contrast, Scrub Jays can cram only three to five seeds into their less flexible throats and mouths, though even this is a considerable improvement over the typically rigid throats of passerine birds other than corvids. Given that an individual can harvest, transport, and cache only 40–50 seeds each hour, natural selection should favor individuals capable of discerning full seeds from seeds containing dried or insect-damaged endosperms (the edible "nut meats"). Pinyon Jays are able to make such discriminations by weighing seeds in their bills and avoiding yellowish seeds, which are usually empty (Ligon and Martin 1974). Few inedible seeds are extracted from cones (one study observed jays extract 197 seeds and discard only 4) and a survey of 655 seeds removed from jays' throats revealed that all were edible (Vander Wall and Balda 1981).

Large flock size may be a behavioral adaptation which allows a greater area to be searched for seeds and reduces predation while utilizing seeds. More jays mean more eyes to spot predators attracted to the concentration of harvesters. Of course, this would benefit group members only if those spotting a predator communicated the information to the rest of the group, and such bold behavior in the face of dangerous predators can be costly. There are numerous reports of scolding birds being eaten by the object of their scorn, and we witnessed one bold Pinyon Jay becoming a Northern Goshawk's lunch during such an encounter! Warning group members of impending danger would evolve when such acts are reciprocated by those warned and/or if some alerted individuals are related to the signaler.

Reciprocity of costly favors may unite several families into permanent troops. Relatively stable membership in such troops would enhance co-operation by keeping relatives together and allowing non-co-operating "cheaters" to be recognized and expelled. Specialized warning behaviors such as sentinel duty could now easily evolve and further increase the safety within a flock. A testament to the efficiency of predator detection by modern Pinyon Jays comes from an unpublished study of accipiter hawk diets by Pat Kennedy. She identified prey remains at Northern Goshawk and Cooper's Hawk nests in the ponderosa forests and pinyon-juniper woodlands of New Mexico. Solitary or pair-living birds such as Northern Flickers, American Robins, Mourning Doves, Scrub Jays, and Steller's Jays made up the bulk of these hawks' diets. Steller's Jays alone comprised 20.5% of Northern Goshawk kills and 28% of Cooper's Hawk kills, but, although Pinyon Jays frequented the study area, not a single one was identified in two years of investigating prey remains. Effective predator detection by a large flock of jays reinforces their social lifestyle as most co-operators can eat, harvest, and cache seeds without scanning for predators themselves. Seed processing is therefore rapid, safe, and efficient.

Reliance during the nesting season on clumped resources, such as seeds cached in restricted areas of the home range and possibly emerging insects, would favor colonial nesting. Commuting costs between food clumps and nests are minimized by nesting in a centrally-located colony (Horn 1968). Moreover, coloniality allows pairs to synchronize their activities, which reduces the periods when a colony is conspicuous and enables parents to leave the colony to forage and recover caches as a group. Foraging success may be increased by group efforts because as more foragers comb an area the chances of one happening upon a rich clump of insects increases. All group members benefit by such discoveries when their locations are advertised. Group members communicate the location of such clumps amongst themselves because insects are a short-lived resource and thus not available for continued exploitation by one individual. Moreover, in a cohesive group the favor may aid relatives and is later repaid when other members discover rich food patches. Visits to the communal caching grounds are group efforts because recovering and handling seeds requires jays to focus their senses on the ground and on seeds, thus reducing their ability to detect potential predators; sentinels in the group increase safety during such distractions. In addition the possibilities of individuals raiding each other's caches is minimized by group recovery as individuals are not alone on the caching grounds.

Cached seeds increase an individual's evolutionary fitness by providing a food source essential to survival through lean winter months. In addition to staving off starvation during the long cold unproductive months caching behavior may have some other proximate benefits. As David Sherry suggests (1985) it allows caching birds to forage for food at one time and eat it at another time. Once food has been safely stored in subterranean caches it can be recovered at the most profitable (least expensive or dangerous) time to do so. This not only puts the food under the control of the bird but also allows the bird to set its own meal times. Another possible proximate advantage of caching food is that it relieves the bird of having to put on large amounts of fat to overcome periods of food shortage (Lima 1986, Rogers 1987) as many animals must do. In birds this is a costly way to store energy. However, evolutionary success depends upon reproduction. Successful seed specialists must rapidly convert cached seeds into baby jays for their specialization to endure over generations. Pinyon Jays do so by breeding early and laying an extra egg when seed crops are abundant.

The fall pine crop can be likened to a large *piñata*, the Mexican statues that are filled with candy and money. At festivals children break open *piñatas* and scramble to fill their pockets with riches. As the pine crop ripens, numerous birds, insects, and mammals scramble to fill themselves and their larders with seeds. This scramble competition has been a major impetus for the evolution of many unique features of Pinyon Jays that allow them to specialize on *piñatas* occurring at unpredictable locations and opening for a few months each year. Most unique is the formation of large, stable flocks that roam undefended home ranges and nest in colonies shortly after bumper crops are harvested. Life in such a large flock is co-ordinated by well-developed and graded communication allowing individual jays to be recognized. Life and death in such a society will be the focus of subsequent chapters, but first we will explore in more detail this jay's unique taxonomic status and mutalistic relationship with pines.

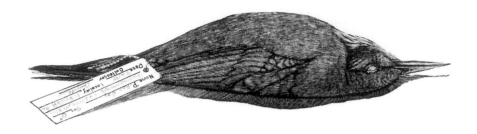

CHAPTER 2

# Birds of what feather?

*The Pinyon Jay "is not a jay at all by any standard except that its plumage is blue in color."*

(J. W. Hardy 1969)

*"the relationships of this peculiar species remain obscure."*

(Mayr and Short 1970)

Maximilian, Prince of Wied, the eminent German naturalist, must have smiled when he discovered the first Pinyon Jay. In the late 1830s the Prince and his party were exploring the remote inner portions of western North America by following the Missouri River. On the Marias River, one of the Missouri's tributaries in northwestern Montana, the Prince first encountered the jay that would bear his name for some 30 years. Maximilian's Jay, as the Pinyon Jay was initially called, was first introduced to scientists in Maximilian's book of travels in North America, published in 1841. However, it was not until 1846 that Mr Edward Kern, a naturalist associated with Colonel Fremont's expedition, collected the first Pinyon Jay specimens for American naturalists to study. Ever since that time, taxonomists (scientists interested in the evolutionary relationships of organisms) have been puzzled by the Pinyon Jay. They have referred to it as a crow and as a jay. Its formal scientific name has frequently been changed, reflecting the jay's uncertain evolutionary relationship. Maximilian referred to Pinyon Jays as *Gymnorhinus cyanocephalus*, but later this was changed to *Gymnokitta cyanocephala*, *Psilorhinus cyanocephalus*, *Cyanocorax cassini*, and *Cyanocephalus cyanocephalus*. Currently we use Maximilian's original Latin name. Apparently, the only thing early naturalists could agree upon was that the Pinyon Jay had a blue head (*cyanocephalus* translates as "blue-headed")! As we will explore in this chapter, the problem of a blue-colored jay that acted like a crow continued to frustrate taxonomists 100 years after Prince Maximilian's trip to Montana.

Pinyon Jays are easily recognized as members of the family of birds known as corvids (more formally, Corvidae) that includes crows, ravens, nutcrackers, magpies, and other jays (Fig. 7). The family is very diverse, but all members share some distinctive features. Their large heads, well endowed with brains and ensuing mental capabilities, powerful bills, and strong, gripping feet with scaly legs are typical of the corvid body plan. Their bold and inquisitive demeanor, exemplified by fearless scolding of potential predators over 100 times their size with loud, harsh "mobbing" calls, is vintage corvid. Although the Pinyon Jay has no blue top-knot or crest like that of its more familiar cousins, it raises and lowers the feathers on the top of its head as if it were extending a crest. Other typical corvidisms exhibited by this drab blue bird include life-long, monogamous pair bonds formed after a prolonged courtship and maintained throughout the year, care of young by both parents, and hoarding surplus food in caches.

It is much more difficult to recognize where Pinyon Jays fit within the corvid family. Seed dependency has produced the unique social organization introduced in Chapter 1 and a unique morphological build that has continually confused scientists searching for the Pinyon Jay's corvid "roots". The quotes beginning this chapter and the fact that Pinyon Jays have been classified in a genus by themselves (itself changing names frequently) underscore their uniqueness and the confusion surrounding their position within the family.

The usual methods for determining the evolutionary relationships among a group of species, such as the corvids, are to quantify many morphological, behavioral, and genetic characteristics of living and fossil members and determine their similarity. Species are then grouped according to resemblances that are derived in common. Groups of species that share derived attributes are suspected to be closely related because the characters they share are presumed to be passed on from a common ancestor. Determining phylogenetic relationships among birds is difficult because the demands of flight impose structural constraints on bird skeletons and muscles resulting in many morphological similarities even among distant relatives. The phylogeny of specialized foragers, like Pinyon Jays, is especially difficult to determine because many behaviors and physical features have been shaped by natural selection favoring efficient exploitation of their predominant foods. Species that specialize on the same foods, regardless of how distantly related they are, can evolve similar actions and appearances that obliterate many useful clues to their true ancestry.

The 115 species within the family Corvidae are categorized into 23 genera (Goodwin 1986). Beginning some 35 million years ago two groups of genera followed different evolutionary courses. One group, the American or New World jays, evolved on the land masses that were then North, Central, and South America. The other group, the old world corvines – crows, ravens, European and Siberian jays, nutcrackers, and magpies in the genus *Pica* – evolved in isolation on a land mass destined to become Europe and Asia. After millions of years of evolution on opposite sides of the Bering Sea, the Bering land bridge formed around 1 million years ago and old world corvines invaded the northern new world, stocking it with ice age crows, ravens, nutcrackers, magpies, and Gray Jays.

Figure 7   *The corvid family tree. A common ancestor gave rise to a variety of jays, crows, magpies, and nutcrackers. The major branches shown here follow work by Goodwin (1986), Zusi (1987), and Hope (1989) which suggest that an early split in the family separated New World jays (the Blue Jay and Pinyon Jay are shown) from Old World magpies, nutcrackers, and crows. Pinyon Jays diverged early from the New World jay line (C.B.).*

It has been a rather simple exercise for taxonomists to sort all modern corvids except Pinyon Jays into either New World jays or Old World corvines. Pinyon Jays, however, have defied classification because many of their defining characteristics are distinctly Old World, whereas others are decidedly New World. Alternative names for the Pinyon Jay – Little Blue Crow or Maximilian's Crow – reflect its crow-like traits. Early work on corvid taxonomy in the 1940s and 1950s by Ashley (1941), Amadon (1944), and Pitelka (1951) classified Pinyon Jays as specialized New World jays because they were blue and lacked nasal bristles (*Gymnorhinus* literally translates as "naked nose" and Pinyon Jays are the only jay completely lacking nasal bristles), and because their bone structure, especially the shape of the end of their humerus (a bone in the wing) was similar to that of other jays. In his revision of corvid classification, Hardy (1961, 1969) dismissed this evidence and concluded that Pinyon Jays had evolved from an Old World ancestor that also gave rise to nutcrackers. (This appears to be one of the few mistakes Hardy made as his classification of other New World jays has stood the test of time.) Hardy's conclusion hung on the observation that, unlike New World jays which hop, Pinyon Jays, crows, ravens, and nutcrackers walk. Ligon (1974b) argued against this conclusion as many Old World and New World corvids both hop and walk depending upon their age and/or their motivation. Pinyon Jays, for example, hop as juveniles and Grey-breasted Jays, which typically hop, walk when engaged in aggressive encounters. Two factors ally Pinyon Jays with New World jays: (1) as nestlings they lack a covering of down which also is lacking in New World jays but is present in nutcrackers, crows, and ravens (Ligon 1974b); and (2) the bones in their lower jaw form a complex found only in New World jays that allows for food (typically acorns) to be opened by pounding upon it with the lower jaw (Zusi 1987). Sylvia Hope is currently retackling the problem of corvid phylogeny and has concluded that Pinyon Jays are indeed New World jays, probably most closely related to Mexican (Grey-breasted) Jays (Hope 1989). Pinyon Jays, along with all the other North American jays, seem to have evolved from an ancestral neotropical jay (Hope, pers. comm.).

A major problem in deciphering the Pinyon Jay's identity has been its superficial resemblance to an old world descendent, the Clark's Nutcracker. Both species have short tails, long wings, chisel-shaped bills, a specialized structure for transporting pine seeds, well-insulated nests, and well-developed seed caching behaviors. These similarities, however, are likely to be the result of convergent evolution, a process whereby natural selection equips distantly related species with similar adaptations to survive similar environmental conditions. The important environmental factor fashioning Pinyon Jays and Clark's Nutcrackers is the periodic, but unpredictable, abundance of nutritious pine seeds. As we argued in Chapter 1, long wings enabling energetically efficient long-distance flights, chisel-shaped bills for opening green cones, and the ability to carry many seeds at once and cache them for future use allow for specialized use of pine seeds. Pine seed specialists may nest in late fall and/or mid- to late winter after a good seed harvest, using stored seeds to supply much of the energy needs of incubating adults and growing nestlings. Fall and winter nesting allows abundant seed stores to be "cashed in" for copies of the harvesters' genes. However, nesting in montane habitats at this time is energetically expensive because temperatures are low. Well-insulated nests may counter this expense.

If similarities between Pinyon Jays and nutcrackers are due to convergent evolution and not due to common ancestry, we should find examples of similar environmental problems solved with slightly different adaptations. Also, it is probable that useful traits present in an ancestral species will not be found to be absent in its true descendents. Both sorts of observations are exemplified by the seed storage adaptations in the throats of jays and nutcrackers. European Nutcrackers and Clark's Nutcrackers have a distensible outpocketing of the throat beneath the tongue known as the sublingual, or gular, pouch (Bock *et al.* 1973). The sublingual pouch holds up to 90 pinyon pine seeds. Pinyon Jays have solved the problem of transporting many seeds with a similar, but less efficient, adaptation. Their throat walls (more specifically their esophageal walls) are distensible when the surrounding muscles are relaxed, allowing up to 56 pinyon pine seeds to be packed into it. This is a beautiful example of the evolutionary principle known as "jury-rigging" (Gould 1980). Natural selection has solved the need to transport many seeds by slightly modifying a pre-existing structure, not by evolving a novel, and perhaps better structure. The result is as expected if convergent evolution has been at work; jays and nutcrackers do not have the same adaptations for seed transport. As far as inherited traits are concerned, it is likely that the sublingual pouch was present in the Old World nutcracker that colonized the New World because both modern nutcracker species, one of which resides in the Old World, possess the pouch. The pouch allows Clark's Nutcrackers to carry nearly twice as many seeds per trip between harvesting and caching areas than the expandable throat allows Pinyon Jays to carry. It is therefore doubtful that Pinyon Jays evolved from an Old World nutcracker ancestor because an efficient structure like the sublingual pouch would not be replaced by a less efficient structure like the expandable throat.

Let us return to the color of Pinyon Jays. Like many New World jays, they are blue. In contrast, there are no completely blue Old World corvines (they are typically black, grey, white, or rusty brown, though the blue magpies and azure-winged magpies have considerable amounts of blue plumage). Pinyon Jays could be blue because they share a common ancestry with New World jays or because they have converged toward them in color. There is no reason to suppose convergent evolution with respect to color; many Old World corvines, such as Clark's Nutcracker and the Common Raven, inhabit the same environments but remain in traditional Old World garb. Blue color may not enhance nor reduce its wearer's fitness and therefore change only by rare, random mutation events. Most descendents of a common New World ancestor should therefore have blue feathers. This is, however, a double-edged sword; perhaps Pinyon Jays are the single rare totally blue mutant Old World corvine. This possibility becomes less likely when we consider other traits, seemingly neutral with respect to natural selection, shared by Pinyon Jays and other New World jays. One such trait may be the Pinyon Jay's rudimentary nasal bristles. All New World jays have relatively short bristles, therefore even if the Pinyon Jay's lack of bristles is an adaptation to reduce pine sap from building up on the beak, their original reduction in length may not have been accomplished by natural selection.

Enough speculation about an event that happened millions of years ago. Suffice it to say that most evidence is consistent with the view that the Pinyon Jay is an extremely specialized New World jay that is superficially similar to Old World nutcrackers simply because they have pursued similar ecological occupations. The quote from Hardy we used to introduce this chapter was intended by its author to persuade others that Pinyon Jays were descendents of Old World corvines. In fact, it may best serve to emphasize this jays' probable affinity with other New World jays.

CHAPTER 3

# A co-evolutionary tale

*The Pinyon Jay's "favorite haunts are the pinon-covered* (sic) *foothills of the minor mountain regions, the sweet and very palatable seeds of these trees furnishing its favorite food during a considerable portion of the year."*

(Bendire 1895)

No story about the Pinyon Jay would be complete without a discussion of its namesake, the pinyon pine. The tree and the bird have a closely linked past that must go back thousands of years. As evidenced by Bendire's quote opening this chapter, the close association between the jay and the pine was obvious to early naturalists. Pinyon pines and Pinyon Jays have probably co-evolved in a type of evolutionary arms race. As the pines evolved defenses to reduce seed predation the jays countered by evolving ways to circumvent these defenses and exploit the seed crop. When jays, insects, or mammals became too proficient at exploiting seeds, pinyon pines evolved new, better defenses. This leads to an escalating arms race as exploiters needed to evolve better ways to counter defenses. Pinyon Jays stepped out of the arms race as they entered a mutualistic co-evolution with pinyon pines. The traits jays evolved became beneficial to trees by providing a reliable way to disperse their seeds. Accordingly pinyon pines evolved many traits that thwarted insect and mammal predators but favored seed harvest by Pinyon Jays. This co-evolutionary tale forms the basis of this chapter. Many of the traits characterizing Pinyon Jays and pinyon pines are the result of their mutualistic association.

## Pinyon pines

In North America there are at present 11 recognized species of "pinyon pines". However, the Pinyon Jay is associated with only two or three of these species. The pinyon pines associated with Pinyon Jays range from northern Baja California to northeastern California throughout most of Nevada, Utah, Colorado, Arizona, and New Mexico. The limits of their ranges extend from southern Idaho to the trans-Pecos of Texas, and east just into the panhandle of Oklahoma (Fig. 1). The two most prevalent species are single-needled pinyon (*Pinus monophylla*) which is the western-most species and Colorado pinyon (*Pinus edulis*) which occupies most of Utah, Colorado, New Mexico and Arizona. The remainder of these trees occur in Mexico, well south of the present distributional range of the jay. Lanner (1981) in his excellent book on pinyon pines states that they dominate over 173,000 square kilometers of the southwestern US landscape.

To the untrained eye, the pinyon pine is anything but majestic. It has been referred to as scrub pine, brush pine, dwarf pine, pygmy pine and nut pine. These names hardly do this fascinating tree justice, but accurately define its posture and notable characteristics. It is a relatively short, stubby, slow-growing tree, that in some years produces huge quantities of large, highly nutritious, and delightful tasting seeds. Branches emerge from the trunk near ground level, giving the tree its "shrubby" appearance.

Although these pines may occur in pure stands, they are most commonly found in association with three species of junipers (native southwesterners often call these trees cedars). These junipers have a growth-pattern similar to that of the pinyon pine thus producing a rather monotonous plant community referred to by plant ecologists as pinyon-juniper woodland.

A woodland is defined as an area of scattered trees separated by enough open space so the tree canopy is open and sunlight reaches the woodland floor. The pinyon-juniper woodland fits this definition well. The spacing of these trees may reflect the scarcity of a critical resource, water. As we discussed in the first chapter, rainfall is periodic and seasonal whereas snowfall is minimal and rare. Summer temperatures are high, consequently soil moisture is rapidly lost to evaporation. Of all North American pines, the pinyon pines are truly best adapted to arid conditions.

CO-EVOLUTION OF PINYON PINES AND PINYON JAYS

In order to understand the special relationship that exists between bird and pine we must briefly examine the evolutionary history of pinyon pine. Much of this information comes from Lanner (1981). The ancestor of the pinyon pines most probably prospered on the Mexican Plateau in what is now southwestern US and northern Mexico. It was a member of a plant community known as the Madro-Tertiary Flora which contained a number of drought-resistant species of trees. About 60 million years ago the climate of North America slowly began to change from a mild, moist one, to one that was warmer and quite arid. The progenitor of the modern pinyon pines was able to withstand this change in climate and even thrived under these arid conditions. Short needles, slow growth, and large seeds probably allowed this species or species complex to move northward into cooler mountain habitats. By the Miocene Era, 25 million years ago, it had moved into southern California and by the Pliocene Era it occupied the Great Basin and northern California. With the coming of the great glacial advances and retreats during the Pleistocene Era the pinyon-juniper woodland showed similar advances and retreats. By the end of the Pleistocene Era (11,000 years ago) pinyon pines had the distributional pattern we see today.

The fossil record documenting the spread of this tree contains an interesting and unique observation. The spread of pinyon pine up and down mountains and mesas, and vast distances in all directions was truly rapid for a tree. As climatic conditions became hospitable and new environments became available the pinyon pines were quick to invade these areas. Most other plants did not show this swift movement across the landscape. Most other pines have small, winged seeds that are dispersed by wind, water, and gravity. These pines generally advance slowly as their agents of dispersal broadcast their seeds randomly over an area, and often only a short distance from the parent tree. The vast majority of these seeds never germinate because they end up in inhospitable places. Paleontologists generally agree that pinyon pines must have been dispersed by some agent(s) other than wind, gravity or water. This is even more likely to be true if ancestors of modern pinyons had the large seed we see in these trees today. In fact, the rapid spread of the pinyons may be attributed to this single morphological structure – a large seed. The paradox is obvious, large seeds are not easily moved by wind, water, or gravity.

Why do pinyon pines possess such a large seed relative to the seeds of most other pines? The answer to this question may be found only deep in the past history of the plant. Earlier we mentioned that these trees lived on the Mexican Plateau at a time when the climate turned warm and dry. Under these arid conditions, sites where seeds could survive, germinate, and grow may have been rare, especially for small seeds that carried little energy and nutrient reserves within them. These small seeds produced tiny seedlings that would have difficulty withstanding dry, hot winds, and would be unlikely to grow quickly enough to establish themselves before all their embryonic materials were consumed in their struggle to survive in these hostile surroundings. One very important means of overcoming these pressures would be to form a large seed that contained the necessary energy and nutrients to give the new seedlings an added boost early in life. This large seed could not only survive in the traditional sites but now would be able to colonize many more sites than a smaller seed because it contained the energy source for rapid root growth. Rapid root growth would enable the developing seedling to trap what little moisture was present in the soil at the germination site. In a hot and dry climate, larger-seeded trees would leave more offspring than those with smaller seeds, thereby enabling natural selection to gradually replace small-seeded trees with large-seeded ones.

The strategy of producing large, energy-packed seeds is not cheap. It has numerous costs, constraints, and pitfalls. The production of these seeds means the tree must somehow capture and store the necessary fats, carbohydrates, and proteins used to build them. If these ingredients were difficult to obtain, the cost of producing large seeds could have placed critical energetic constraints on the trees. These trees would have had only two alternatives: either produce only a few, high quality seeds year after year, or store up the necessary materials and then produce a bumper crop of seeds at intervals greater than one year. The first scenario was possibly the ancestral one.

The second major problem with the evolutionary decision to build large seeds was the ever-present search of seed predators to find enough food to survive and reproduce. Large seeds are easily detected and efficient to harvest. Most certainly, efficient seed predators quickly caught on to the presence of these energy packets and took full advantage of them. As cones ripened and seeds fell to the ground under the trees, mammals and birds concentrated their search there, possibly eating every seed produced. The conifer seed-eating birds and mammals could have decimated small seed crops year after year. Obviously, the program of producing a small yearly crop of large, nutritious seeds would have been disastrous for the pinyon pines, as now no seeds (or very few) would escape to germinate as seed predators took full advantage of this small but predictable supply of energy. The rapid colonization of new habitats, so striking in the fossil record of pinyon pines, could not have taken place under these conditions, and thus, it is necessary to add another twist to this intriguing story.

If the trees could develop a means of capturing and holding on to the essential ingredients used to form a pine seed for some length of time, then large, bumper crops could be produced in some years and in other years no seeds would be grown. The effect on the seed predators would have been dramatic. In years when no seeds were produced these predators would have to look elsewhere for food, thus removing the element of predictability from their normal foraging pattern. They could no longer rely on a ready supply of food each year from the pines. In the most extreme case many seed predators may have starved to death. Thus, when the trees produced their next large crop of seeds there would be few predators available to capture them. This would be especially true for relatively short-lived insects. However, most vertebrate seed predators are generalists that can take advantage of many different kinds of seeds. The presence of massive numbers of seeds could have had another effect; their sheer numbers may overwhelm predators so that all the seeds could not possibly be eaten. Thus, some seeds would have escaped predation in "safe" sites where they could germinate. Swamping potential

seed predators can be a complicated business; in order for it to operate efficiently all trees in the population must be on the same yearly cycle, as a tree
that produces its seeds out of synchrony with its neighbors will most probably
have all its seeds consumed by seed predators. The timing of seed production
had to be synchronized among all trees in an area if the strategy was to be
successful. Trees that were out of step woud have been at a severe disadvantage and would have been gradually replaced by trees that used the same
timing mechanism (Ligon 1978).

We can speculate that the interaction of a harsh environment and persistent
seed predators was responsible for the initial pattern of seed production seen in
pinyon pines. One important piece of this puzzle is still missing. As mentioned
earlier, large seeds are not transported easily by wind, water, and gravity. A
dispersal agent is still missing if our reconstruction is to work. At this point we
can return to the central character of our tale, the Pinyon Jay.

At first, conifer seed-eating birds simply took advantage of alternate foods.
Pinyon Jays probably occurred in the same areas where these pines lived, and
probably fed off the bumper crops as did many other seed predators. All
corvids are, by nature, inquisitive creatures that spend time exploring and
probing in what appear to be unlikely locations. Faced with a super-abundant
supply of pine seeds these birds may have simply started to manipulate them
in their bills and on occasion put them in concealed locations. (Most captive
corvids hide objects in their cages, even when very young.) Such behavior
may have led to the habit of hiding pine seeds in concealed locations and
recovering them days, weeks, or even months later. If seed availability
declined rapidly in autumn and winter these hidden seeds would have been a
valuable addition to a bird's diet. We can speculate that at first these hidden
seeds were located by random search and accidental hit or miss probing. Birds
that remembered the locations of seeds they hid would have a decided advantage over birds that forgot the locations of their hidden seeds. These seeds
could make the difference between life and death during a cold, unproductive
winter.

An important component of this behavior would be the type of sites chosen
by birds for hiding seeds. At first, birds probably simply stuck seeds in cracks
and crevices near where they picked them up. If a bird took seeds from an open
cone it may have simply put the ones not consumed in an arboreal location
such as a crack in the bark of the tree or in a crotch of the tree. These exposed
seeds would then be available to the next bird that happened to search through
this area. The use of inconspicuous locations would help ensure that seeds
would not be found by other birds, but at the same time they may not have
been evident to the jay that hid them. The habit of using subterranean cache
sites was then probably favored as seeds could easily be buried and hidden
from view of other jays and other seed predators. Burying seeds in the soil was
probably closely linked with the use of memory to find them as visual search
would not reveal their location. The use of inconspicuous sites and the use of
memory to find these seeds at some later time could have led to a dependence
on these cached seeds during the non-productive time of year.

Now the stage was set for the jays truly to manage the resource. Many seeds
would be eaten immediately upon harvest, but after satisfying their daily

needs birds now could provision larders with seeds for future use. Thus, a bumper crop of pine seeds could provide the birds with a guaranteed supply of food for many months thereafter. Man does the same by growing agricultural crops during the summer season and storing most of the harvest for use during the next winter.

A major problem with using subterranean cache sites in the pinyon-juniper woodland may have been the presence of competitors for these cached seeds. One would expect that even buried seeds could be found by hungry predators such as the modern pinyon mouse that rely on smell to locate food. Modern day pinyon seeds have a strong odor that these mice presumably home in on. This is puzzling as any clue to a seed's location should have been eliminated in the evolutionary process as it endangers the seed's survival. Maybe the odor is associated with some critical nutrient the plant must have and cannot synthesize in an odorless form. In all probability, Pinyon Jays do not use this odor to locate their hidden seeds, as they have a rather poorly developed olfactory system (Larry Clark, pers. comm.).

The need to overcome intense pressure from these specialized seed predators may have been the very reason pinyon pines rapidly colonized new areas. The birds may have simply transported the seeds out of the pinyon-juniper woodland and into habitat types where specialized predators did not exist, or where they lived at low densities. The spread of these pines would now be paced by the ability of birds to transport seeds.

One last major question needs to be answered. Why would these trees engage in a relationship with the jays if all the benefits went to the jays? Some seeds at least had to somehow be carried off and buried by the birds and then escape their notice. Some of the seed caches may have remained unrecovered because the jays either forgot their location, or simply could not utilize (or did not need) all the caches they created. Some jays certainly perished before harvesting all their caches. These seeds could then have germinated and grown to maturity. It is possible that the only seeds that contributed individuals to the next generation were those that were stored by the jays in subterranean caches and then escaped detection. Thus, the trees paid a heavy price to have their large seeds dispersed out of reach of waiting seed predators. However, once the large seed strategy evolved, few alternatives were available. One such alternative would have been to produce a seed hull so thick that no animal could open it. This would have been an additional energy drain on the tree but it might have produced the needed results. In Mexico there is just such a pinyon pine, but at present little is known about its biology.

## MODERN INTERACTIONS BETWEEN PINYON JAYS AND PINYON PINES

Up to this point we have proposed how the evolution of the pinyon pine and the Pinyon Jay were intertwined with one another. However, we can only theorize, as we have no proof or documentation from the fossil record that things happened as we proposed. Others, including Ligon (1978), Smith and Balda (1980), Lanner (1981), and Vander Wall and Balda (1981) have made similar suggestions.

Recently, Tomback and Linhart (1990) proposed three different scenarios to explain the evolutionary sequence of events which ultimately resulted in the bird-dependence in this special group of pines. Our above scenario is very similar to one of their models, which can be termed the adaptationist model. They present three caveats, however, for this model which may be relevant: (1) cone and seed traits of the trees may not be the consequence of directional selection by only the birds, (2) phylogenetic constraints may limit the response of plants to the selection pressures of the dispersal agents, and (3) directional selection by the birds does not always result in continuous changes by the plant because of various genetic constraints and adaptive compromises. They rightly point out that seed dispersal by birds may result in small, isolated populations of these pines. Small populations are far more prone to inbreeding and random genetic drift. Directional selection, as outlined above, coupled with the consequences of small population size probably caused the process to proceed rapidly.

One way to give some credence to our suggested evolutionary pathway is to examine the present day interactions of the tree and the bird in an attempt to determine the mechanisms that led to the successful reproduction of both.

*Pinyon pines*

The pinyon pines of the southwestern US have many novel characteristics that they use in their race to attract jays to harvest their seeds. We refer to this suite of traits as "enticer traits".

An important enticer trait which we have discussed is the size of pinyon pine seeds. They are the largest seeds of any southwestern conifer (Little 1950) and their energy value is quite high (31kJ/g; Botkin and Shires 1948). High energy content results because pinyon seeds are comprised primarily of fat (61% fat, 16% protein, 16% carbohydrate (Botkin and Shires 1948)). Large, nutritious seeds probably evolved to maximize germination rate in arid conditions, but they also served to attract the jays.

The seed hull of the pinyon pine also has some interesting attributes. In conifers there is a clear relationship between the size of the seed and the thickness of the hull; larger seeds have thicker hulls. However, pinyon pine deviates from this pattern as it has a thinner hull than expected based on seed size (Vander Wall and Balda 1977). This makes it relatively easy for jays to open. Healthy, edible pinyon seeds have different colored seed coats than empty, aborted seeds. Edible seeds have a rich or deep, dark brown color whereas aborted seeds have a yellowish, mottled appearance. Thus, the seeds are effectively labeled for easy identification.

Pinyon Jays in the wild and in laboratory conditions discriminate between these two seed types and avoid the yellow ones. Some dark brown seeds, however, are also hollow, but these are picked up by birds and readily discarded. Sophisticated seed assessment techniques may allow jays to pull off this trick. After extracting a seed from a cone, jays point their heads slightly upward and rattle the seed in their bill. Ligon and Martin (1974) refer to this behavior as "bill clicking" and believe it is a means of assessing seed quality. Another seed

*A Pinyon Jay probes into a seed-filled pinyon pine cone to extract a dark brown, viable seed.*
*(J. Marzluff).*

assessment technique used by jays is to just hold the seed momentarily in the bill as if to weigh it. Inedible seeds that were clicked and weighed were quickly discarded. In fact these researchers fooled jays into accepting "empty" brown seeds by putting a small lead weight into the seed so it felt normal!

Pinyon pine seeds are wingless, a rare trait for a pine. The wingless seeds are easy to handle as wings must be removed before a seed can be opened. This structure is not totally lost in the pinyons, however, as it is embryonically retained in the form of a special ridge that acts to hold the seed on the cone scale in the opened cone. Wind dispersed pines are quick to lose their seeds before seed predators can locate the open cones. Pinyon pines use just the opposite tactic and hold seeds in a conspicuous cone for jays to find them. Seeds may be held in the cones well into the fall and usually start dropping to the ground in late October and early November after the first frosts. Thus, pinyon pine seeds ride on the wings of Pinyon Jays, rather than on their own wings.

Pinyon pine cones also show some interesting enticer traits. The cones of most pines are armed wih sharp, pointed spines that act as efficient deterrents to seed predators bent on robbing them. This armor is notably absent in pinyon pines, thus seeds can be plucked from the cone with relative ease.

The cone scales that provide the support structure for the seeds also act to shield the seeds from view in wind-dispersed pines, thus hiding them until they are released from the cone. In pinyon pines, these scales are short relative to seed length so seeds stick out farther than in wind dispersed pines and can be more easily seen by dispersal agents. These cone scales are also arranged in a unique pattern on the cone so that seeds are conspicuous to the birds. Wind-dispersed pines have their cone scales radiating out from the central axis of the cone at many different angles thus obstructing a clear view of a seed. Pinyon pine cones hold their scales out at a more uniform angle making sighting a seed much easier. Thus, the cone itself has a group of characteristics that collectively serve to make the job of locating edible seeds and extracting them fast and efficient.

The pattern of placement of the cones on pinyon pines may also make it easier for birds to harvest seeds. The cones of wind dispersed pines typically orient downward so the seeds can be swiftly and efficiently released from them, but this also conceals the seeds from view. In contrast, the cones of pinyon pines tend to be oriented outward and upward, thus holding the seeds longer and increasing their visibility and accessibility.

In the Southwest, the production of pinyon pine seeds is synchronized, irregular, and unpredictable. Large crops occur every four to seven years. The boom or bust pattern of cone production was evident when we examined the seed harvests of Native Americans in our study area (Fig. 4). During the period from 1972 to 1985 there were five years (1972, 1973, 1975, 1979, 1985) when few pine seeds were harvested by Native Americans. In contrast, in 1974, 1976, 1980, and 1983 bumper crops were harvested. In the other years, crops were intermediate in size and usually widely scattered over many hundreds of miles. That is, huge areas of the woodland bore large crops, but neighboring areas were barren. Intermediate years rarely resulted from a general low level of production across the entire woodland.

One has to be impressed with this temporal and spatial pattern of variability. In one year virtually every tree may bear hundreds or thousands of cones and the next year one can walk for miles without finding a single tree with cones. In years when cone crops are low and patchy one can simply follow converging flocks of Pinyon Jays to these "hot spots". In years when cone crops are low to intermediate the entire crop vanishes by December and not a single edible seed can be found in cones or on the ground. In years of abundance, seeds can easily be found in cones and scattered on the ground until the snow flies. However, even in years of the largest crop, all seeds have disappeared by mid-winter.

It is hardly conceivable that pinyon pine trees show all of these unique characteristics for reasons other than the selection pressures placed on them to build a large seed to withstand the harsh environment and then avoid complete annihilation of these prized seeds. This group of characteristics would probably never have appeared under any other set of conditions. Pines in Europe and Asia that are bird dispersed also show some or all of the traits discussed above.

*Conifer seed-eating corvids*

Now we are prepared to discuss the behavior of the birds that inhabit the pinyon-juniper woodland and capitalize on the cone crop provided to them by these remarkable trees. In northern Arizona and New Mexico four species of corvids harvest, transport, cache, and recover the seeds of the pinyon pine. Each is somewhat unique in the technique that it uses to gain its share of the seed crop. We begin our discussion with the Pinyon Jay and then compare its behavior to that of the other three corvids.

The response of Pinyon Jays to the presence of pinyon pine seeds is dramatic and instantaneous. When we presented seeds removed from cones to wild jays they immediately ceased all other behaviors and concentrated solely on eating and gathering these seeds. They quickly assessed the seeds using bill-clicking and bill-weighing. From our observation post we saw only dark brown seeds being gathered and after the birds departed we found only yellow-hulled, inedible seeds at the feeder. Flock members gathered at the concentration of seeds, often feeding shoulder to shoulder with few hostile interactions as they literally vacuumed up the seeds. After opening and feeding on seeds at the feeder, the jays began to pick up seeds and hold them in their throats. On average, jays are able to hold about 38 seeds as the esophagus usually expands to a volume of 17 ml. (We employed a sophisticated scientific technique to measure the volume of a jay's esophagus: we sealed off one end, attached the other end to a water faucet, and turned on the water until the esophagus expanded no further. The volume of water indicates the volume of the esophagus. Of course this procedure works only on dissected material!)

We followed these jays when they left the feeder to cache their bounty of seeds. They flew as a group up to 7 km and cached in concealed locations. When native cone crops were light the members of this flock visited our feeders and left with pine seeds up to seven times in a day.

In years of bumper seed crops in the pinyon-juniper woodland, Pinyon Jay flocks may become sedentary for up to an hour or more as members extract seeds from cones. In early autumn when the green cones are still tightly closed and laden with copious amounts of pitch, birds remove these cones from branches by either twisting the cone until the pedicel breaks, or hammering at the pedicel with strong blows of the bill. Once a cone is detached it is carried in the bill to a nearby crotch where it is tightly wedged. Now the bird strikes the cone with strong, forceful jabs of the bill accurately directed at the junction between the cone scales. With repeated jabs of the bill the cone scales are ripped apart or shredded and the exposed seed extracted, if it is dark brown in color. A bird may spend up to 15 minutes extracting seeds from a single cone. Often the same "anvil" crotch is used repeatedly and a pile of cones accumulates on the ground beneath it. The sharp, long, tapered bill of Pinyon Jays serves them well to chisel through the tough green cones, and allows them to probe deeply within the cone to remove seeds without the facial feathers contacting the sticky pitch that covers the cone (Fig. 8). As we mentioned in the second chapter, Pinyon Jays also lack the typical corvid feathers that cover the nostrils, which may also minimize contact between feathers and pitch.

Later in the season when the cones have opened and have lost most of their pitch, the jays hop through the branches visually scanning each cone in search of edible seeds. When one is located, the bill is used to extract the seed from the attached cone. Now, the flocks move rapidly through the woodland and birds quickly inspect open cones. Pinyon Jays are just as selective about the quality of the seeds they remove from cones as they are at feeding stations. Yellow-

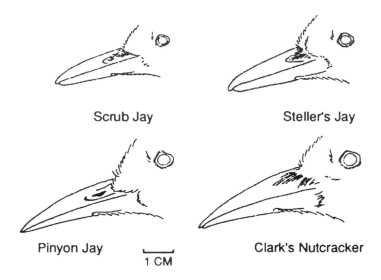

Scrub Jay                                    Steller's Jay

Pinyon Jay        |———|        Clark's Nutcracker
                   1 CM

Figure 8    *Bill profiles of four conifer-seed caching corvids. The least specialized (Scrub and Steller's Jays) have short, blunt bills in comparison to the long, chisel-shaped bills of the specialists (Pinyon Jays and Clark's Nutcrackers). Note that Pinyon Jay's bills are functionally lengthened because they lack feathers covering their nasal openings. From Vander Wall and Balda (1981).*

hulled seeds are left in the cones. We collected 655 seeds from the expandable esophagi of Pinyon Jays and all were edible!

After some period of harvesting seeds, loud *Kaws* are given by a few birds and other flock members ascend to the upper reaches of the trees. More *Kaws* are given as the flock assembles and then departs as a group. Pine seeds may be transported up to 20 km before they are cached. These long flights take place above tree level where birds may roll and swirl as they fly out of sight. Caching areas, about 4 ha in size, are discrete and located on a flock's home range (see Chapter 7, Fig. 45). Each flock may have seven or more such caching areas that are visited regularly during the fall caching season. In years of a bumper crop when birds do not have to travel far to encounter cones, a single flock of birds can make up to six round trip visits in a day. For a single bird carrying 50 seeds per trip, that amounts to 300 seeds cached per bird per day. In a 90-day caching season a single bird could cache over 25,000 seeds. J. David Ligon (1978) estimates that a single flock of birds in central New Mexico caches up to 4.5 million seeds in a bumper year!

In years when cone crops are patchy or generally low, far fewer seeds are cached and most of the crop is consumed by jays and other seed-eaters. Flocks spend more time moving through the woodland in search of the rare tree that has cones. We have seen as many as 15 jays in a single tree that happened to produce cones in a bust year. Chances of leaving a single progeny from such a cone crop are indeed slim. When a "hot spot" is located, numerous flocks converge in the area and it is soon depleted of seeds.

Caching areas are often sparsely vegetated, with patches of bare, exposed soil, rocks, and cinders indicative of good drainage. These sites often are snow free in winter because of the direct sunlight and wind. Birds may also prefer to cache close to the trunks of trees where less snow accumulates and snow melt is rapid once the sun reappears.

What happens to all these buried seeds? The hundreds of hours spent by the birds each fall caching seeds would be wasted if there was no return on their effort. The fact is, the flock returns periodically to their caching areas to search for their hidden treasures of fat, carbohydrate, and protein. After surveying the area from high in surrounding trees the birds descend to the ground and move as a blue wave in a uniform direction. Birds walk slowly, intently peering around and under objects and in the litter. Occasionally a bird will make a deep probe with its bill and extract a pine seed from the soil. We have often attempted this on our hands and knees and never have we successfully located a single seed. The seeds are buried in the top soil just below the litter layer out of sight. A bird with a seed proceeds to a suitable perch where the seed can be placed between the toes and cracked open with jabs of the bill. Often birds carry recovered seeds in the esophagus where they are hidden from view, as a bird carrying a seed visibly in its bill often must contend with other birds attempting to take it away.

How do birds locate their hidden seed caches? One could argue that random or trial and error search are the only techniques available to the birds. Random techniques would be very inefficient and inaccurate, resulting in thousands of unsuccessful probes relative to successful ones. Such is simply not the case. We have seen the birds make too many successful probes. In order to study this phenomenon in greater detail and with higher resolution, we brought some Pinyon Jays into the laboratory where we conducted cache-recovery sessions under rigid experimental conditions (Balda and Kamil 1989).

We constructed a raised wooden floor with 180 holes in a 3.4 × 3.4 m room. Each hole held a sand-filled cup or a wooden plug. The room contained many landmarks and a centrally located feeder on which pine seeds could be placed. The feeder had a false bottom that was controlled electrically so that seeds remaining after the prescribed number of caches had been made could be moved out of the bird's reach. Birds were given ample time to become accustomed to the room. At first Pinyon Jays isolated in the room appeared nervous and over-active. This problem was resolved by placing a caged Pinyon Jay outside the door of the experimental room. The vocalizations of the caged bird had a calming effect on the experimental bird. Birds eventually began caching seeds in the sand-filled cups.

Each bird was allowed to cache and recover seeds under two diferent sets of conditions. In one situation 90 holes (half the total number) were opened for caching. In the other, only 15 randomly selected holes were opened. Birds were allowed to make only eight caches under both sets of conditions. After each caching session the bird was removed from the room and held in its home cage for eight to ten days. Then the bird was returned to the experimental room to search for its caches. While the bird was out of the room all signs of digging were removed so no clues existed in the room to aid the bird in its search for seeds.

Before proceeding, we should explain the rationale for this experiment. A number of alternative explanations are available to explain how birds find their caches. For example, birds may have an inherent preference for certain sites or locations, always caching food in such sites when presented with them. Or, birds may have a rule that they use for distributing their cache locations. Under both of the above conditions birds simply have to know their preferences or remember their rule to accurately recover caches they previously made. In the 90-hole situation birds could put a rule or a set of preferences into play when making their caches, because so many potential cache locations were present in the room. Then, when given the opportunity to recover the caches their accuracy should be very high – that is, they should not make many probes in holes where they did not cache seeds. On the other hand, just the opposite should occur in the 15-hole condition as now the holes are few in number and widely spaced in the room. The chances are high that preference or rule holes are not present (or few in number) in this sample of open holes. Now the bird must use some means other than rules or preferences to locate caches, and we predict that recovery accuracy will therefore be much lower than in the 90-hole condition.

The results of this rather simple experiment are interesting and exciting. First, the Pinyon Jays performed well above chance levels under both sets of conditions. We conclude from these data that the birds were using some type of spatial memory to locate their caches. However, in the 90-hole situation their performance was spectacular. Their accuracy was close to 90%, a figure we have seldom seen in our many studies of this type. A closer look at the actual placement of caches revealed that Pinyon Jays placed their caches almost as close together as was possible in the 90-hole situation. Thus, they spatially arranged their caches in clumps. When searching for their caches they simply went directly to the place where they had created a clump and searched all the holes at that site. Because their caches were so close together there were few holes available to probe that did not contain a seed. Further study by Nancy Stotz (1991) confirms that Pinyon Jays are more accurate at recovering seeds placed close together than seeds that are widely dispersed.

In nature Pinyon Jays cache while in a flock with all birds slowly walking in the same direction. They may stop periodically to create a clump of caches or they may string them out in a line. In both situations the bird would be aided in its search if it could remember the starting point of its caching-walk. At present we know very little about the spatial distribution of caches in the wild, but work has started on this question.

Up to now we have concentrated on the caching behavior of the Pinyon Jay alone. Now we turn our attention to the other three seed-caching corvids in northern Arizona. We begin with the most specialized, the Clark's Nutcracker, and work through the less specialized Steller's and Scrub Jays.

Clark's Nutcrackers inhabit the high alpine areas of the western United States. They specialize on the harvest of high alpine species of pines, but if pinyon pines grow near their haunts they are quick to take advantage of this rich food source. Of all the species, this bird shows the greatest morphological specializations for this behavior. It has a long, stout, sharp bill (Fig. 8) and very powerful jaw muscles allowing it not only to hammer open green pinyon cones but to crush seeds open with its mandibles. This latter behavior led to the name "nutcracker". Nutcrackers, like Pinyon Jays, extract only dark brown seeds from cones. These seeds are held in a unique morphological structure known as the sublingual pouch (Bock *et al.* 1973) which is located on the floor of the mouth and is only evident when it is packed with seeds. Nutcrackers pack this pouch with up to 90 medium sized pinyon pine seeds; quite a mouthful!

Nutcrackers are strong fliers and have been observed carrying seeds up to 22 km from the site of harvest. Because these birds live and cache above the normal modern day elevational limit of the pinyon pine they may be responsible for much of the extremely long distance dispersal of pinyon pine seeds. In the San Francisco Peaks of Arizona there is a stand of 100-year-old pinyon pines at 1400 m, well above their usual elevation. Nutcrackers still use this site to cache pinyon pine seeds each fall, and we believe they are responsible for planting this stand of trees.

Field and laboratory studies reveal that Clark's Nutcrackers have exceptional spatial memory abilities that are about equal to those of the Pinyon Jay. Birds have been observed recovering cached seeds through a thick layer of snow. In the laboratory experiment discussed above, Nutcrackers did not clump their caches but spaced them out over the entire room. Thus, spatial memory may be more highly developed in this species and rules may be used less.

Like Pinyon Jays, Clark's Nutcrackers are among the earliest nesting bird species in North America. In years when an ample supply of pine seeds is harvested in the fall, Nutcrackers can be found sitting on eggs in late February when the snow level is deep and snow storms common. Both male and female Nutcrackers possess brood patches and incubate. This is possibly because they must often fly long distances to obtain their cached seeds; under these conditions it would be uneconomical for the male to provide all the food to the female and brood of young. When both birds share in this duty, one of them is on the nest incubating or brooding nearly continuously until the young are able to regulate their own body temperatures.

The Steller's Jay is a common species of the ponderosa pine forest. This jay spends much time harvesting and caching acorns but is relatively ill-adapted for harvesting pinyon seeds. Birds that live on the edge of the pinyon-juniper woodland, however, descend to harvest pinyon pine seeds. Steller's Jays' bills are not sharp or pointed and consequently they cannot open closed cones (Fig. 8). Thus, the only seeds they can capture are those available after cones open. In some years the entire crop has vanished by this time. Steller's Jays also have an expandable esophagus. However, it is not expandable to the degree as that of the Pinyon Jay. Steller's Jays only carry from nine to 18 pinyon pine seeds per trip and they fly only about 3 km per trip. Steller's Jays also are not as discriminating as Pinyon Jays as we have seen them carry off yellow-hulled seeds from our feeding stations. This species does not nest early and apparently relies on other foods for its reproductive energy. They may be partly responsible for the movement of pinyon pines upward into the ponderosa pine forest.

Another generalist that consumes and caches pinyon seeds is the Scrub Jay.

This jay is a common and relatively sedentary inhabitant of the pinyon-juniper woodland. The Scrub Jays' bills are incapable of opening green cones (Fig. 8) so they harvest seeds only from open cones. They cache seeds but they do not appear as industrious as the other three species. Instead of harvesting and caching seeds full time, we have watched them simply "loaf" in the midst of a bumper crop of pine cones. Our laboratory experiments indicate that Scrub Jays use spatial memory to locate caches, but that their accuracy is well below that of Pinyon Jays and Nutcrackers.

Scrub Jays may not be highly specialized to harvest pinyon pine seeds, but they are ingenious. On numerous occasions we have observed them sitting quietly near a Clark's Nutcracker as it opened a green cone. The instant a seed was extracted the Scrub Jay flew aggressively at the Nutcracker, startling it and causing the cone to drop to the ground. The Scrub Jay then flew to the partially opened cone and extracted a few seeds. Scrub Jays may be responsible for the bulk of the regeneration of the woodland it inhabits.

The four species discussed above all seem to play some role in the active dispersal of pinyon pines. In evolutionary time, the Pinyon Jay (or its ancestors) and the pinyon pine (or its ancestors) may have initially established this close relationship, but at present other species have learned to take advantage of the large seeds. Most likely, the Clark's Nutcracker originally adopted its seed caching habits with some alpine species of pine and later extended this behavior to pinyon pines where its geographic range overlaps with that of the pinyon pine.

CHAPTER 4

# Vocal communication

Pinyon Jays are rarely quiet. They give soft *Near-er* or *Kaw* calls as they fly,
forage, or approach their nest. Occasionally they erupt into a harsh *Krawk-
Krawk-Krawk* as they encounter a predator. Pairs often sit side by side quietly
cooing as they allopreen each other. Suddenly the female may interrupt their
intimacy with a descending *Brrrt* as the male sidles too close. These are not
random bursts of noise. Instead, they are examples of a sophisticated communication network that Pinyon Jays use to convey information about their inter-

nal and external states. As our opening quote suggests, the intricate nature of this communication network impressed naturalists over 100 years ago.

The vocabulary of Pinyon Jays has likely been shaped and expanded by natural selection. Jays that are able to vocalize accurate, timely, and specific instructions to nestlings, mates, and flock members should live longer and fledge more young than jays possessing lesser vocabularies. Take the following simple example. A male jay returning to the nesting colony to feed his mate would be at an advantage if he produced individualized calls to inform his mate that he is approaching with food. The female can then rise off the nest and prepare to be fed. Recognition of these calls would allow the female to respond only to her mate and to ignore the other returning males who will not feed her. Jays that tune in only to their mates reduce unnecessary interruptions in brooding and incubation. Imagine now a male that is not only able to say "here I come", but also able to say "the coast is clear" or "a raven is circling above the nest, don't move a feather". Clearly, vocabulary enrichment can increase fitness by reducing costs of incubation and lowering the chances of exposed eggs freezing or being eaten by predators.

Natural selection has equipped adult Pinyon Jays with a repertoire of at least 15 distinct calls. This number is fairly modest in respect to other corvids. Adult Rooks regularly use 20 calls (Røskaft and Espmark 1982) and Common Ravens, the champion corvid chatterer regularly use 30 or more calls (Gwinner 1964, Bruggers 1988). At least two factors may explain the verbosity of corvids. First, all corvids studied to date form lifelong, monogamous pair bonds. A large variety of calls may help maintain these pair bonds and coordinate the partners' actions as they perform all the behaviors associated with successfully rearing young. Second, all corvids spend at least part of their lifetimes in social gatherings (even typically territorial species like Steller's Jays and Common Ravens spend a portion of their pre-breeding lives in flocks (Brown 1963, Heinrich 1988)). Sociality favors a variety of calls in order for individuals to keep in contact with their flock members, identify themselves, demonstrate their status and motivation during encounters, and rapidly receive information about their surroundings. In other words, a corvid without a diverse vocabulary is like a gagged human who is expected to maintain a marriage, raise children, master the family dog, and interact with friends and colleagues by only mumbling.

Corvids are able to do more than just communicate with a large vocabulary. They pack information about urgency into many calls by varying intensity and pitch just as we yell louder and in a higher tone as we get more excited. Additionally, the individuality of corvids' voices adds a signature to calls so that listeners know not only what was said but who said it. This may enable Pinyon Jays and other corvids to temper their responses depending upon who is calling. Perhaps they respond quickly to their mate or a relative, but dally in response to a juvenile who often "cries wolf".

This chapter is like an introductory course in the language of Pinyon Jays. We will first provide a dictionary of the major calls used by jays. Second, we will discuss how several calls may enable jays to recognize their friends, relatives, and mates. Lastly, we will use the Pinyon Jay's alarm call to illustrate the wide variety of information that can be packed into a single call.

A Pinyon Jay dictionary

Lyle Berger and J. David Ligon kept six pairs of adult Pinyon Jays captive in order to study the variety of calls they uttered and the situations evoking these vocalizations (Berger and Ligon 1977). The 15 calls they recorded and named are described below. Behavioral ecologists avoid naming calls with terms indicative of their perceived function (e.g., Distress Cry or Mating Call) because often a call serves many functions or our perception of its function may be preconceived. Berger and Ligon have appropriately avoided such names, instead using onomatopeic ones.

*Rack or Krawk* (Fig. 9A)

If you are lucky enough to watch a foraging flock of Pinyon Jays you will hear the nasal-sounding Rack or Krawk. This short note is given singly or may be combined to form a series of variable length. Single and double racks probably serve as contact calls between foraging birds and between members of a pair as they construct their nest. Racks, therefore, are a form of vocal cement that keeps individuals in the flock together and coordinated as they move in and out of visual contact with their mates and other flock members. It is likely that the Rack is a non-aggressive signal that allows individuals to approach each other passively. When cages containing yearling jays are placed together the birds give racks immediately. The din is loud, noisy and continuous. Racks may therefore also be greeting calls.

*Near or Near-er* (Fig. 9B)

Pinyon Jays do not wear name tags but several calls, most notably the soft, nasal call that sounds like "near" or "near-er", function like identification badges because each individual has a unique rendition of the call. The Near is most commonly used between mates while they are courting, nest-building, and raising young. It is this call that mates use to identify each other in the crowded colony and that nestlings and fledglings use to identify their parents at feeding time (see Table 11 and Chapter 8).

*Kaw* (Fig. 9C)

Kaws are louder, longer and more quavering in pitch than Nears. The Kaw is the typical flight call that most people find characteristic of Pinyon Jays. It must also encode some degree of jubilation as Berger and Ligon heard this call given by both members of the pair after copulation!

*Rick* (Fig. 9D)

Pair members often use soft, short Ricks when in close contact during nest construction, preening, Swagger Walking, Sidling, or "Silent" Sitting. Ricks convey a non-aggressive message quite possibly one indicating approval with a situation thereby fostering courtship and pair-bonding.

### Buzz, Burrt, or Brrack (Fig. 9E)

Pinyon Jays in close contact often give a complex call composed of many very short notes given in rapid succession, called a Buzz. The Buzz appears to function just like the near except that its soft quality would restrict its use to

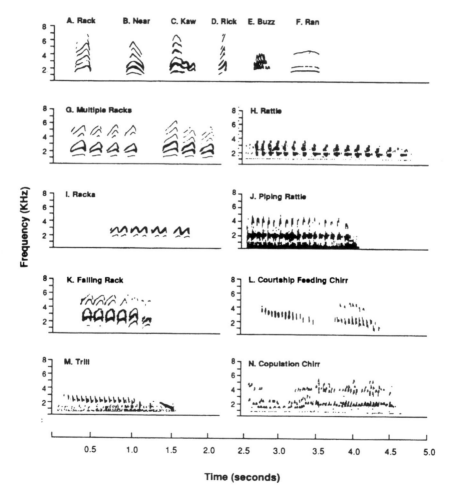

Figure 9    *Sonograms of the major Pinyon Jay vocalizations. A) Rack, B) Near, C) Kaw, D) Rick, E) Buzz, F) Ran, G) Multiple Rack, H) Rattle, I) Racka, J) Piping Rattle, K) Falling Rack, L) Courtship Feeding Chirr, M) Trill, N) Copulation Chirr. From Berger and Ligon (1977).*

times of close association. Accordingly it is very common between "Silent" Sitting, Sidling, and foraging mates. It also is used by males in the precopulatory display. Buzzing jays are apparently somewhat tense as they often flick and spread their tails which indicates an aroused or aggressive state. Buzzes are quite variable. They may inflect upward to sound like Burrt or end in a Rack to sound like Brrack. These variations may not encode different messages but may provide clues to the motivational state of the caller.

### *Ran* (Fig. 9F)

Frustrated male Pinyon Jays give a soft, vibrato, nasal call that sounds like "ran". This is especially common when males Sidle with their mates and their mates ignore or Rattle at them. Isolated males may also give Rans.

### *Multiple Rack* (Fig. 9G)

A Pinyon Jay that encounters a fox, cat, coyote, hawk, owl, or human invariably gives a series of Racks called a Multiple Rack. Each series is separated from the next by a short pause. Jays giving Multiple Racks are usually perched in the exposed upper branches of a tree and are often quite close to and/or approaching the source of concern. This call may send flock members flying for cover, freeze nestlings or parents near the nest, or quickly assemble a mob around the perceived threat. Great-horned Owls are especially likely to

*Adult Pinyon Jay in ponderosa pine tree scolding the photographer as he gets near its nest.*
*(J. Marzluff).*

assemble a mob of scolding jays. We have seen jays mob owls continuously for several hours before returning to their daily routine. One time jays foraged and napped for an hour after such a mob scene and suddenly returned to the spot where they left the owl to dish out a further verbal bashing.

The use of Multiple Racks as alarm or mobbing calls means that occasionally they are the last call uttered by a bold (or careless) jay. Scolding predators is risky and we observed a Northern Goshawk snatch and eat one jay that edged too close and uttered one too many Multiple Racks!

Why should a jay that has spotted a predator give a conspicuous, easy to locate call like the Multiple Rack? Apparently the major intent of a scolding jay is to draw attention to the predator which may then be evicted from the flocks's nesting colony, foraging area, or roost site as many jays cease all other activities and approach the potential predator. Thus, Multiple Racks serve as a rally cry. Additionally, Multiple Racks alert flock members to a potential threat and they inform the predator that their element of surprise, often the key to making a kill, has been negated. It seems obvious that these functions must, on balance, enhance the fitness of the jays giving and receiving alarm calls or else the clearly detrimental aspects of these calls would erase them from the language. Alarm calls would be especially beneficial during the nesting season when predators can wipe out a pair's reproductive efforts. As expected,

*A Northern Goshawk prepares to eat a freshly captured Pinyon Jay. (C.B.).*

Multiple Racks are given most frequently as daylength increases signaling the onset of the breeding season (Cully and Ligon 1986).

Jays giving alarm calls may also be purposefully drawing attention to themselves. The jay that was killed in the act of scolding a goshawk was one of the most dominant birds in the flock. Apparently dominants must continually show their power by taking the lead in times of danger. Perhaps this dominant male was showing his fearlessness to his flock mates and potential rivals by going where none of them dared to go. Dominant male Black-billed Magpies also appear to show off by edging close to predators (Maholt and Trost 1989).

*Rattle, Trill, and Piping Rattle* (Fig. 9H, J, M)

All female corvids studied to date have Trills and Rattles or Knocking calls that males do not possess. Female Pinyon Jays have at least three distinct calls not used by males. The Trill is a series of short notes ending with a long descending slur. Trills appear to be individually distinct and may provide another way for mates to recognize each other. Trills are especially common during the spring and summer and nearly exclusively given by females to their mates. Agitated females Rattle. The Rattle is similar to the trill except that its harmonic and unharmonic structure give it a very harsh tonal quality. Females Rattle in response to approaching predators, when other jays give alarm calls, or when their mates sidle too near. Rattles are apparently the female Pinyon Jay's way to say "back off" as a male sidling close to a female usually stops in his tracks when she rattles. Sometimes females use another form of a Rattle with higher-pitched piping notes known as the Piping Rattle. This Rattle also carries an aggressive message and possibly an escape message as well. This series from Trill to Rattle to Piping Rattle is a good example of a graded response. All calls are structurally similar but they replace each other as females swing from complacent to aggressive to fearful moods.

*Racka* (Fig. 9I)

Most birds have at least two alarm calls. One, like the Multiple Rack, draws attention to a predator; the other is a high, thin whistle that is difficult to locate and simply means "take cover immediately" (Marler 1955). Pinyon Jays lack a "take cover" alarm call but use the Racka to fortify the "assemble to mob" message of the Multiple Rack. Rackas usually precede Multiple Racks and quickly assemble a mob.

*Falling Rack* (Fig. 9K)

Lone jays are naked, having been stripped of their highly adaptive clothing of flock members. Given the importance of being in a flock, it is not surprising that jays isolated from their flock members give a unique series of descending Racks termed a Falling Rack. We have heard males that recently lost their females give this call endlessly from high in a tree. The Falling Rack appears to be a Pinyon Jay's vocal version of a signal flare.

### Courtship Feeding Chirr (Fig. 9L)

Adult female corvids receiving food from their mates vibrate their wings and tails, and utter prolonged chirring calls similar to juvenile begging calls. Females Chirr when their mates feed them prior to nesting (courtship feeding) or while she is on the nest incubating or brooding. Females may also Chirr when their mates present sticks during nest construction and occasionally as they chase their mates to solicit food. Chirrs are loud and easily located. We often wait near a suspected nesting area, listen for Chirrs and then dash in their direction in the hopes of finding an elusive nest.

### Copulation Chirr (Fig. 9N)

During copulation and while presenting themselves in precopulatory position females give an extended version of the Courtship Feeding Chirr known as the Copulation Chirr. The structural similarity between these two calls led Berger and Ligon to suggest that the Chirring during courtship feeding may be a form of vocal cement that strengthens pair-bonds.

### Kack

Jays in extreme distress give multiples of quiet, sharp calls with harsh tonal qualities known as Kacks. Berger and Ligon heard this call only while chasing jays at close quarters. It is so quiet that it may rarely provide a signal to other jays and may just be a call of exasperation.

### Sub-song

Corvids rarely sing, but many young corvids and isolated Pinyon Jays often sit quietly and pour forth a mixture of sounds that may last for many minutes. We once recorded 20 minutes of continuous Sub-song given by a single bird. Many sounds in the Sub-song are recognized as typical calls (e.g. Nears, Trills, Kaws, Buzzes, Ricks and/or Rans are usually included), but many others are conglomerates of call notes and novel sounds (described by Berger and Ligon as "toots, whistles, and hollow-sounding notes reminiscent of a woodwind instrument, plus clicks and grating sounds"). No two Sub-song sessions are alike and within a session the song appears to be composed spontaneously. It is hard to understand what message singing jays are broadcasting and for whom the song is intended. Like a person in the shower, they may sing because they lack any external stimuli, to try out new noises in private, to practise vocalizations they will use later in life, or simply to hear their own voices.

## *Juvenile Begging Chirr*

Nestlings and fledglings still dependent upon their parents for most of their food use one primary vocalization, the Begging Chirr. The chirr develops from a combination of quiet squeaks into a loud stereotyped call as young jays enter their second week of life. Pat McArthur (1979) studied the development of this call and determined that parents use the individualistic properties of this call to identify their young just before they fledge from the nest. This call is given before and during food transfer as the juveniles crouch and shake their wings. This display functions to appease the dominant parents and secure food. As in most corvids, young jays stop begging after they are weaned from parental care, but continue to use the superficially similar Courtship Feeding Chirr later in life when they receive food from their mates.

## *Bill Clack*

Many corvids, most notably Common Ravens (Bruggers 1988), make a clacking noise by snapping their bills shut. Bill Clacking may be another non-aggressive signal used between mated Pinyon Jays, but it has been observed too infrequently to be interpreted.

### RECOGNITION OF FLOCK MEMBERS BY VOICE

When we answer the phone we quickly gather a wealth of information about the person calling us. Just by listening, we are often able to identify the caller's sex, country and perhaps region of origin, and determine their individual identity. Pinyon Jays are also able to identify individuals by voice alone. We have been able to determine this by "telephoning" jays and monitoring their responses to our calls. Of course, we didn't use a telephone, but we used an analogous process known as a "playback" experiment. During an experiment we play recordings of two different callers to a captive or free-ranging jay. If the jay calls or moves more after hearing one recording compared to the other recording this suggests that the listener recognized a difference in the calls. We have contrasted jays' responses to calls of their mates versus neighbors in the nesting colony, their responses to parents and offspring versus unrelated flock members, and their responses to randomly selected flock members versus members of other flocks. In all cases, the responses of jays suggest they are able to discriminate between individual voices. Here we illustrate this recognition ability by discussing a set of playback experiments aimed at determining whether or not flock members are recognized. In later chapters we will show that individuals are capable of the finer discrimination necessary for mate (Chapter 8) and kin (Chapter 9) recognition.

Social life and recognition of individuals within a society may be linked by positive feedback. In an escalating spiral, increasing sociality favors greater development of recognition which may foster greater development of sociality which favors still greater recognition and so on. Let's investigate the reasons that sociality and recognition may be caught up in a spiral of positive feedback. Daily interactions between members of a society could be more efficient if group members could be recognized and discriminated. Discriminating individuals could aid and learn from their relatives, reciprocate altruistic acts, reduce their aggression when in familiar company, and guard their mates more effectively than non-discriminating individuals. This discrimination ability would foster sociality because forces favoring group life such as kin selection, nepotism, reciprocal altruism, and parent-offspring manipulation (parents coercing their offspring to co-operate) are more efficient if group members recognize each other. Sociality does not depend on an ability to recognize potential or past co-operators (Brown 1987), but recognition may allow sociality to evolve more rapidly and become more complex, especially when many related and unrelated individuals live together. This positive relationship between sociality and recognition may explain why recognition among social organisms is common, if not universal (for many examples, see Colgan 1983).

### Playback experiments

From 1984 to 1986 we captured 44 Pinyon Jays (26 adults and 18 yearlings) and held them from one to three months in aviaries and cages. These captive "test birds" were housed in groups of two to six that included others from their flock and from different flocks. Groups could hear, but not see each other. These housing conditions are important because they may have influenced our results. For example, because our jays could hear flock members and birds from other flocks their ability to distinguish between flocks might have been affected. A logical prediction is that being housed in a captive "flock" which included jays from several natural flocks would blur a jay's distinction between flock and non-flock members thus reducing its ability to respond strongly to members of its own flock during our tests. If this were the case, our claims of recognition are conservative. Alternatively, exposure to several flocks during captivity may have allowed jays to learn any differences between flocks which would increase their ability to disciminate between calls during our tests. If this occurred, our claims that jays recognize flock members would be weakened, but they still should not be dismissed. Discrimination would still indicate that jays know their flock members' calls and choose to respond more strongly to them than to the calls of non-flock members. Jays encounter other flocks in nature and comparisons during times of flock mixing may be the way they learn to recognize members of their own flock. We minimized these problems by testing jays only with calls of birds from groups with whom they were not housed.

On the day of an experiment we placed a test bird in the central part of an experimental chamber and played it calls from two birds through speakers, located in two, opposite peripheral chambers (Fig. 10). We conducted two types of playbacks. In "choice experiments", one of the tape recordings was of a bird from the test bird's flock and the other was of a bird from another flock. In "control experiments" both recordings were of birds from other flocks;

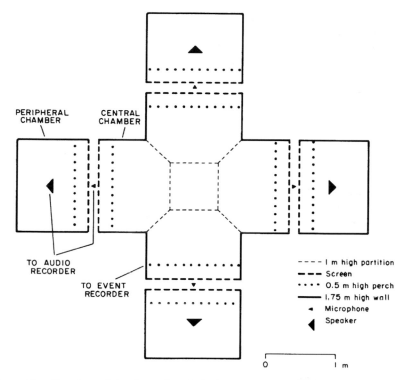

Figure 10   *Experimental chamber used in playback experiments. View here is from above as seen by the closed circuit television.*

neither recording was of a flock member. These experiments allow two ways to assess whether calls signal the flock identity of the caller. First, we can compare the test birds' responses to flock members versus non-flock members within a choice experiment. Second, we can compare the responses of test birds in choice experiments that contain flock members' calls to responses in control experiments that do not include flock members calling.

We broadcast four types of calls to our test birds. Three were single call types given by single birds in the laboratory or in the field. These included Kaws recorded from ten jays, Nears recorded from 22 jays, and Multiple Racks recorded from 20 jays. The remaining playback used an amalgamation of call types recorded as wild flocks foraged and flew around their home ranges. We recorded these "general flock calls" from four flocks in 1984.

Previous work on Pinyon Jay vocalizations, which we summarized in the dictionary section of this chapter, allowed us to make several predictions. If flock members are recognized we expected jays to: 1) give more calm calls (Nears, Kaws, Ricks), give fewer alert calls (Multiple Racks, Rattles and Rackas), and move more while hearing general flock calls, Nears, and Kaws of their flock than while hearing these calls from members of another flock, and 2) give more alert calls, give fewer calm calls, and move less while hearing Multiple Racks of their flock members than while hearing Multiple Racks from members of other flocks. We tested these hypotheses by watching our test bird on closed circuit television from another room and counting the number and type of vocalizations it made and the number of times it moved (changed perches or changed orientation on a perch) during a playback experiment.

When we study animals we are always hindered by an inability to talk with our subjects. Playback experiments are as close as we can come to bridging this communication gap. Unfortunately we must use an inanimate interpreter, a tape recorder, to carry on our conversation. Therefore we never know exactly what we said and we must infer from the birds' reactions what type of message we sent. It was obvious we were saying something relevant to our test birds during playback because their behavior changed dramatically when the calls were played. We allowed birds to acclimatize for 15 minutes in the experimental chamber before beginning a playback. During this time birds perched silently in 49% of 204 trials. On average, acclimatizing birds moved four times, gave 2.2 Nears, 1.95 Kaws, 1.64 Multiple Racks and Near-Multiple Racks (Multiple Racks preceded by a Near, see next section), and 0.24 Rattles. Two acclimatizing jays sang soft Sub-songs. In contrast, when we "spoke" to our birds with the tape recorder they immediately became alert and moved and called five to ten times more frequently than they had during acclimatization. We always got a thrill when we could command a bird's attention by playing it calls.

Our playback experiments strongly suggest that jays recognize the calls of their flock mates and use this information to alter their response to a call. We provide data to support this claim in Figures 11 and 12. Not surprisingly, the more natural "general flock calls" elicited the greatest response from jays (Figs. 11A, 12A). Jays moved and called frequently when we played these recordings and they responded significantly more strongly to calls from their own flock than to calls from other flocks. They moved more and called more, especially using Nears and Kaws, when they heard their flock than when they heard another flock. Adult males, adult females, and yearling males gave significantly stronger responses to their own flock's calls than to another flocks' calls. Yearling females did not discriminate between flocks. Comparing responses during choice tests where one recording was of a jay's home flock to responses during control tests where neither recording was of the home flock allows us to sort general responses from responses specifically triggered by the sound of familiar birds. It appears that vocalizations of any flock stimulate a lone jay to increase its calling, especially calm contact calls like Nears and Kaws. However, calls from a jay's home flock are needed to stimulate a jay to move (notice that jays move more when they hear their home flock and move twice as much in choice versus control [Fig. 12A] tests). Isolated jays seem quick to make vocal contact with any passing jay, but they are especially quick to fly to investigate and join their home flock.

Jays are also able to discriminate between callers giving particular types of calls. Kaws (Figs. 11B, 12B) and Nears (Figs. 11C, 12C) elicited greater responses than Multiple Racks (Figs. 11D, 12D). Our captive jays responded to these calls just as we expected based upon their use in nature. As we sug-

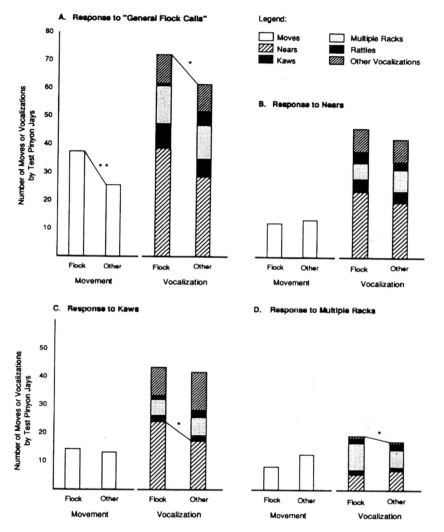

Figure 11 *Average responses of test birds to calls of flock members (flock) versus calls of birds from other flocks (other) during "choice" playback experiments. A) Comparison of responses to flocks foraging in the wild (N = 38 experiments). B) Comparisons of responses to single birds Nearing (N = 26 experiments). C) Comparison of responses to single birds Kawing (N = 28 experiments). D) Comparisons of responses to single birds giving Multiple Racks (N = 28 experiments). Responses that were significantly different are indicated by stars (\* = P < 0.05, \*\* = P < 0.01, \*\*\* = P < 0.001, MANOVA test on paired responses of individuals in each experiment).*

gested in the dictionary, Nears and Kaws are contact calls which are responded to by others Nearing and Kawing, but Multiple Racks are alarm calls which elicit mobbing or freeze jays silently in place. Jays in the laboratory Near, Kaw, and move when they hear Nears and Kaws, but they move less

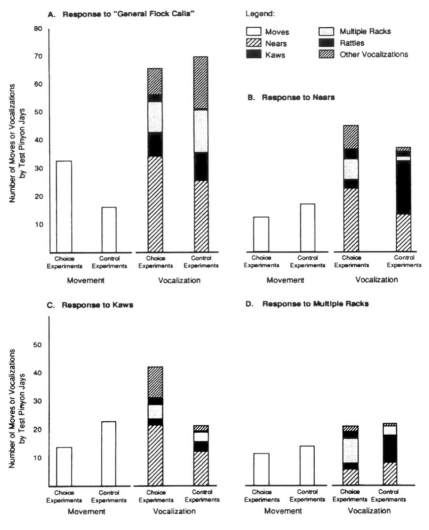

Figure 12    *Average responses of test birds in* choice *versus* control *playback experiments. Sample sizes for choice tests are given in Figure 11. Responses to flock and non-flock are averaged here. There were 16 control experiments using general flock calls, 12 using single birds Kawing, 15 using single birds Nearing, and 13 using single birds giving Multiple Racks.*

frequently and primarily give Multiple Racks when they hear Multiple Racks. All three of these calls encode the above messages *plus* the flock identity of the caller. Jays gave significantly more Nears when their flock members Kawed than when birds from other flocks Kawed. Kaws appear to contain the strongest "flock signature" of any single call we investigated; jays discriminated between Kaws in choice tests (Fig. 11C) and called twice as frequently in choice tests compared to control tests (Fig. 12C). Adult females and yearling males showed the greatest discriminatory response between Kaws of flock members and Kaws of others. Nears, although excellent indicators of individual identity (see Chapters 6 and 10), appear to have less flock specificity. Only adult test birds discriminated between flock and non-flock Nears with males Nearing more and females Kawing more to the Nears of flock members than to the Nears of birds from other flocks.

Our playbacks using Multiple Racks suggest that flock members come to each other's aid more than they come to the aid of a bird from another flock. Jays responded with Multiple Racks at twice the frequency when their flock member called than when a bird from another flock called. Jays also gave these alarm calls three times as frequently in choice tests (Fig. 11D) compared to control tests (Fig. 12D) suggesting that it is the sound of a flock member scolding that stimulates a jay to join in and give scolding vocalizations. Adult males, adult females, and yearling males all responded more strongly to the Multiple Racks of their flock members than to the Multiple Racks of other jays, but adult males showed the strongest and most consistent discriminations. Yearling females did not discriminate and rarely scolded along with the recordings. When the chips are down flock members, especially adult males, rally together!

### Discussion of flock recognition

Our experiments were designed to test the ability of jays to recognize the voices of randomly selected flock mates. Our results suggest that such discrimination is possible for a variety of contact and alarm calls and that Pinyon Jays adjust their response to the identity of the caller. Adult males, adult females, and juvenile males often discriminated between our calls. Juvenile females rarely did so. Why were young females so coy? It is possible that they had not learned the voices of their flock members by the time we captured them, which reduced their ability to discriminate. Juvenile females are the dispersing cohort (Chapter 11) and it is likely that many of the females we tested had just immigrated into the flock with which we caught them. In their young lives they had already experienced the vocalizations of more than one flock. This would reduce the time they had for learning any flock-specific peculiarities in voice. Males rarely disperse so they have spent several more months in the company of their flock members than females have. Not only may this give them an advantage in learning flock members' voices, but they have probably also established stronger social bonds with their natal flock members than have juvenile females. Juvenile males are still in the company of their parents and a few more distant relatives (see Fig. 34), but many females are in a sea of strangers.

What type of recognition system allows Pinyon Jays to discriminate flock members from other jays? Individual recognition is not needed for this discrimination; jays need only to place a call into a flock or non-flock category. Strong discrimination between Kaws suggests that these calls have flock specific features (perhaps tone, vibrato, or duration) that enable quick classification. They may also have individually distinct properties that allow finer discrimination, but this has not been investigated. Nears are individually distinct (Chapter 8) and this complexity appears to hinder flock classification, especially by juveniles. It seems that adults have learned their flock members' Nears and use individual recognition to sort flock members from other jays. Juveniles have not learned all individuals in their flock and rely more on calls like Kaws with properties indicative of the flock as a whole to distinguish flock members from other individuals.

Individuality in calls does not always preclude flock recognition by juveniles. Multiple Racks are individually distinct (see below) and easily sorted by flock (Figs. 11D, 12D). These calls appear to have a flock signature in addition to their individual signature, as we discuss in detail in the next section. Why should alarm calls have a flock signature, while contact calls like Nears do not? Natural selection has probably shaped these calls, favoring individuality in Nears because jays use these calls to keep in touch with their life-long mates and relatives. There is no need to have strong flock-specificity in these short-distance, intimate calls. Selection also favored individuality and to a lesser extent flock-specificity in Multiple Racks because these convey a message of danger over long and short distances. Long distance alarm calls need to be simple and unambiguous in structure so the message is clear and a rapid response is evoked. However, selection also favors strong responses to calls that indicate your fitness is at stake (calls from mates, kin, and neighbors in the nesting colony) and such a response necessitates individual recognition. Selection may also favor differential responses to flock members because they are potential reciprocators. It would be rare for birds living in different flocks to have opportunities for reciprocation, therefore flock recognition would be favored.

## WHAT ARE ALARMED PINYON JAYS SAYING?

### *Lying and deceit*

A few biologists have recently suggested that animals willingly withhold information from each other (Krebs and Dawkins 1978). That is to say, animals lie! As a Pinyon Jay that has come upon a hunting Cooper's Hawk, you could fly to safety in a nearby tree and unleash a barrage of Multiple Racks or keep tight-billed. A lying jay would keep quiet, or worse still, confuse the situation by giving an all clear signal like a Near or a Kaw. An honest jay would scold the predator so that all the flock members nearby could react and avoid the hawk's talons. It is easy to see how lying and deceit could evolve. If unrelated birds are killed or have their reproduction reduced by a liar's actions, then the liar's fitness will increase and the propensity to lie will spread through the population (because fitness is a measure of *relative* success among individuals in a population). The evolution of lying is more difficult to imagine in a permanent group like a Pinyon Jay flock because lying may endanger relatives and may eventually take a direct toll on the liar. "You will reap what you sow" is an appropriate human adage in this respect. In a permanent group, liars will be lied to and eventually must pay the consequences.

Do Pinyon Jays lie? We addressed this question with the help of an eager undergraduate student, Mike Munoz. Mike built a Pinyon Jay "lie detector" in order to test whether Pinyon Jay alarm calls are deceitful or, conversely, rich in information. The lie detector was a 2.5 m × 1.5 m × 2 m aviary that contained an elevated box at one end and three perches at 0.5 m intervals from the box. Mike placed a jay in the aviary, put a predator (a live cat, stuffed and mounted Cooper's Hawk, American Crow, Abert's squirrel, or a Pinyon Jay as a control) in the box, hid from the jay's view and opened the box. Liars and cheats should stay quiet, give misleading calls, or give calls devoid of much information. Honest jays should give Multiple Racks rich with information about the type and severity of the predator.

In order to assess a jay's response to the predator Mike counted the jay's movements in the aviary before and after he opened the box. He also determined if the jay approached or withdrew from the predator and recorded all calls on a tape recorder. In 1985 we repeated this experiment in the field by presenting the same mounted "predators" to nesting jays. We carried the predator in a paper bag to a spot 1.5 m from a nest with an incubating female. (We did not use the cat in these experiments because, as you all know, cats are easy to "let out of the bag", but less easy to get back into a bag!) We removed the predator, placed it in plain sight of the female and monitored the reaction of males returning to feed their mates (females did not leave their nest until their mates arrived). We were able to test the mobbing responses of 24 jays in the lab and 13 pairs in the field.

When Mike ran his first lie detector test it was obvious that Pinyon Jays were not blatant liars. Upon seeing a Cooper's Hawk or a cat the jays scolded vigorously, moved often, and retreated from the predator (Table 1). Reactions to Abert's Squirrels and Pinyon Jay models were subdued. Similarly, nesting jays vigorously scolded the Cooper's Hawk, recruiting an average of six others to mob this intruder. In the field and lab reactions to crows were strong but not as strong as reactions to hawks and cats. Differential response strength was correlated with the threat posed by the predator. Hawks and cats pose the most serious threat because they eat adults and their nest contents. Crows and squirrels pose less of a threat because they eat only nest contents. Pinyon Jays pose little threat to adults and nest contents.

It would be difficult for a jay to watch another jay scolding and determine the object of concern based upon the scolder's general behavior. There are clearly *relative* differences in scolding that are predator dependent, but there are no *absolute* rules evident in Table 1. Rather, scolding behavior is affected by the condition of the scolding jay. For example, captive jays were more subdued in their mobbing than wild jays (Table 1) and scolding is especially vigorous during the breeding season (Cully and Ligon 1986).

*Variability in alarm calls*

In order to determine if there were any reliable cues to predator identity

**Table 1.** *Responses of Pinyon Jays to predators placed near their nests in the field and with them in an aviary. Predators were mounted specimens, except cats which were alive and restrained just outside the aviary. A Pinyon Jay model served as a control since jays do not prey upon each others' nests. The column headed "change in position" refers to the movement of jays in the aviary before, compared to after, the stimulus was unveiled. Negative values indicate that jays withdrew from the stimulus; positive values indicate that jays approached the stimulus; and zero values indicate no change in a jay's position when a model was introduced. Larger values indicate stronger responses. Starred responses are significant (\* P <0.05, \*\* P <0.01, Wilcoxon Matched-pairs, signed-ranks test comparing average position before and after presentation of stimulus). Last rows indicate results of one-way ANOVAs testing the differences in mean responses across predators. N equals sample sizes for ANOVA tests.*

| | | Response to stimuli near nests | | | | Response to stimuli in aviary | | |
| --- | --- | --- | --- | --- | --- | --- | --- | --- |
| Stimulus presented | N | Number of multiple racks | Number of times jays dived at stimulus | Number of jays scolding stimulus | N | Number of times jays moved | Number of multiple racks | Change in position |
| Cooper's Hawk | 13 | 48.4 | 3.9 | 6.3 | 46 | 202.5 | 9.6 | −1.7* |
| Cat | 13 | | not used | | 48 | 261.7 | 4.7 | −0.8* |
| American Crow | 13 | 17.4 | 2.5 | 3.2 | 33 | 161.8 | 1.5 | −1.0** |
| Abert's Squirrel | 13 | 7.0 | 3.7 | 0.4 | 47 | 50.4 | 1.3 | −0.2 |
| Pinyon Jay | 13 | 3.2 | 0.1 | 0.1 | 47 | 44.0 | 1.0 | 0.0 |
| *Analysis of Variance* | | | | | | | | |
| F | | 7.3 | 2.1 | 9.2 | | 5.9 | 8.9 | |
| DF | | 3, 51 | 3, 51 | 3, 51 | | 4, 220 | 4, 220 | |
| P | | <0.001 | 0.12 | <0.001 | | <0.001 | <0.001 | |

contained within calls we looked in detail at the structure of Pinyon Jay Multiple Racks. We played our tape recordings into a "real time analyzer". This machine produces a voice print or "sonogram" of the call which is a plot of the sound's frequencies (pitch) against time (Fig. 9 is a collection of sonograms). Armed with a scaled picture of a call we could measure calls and compare them between individuals and between situations. If calls contain little information except a general message of alarm then they should look the same regardless of what predator elicited them or what jay gave them. Conversely, if they are rich in information, such as the identity of the predator, then all calls given to hawks should be similar, but different from all calls given to cats or squirrels.

There are two levels of variation in alarm calls that we measured. We term the first "macrovariation". This includes variants which are easily distinguished by humans such as differences in the number of Racks per Multiple Rack, the number of frequency bands present in each rack, and combination of other calls with Racks. We term the second "microvariation" because this variation is not easily perceived by the human ear. This includes subtle differences in Racks resulting from variance in call duration, shape, and tonal quality.

*Macrovariation*

Most Multiple Racks consist of only three to five Racks, but occasionally alarmed Pinyon Jays act like broken records and string out a stupendous series of Racks seemingly without pause to breathe (Fig. 13). Our record racker gave a Multiple Rack consisting of 70 Racks! Berger and Ligon noted this variation and speculated that it was an example of a gradation where each increase in the length of Multiple Racks translated into an increase in perceived danger.

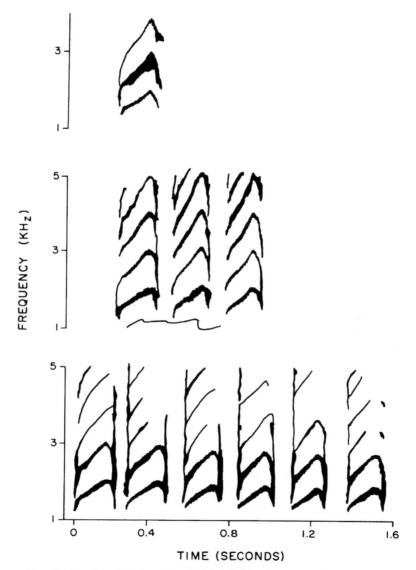

Figure 13   *Combination of Racks to form Multiple Racks of various lengths. The repeated bands of sound at different frequencies appear to be harmonics (multiples) of the fundamental (lowest) frequency.*

The variation in length of Multiple Racks we documented does not appear to function in this way. In our field experiments jays become more motivated and mobbed most intensely two to four minutes after they discovered a predator. (Fig. 14A). If Berger and Ligon are correct then we should expect to see a similar change in the length of Multiple Racks, but this was not the case (Fig. 14B).

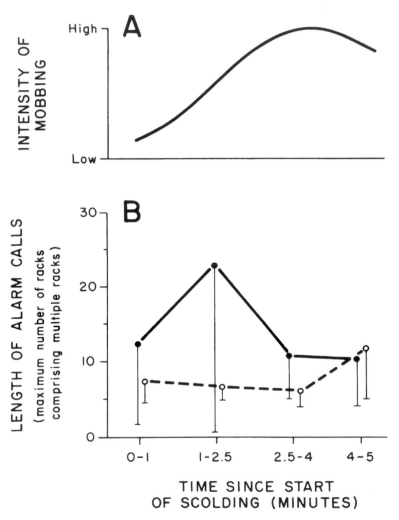

Figure 14   *Changing behavior of scolding jays during the course of field experiments. A) The intensity of mobbing (indexed by the number of birds mobbing, the tendency to dive at and approach predators, the rate of calling, and the volume of calls) increases through 75% of the experiment and then begins to wane. B) The length of Multiple Racks was constant through the course of an experiment. Average length minus 1 SD of Multiple Racks given to Cooper's Hawks (solid line) and American Crows (dashed line) are shown. Kruskall-Wallis tests indicated that Pinyon Jays did not give longer Multiple Racks as they became highly motivated (hawk: P = 0.67, crow: P = 0.54).*

Moreover, if relative danger is coded by the number of Racks then we expect laboratory birds to give especially long Multiple Racks to cats and hawks. In contrast, all predators and even control situations elicited Multiple Racks of similar length (Fig. 15).

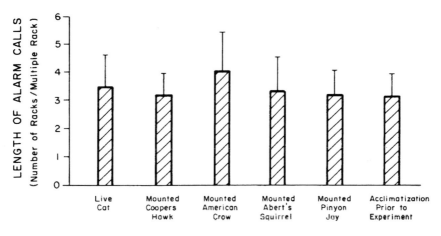

STIMULUS PRESENTED IN LABORATORY

Figure 15    *Length of alarm calls given to various stimuli in the laboratory. Predators and control situations (Pinyon Jay model and acclimatization before models were presented) elicited alarm calls of similar length (F = 0.92, DF = 9, 49, P = 0.52; one-way ANOVA with calls by same birds to same predators as repeated measures). Mean length of call plus 1 SD are shown. Means are averages for 19 experiments with cats, 14 with hawks, ten with crows, six with squirrels, five with jays, and 46 acclimatization periods.*

The number of frequency bands in each Rack may represent a graded signal indicative of the caller's motivation. (These bands include harmonics of the fundamental frequency as well as non-harmonic bands that may result from frequency modulation, Marler 1969). In the field we noticed that Racks were composed of two to eight distinct bands (five are visible in Fig. 13). In 11 experiments jays mobbed vigorously in the later minutes of exposure to the predator and in all cases the number of frequency bands per Rack was greater during these periods of intense mobbing than during the initial weaker mobs (median number of bands at start of experiment = 2, median at end = 4, Mann-Whitney U-test, P = 0.004). Our ability to record a variable number of bands from the same birds scolding in the same locations suggests that highly motivated jays give louder, harsher, and higher-pitched calls than less motivated jays.

The inclusion of other call notes into a scolding sequence may further indicate a jay's motivation. We commonly recorded Multiple Racks preceded by Nears or Rackas, interrupted by Ricks, or ending with feeding Chirrs (Fig. 16). Nears, Rackas, and chirrs were most often included with Multiple Racks given

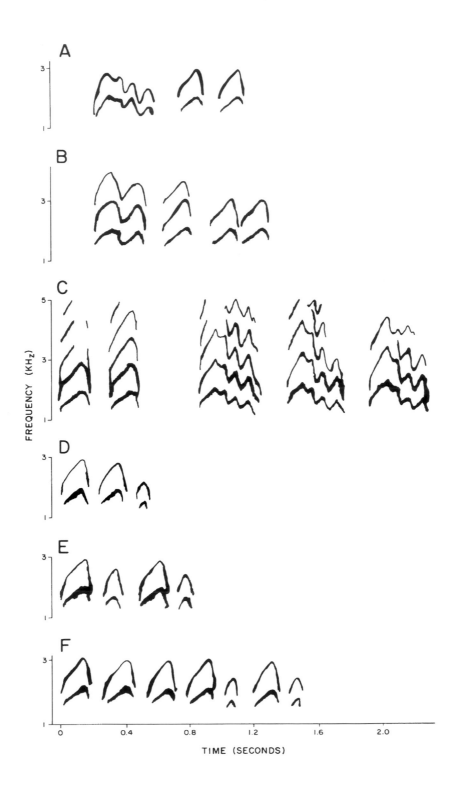

by very excited jays scolding the most dangerous predators (Table 2). Courtship Feeding Chirrs (or begging) were only performed by the most agitated females, typically when scolding a Cooper's Hawk. Balda and Bateman (1973) reported female jays begging from stuffed Great-horned Owls and females occasionally beg from us when we band and weigh their nestlings. Why should predators stimulate a female to perform a behavior usually reserved for appeasing their mates? It is possible that females are caught in a conflict over whether they should attack or flee from the predator. Behavioral conflicts like this often result in performance of seemingly inappropriate behaviors known as "displacement activities" (Lorenz 1981).

**Table 2.** *Combined calls given by captive and free-ranging Pinyon Jays. Experimental protocol and combination of calls are described in the text.*

| | Number of each call combination uttered | | |
|---|---|---|---|
| Stimulus presented | Near-Rack | Racka-Rack | Rack-Chirr |
| *Free-ranging birds* | | | |
| Cooper's Hawk | 8 | 12 | 10 |
| American Crow | 4 | 1 | 1 |
| Abert's Squirrel | 0 | 0 | 1 |
| Pinyon Jay | 0 | 0 | 0 |
| *Captive birds* | | | |
| Cooper's Hawk | 3 | 2 | 0 |
| Domestic Cat | 0 | 1 | 0 |
| American Crow | 0 | 6 | 0 |
| Abert's Squirrel | 0 | 0 | 0 |
| Pinyon Jay | 2 | 4 | 0 |
| Before presentation of stimulus | 5 | 1 | 0 |

Interspersing Ricks with Racks bestowes a laughing quality on Multiple Racks (e.g. hah-hah-hah-hah-uh-hah-uh for the sequence in Fig. 16F). Rick-racking was not correlated with a jay's motivation as its sound suggests. Instead, it may be a habit held by only a few jays. Rick-racks were given in all types of experiments but only eight of our lab birds and seven of our field birds ever Ricked. In the lab, these birds included Ricks in 39% of their Multiple Racks. For an unknown reason it seems that some birds commonly mix Ricks with Racks while others rarely do.

*Microvariation*

We quantified microvariation in Multiple Racks by measuring six frequency variables, six time variables, and four shape variables (Fig. 17). Racks within Multiple Racks are essentially carbon copies of each other so we saw no

Figure 16   *Inclusion of other calls into sequences of Multiple Racks. A) Multiple Rack preceded by Near. B) Multiple Racka preceded by Racka. C) Multiple Rack ending with harsh feeding Chirr. D, E, F) Inclusion of Ricks (shorter, lower frequency calls) with Racks.*

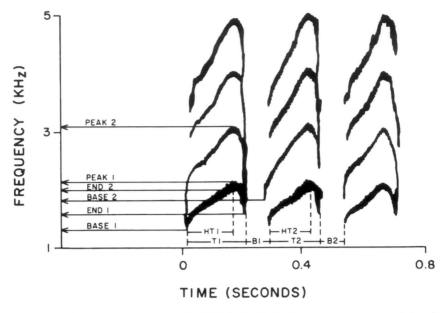

**TIME (SECONDS)**

Figure 17   *Sonogram, or voice print, of a Multiple Rack indicating measurements we made in order to determine its information content. The six frequency and six time variables measured to assess microvariation in racks are indicated. Four shape variables were computed by combining these variables as follows:  %Peak1 = HT1/T1, %Peak2 = HT2/T2, Slope1 = (Peak1 − Base1)/HT1, Slope2 = (Peak2 − Base2)/HT2.*

reason to measure beyond the second Rack. (All canonical correlations between suites of variables measured on each of up to eight Racks are significant with $R^2$ ranging from 66% to 98%.)

We were interested in determining if jays varied the microstructure of their calls in a way that could be used to predict the type of predator they were scolding. To do this we needed recordings of many jays scolding a variety of predators. We could then measure several calls by each jay to the same predator in order to tell how different calls are even when the predator stays the same. Most of this variation is probably due to differences in individual jays' voices. Then by measuring variation in calls by the same jay given to different predators we could tell how much variation in call structure is induced by changing the predator. If the latter variation is more than the former we can feel justified in concluding that jays have specific call features that identify predators. Take base frequency of the first Rack as an example. If individuals give calls with base frequencies of 1300–1320 Hz when they see a hawk, 1350–1370 Hz when they see a crow, and 1380–1400 Hz when they see a cat, then we can conclude that each predator elicits a unique call. On the other hand, if calls in reaction to every predator spanned the range 1300–1400 Hz then we could not conclude that specific predators are greeted with unique calls.

In order to assess variation within individuals and between predators for a suite of call features simultaneously we have to rely upon a complex statistical procedure known as discriminant function analysis (DFA). This procedure gives each call a co-ordinate score on a set of axes just like a normal X–Y graph. If all calls for each predator form distinct clusters, regardless of which jay gave them, then we can conclude that alarm calls give reliable information about the predator eliciting them. If calls for each predator do not segregate on the discriminant function plot we might just as well flip a coin to tell what predator stimulated a call. DFA compiles a useful summary of the ability to discriminate amongst calls given to different predators. It creates a classification function which combines each measure of microstructure into a formula to predict the most likely predator causing the alarm. It then takes each call and, based on its classification function score, assigns it to a predator. Lastly it checks to see what predator really incited the call. If this classification is correct most of the time it suggests that there are reliable rules with which to categorize calls which of course is only possible if each predator elicits a distinct call.

How did our calls fare under the scrutiny of discriminant function analysis? Not well. There were no distinct clusters of calls corresponding to predators (Fig. 18) and we did not gain much improvement in classification ability by using DFA (Table 3). We recorded calls given to three predators in the field so by random chance we would be able to pick the predator stimulating a given call one out of every three times. Using the classification functions created by DFA, we can use the microstructure of a call to help us infer which predator stimulated it. This only improved our ability to assign a predator to a call by 15% (Table 3). Similarly, among calls given in the lab we only gained a 10% improvement in classification ability by knowing the microstructure of a call.

Calls in the lab given to cats and crows were steeper (quicker rise from base to peak in the fundamental frequency of the first syllable) than calls given to other predators or under control conditions (see MANOVA results in Table 3). This increase in pitch is likely to indicate that a serious predator has caused the caller considerable distress. European Starlings exhibit a similar increase in the slope of their alarm calls when they become distressed (Aubin 1989). Some individual jays showed similar pitch shifts when they detected very serious predators. We recorded ten birds in the lab as they scolded different predators. By comparing three scolds by the same individual in reaction to each predator we could tell if the same bird changed its voice depending on the type of predator it encountered. Four birds showed significant increases in peak frequency and another showed an increased slope that corresponded to increasing danger of the predator (cats and hawks received the highest frequency, steepest scolds, crows were intermediate, and control calls were lowest and flattest). Two of these birds showed significant and opposite trends for slope or frequency. Four birds showed no significant changes. Again we lack consistent evidence that a Pinyon Jay's voice may indicate to others the type of predator.

A Pinyon Jay's voice may not tell about the source of danger but it does reveal the caller's motivation. We recorded 11 individuals scolding for protracted sequences in the field. As we noted earlier, motivation increased from start to finish of an experiment (Fig. 14A). The peak frequencies of the first and second Racks and the ending frequency of the first rack shifted signifi-

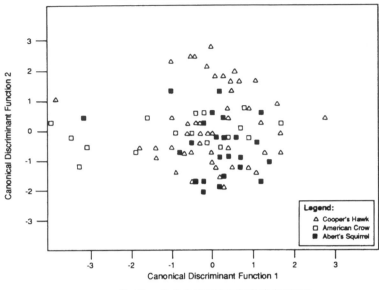

**A. Alarm Calls to Predators Placed Near Nets**

Legend:
△ Cooper's Hawk
□ American Crow
■ Abert's Squirrel

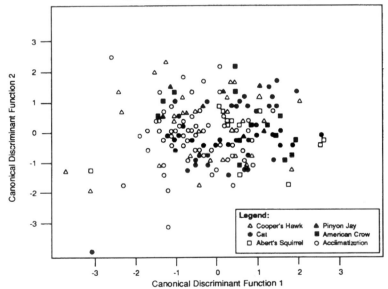

**B. Alarm Calls to Predators In the Laboratory**

Legend:
△ Cooper's Hawk      ▲ Pinyon Jay
● Cat                ■ American Crow
□ Abert's Squirrel   ○ Acclimatization

Figure 18    *Lack of consistent microstructure in calls stimulated by predators in the lab (A) and field (B). Each call is plotted in two dimensions (the only ones that were significant) calculated by discriminant function analysis. Each axis (known as a canonical discriminant function) is a weighted combination of all the properties we measured on calls. Each axis taps a different combination of call features. If each predator receives a special type of call then calls to each predator cluster together and rarely overlap with calls to other predators. This is not the case here as calls to crows, cats, squirrels, hawks, Pinyon Jays, and during acclimatization overlap widely. Numbers of calls per predator in the lab tests were: crow = 18, cat = 50, squirrel = 16, jay = 11, hawk = 43, acclimatization = 61. In the field, sample sizes were crow = 30, squirrel = 17, and hawk = 59.*

**Table 3.**  *Statistical analysis of microvariation in Multiple Racks. Calls are grouped by a variety of factors and then analyzed with discriminant function analysis (DFA) and multivariate analysis of variance (MANOVA) to ascertain the significance of each grouping factor. Factors that greatly improve our classification ability and have significant MANOVAs are likely to be encoded in multiple Racks and may be perceived by listening jays. N refers to the number of groups defined by each factor. Correct classification by chance equals 1/N. Improvement in classification equals DFA classification minus chance classification. Two MANOVAs were done: One used time and shape features of calls. The other used frequency measures. Features of calls that were measured are shown in Figure 17. MANOVAs for all factors except predators presented in the field utilize three calls per individual as repeated measures. MANOVAs for field predators use six calls per individual as repeated measures. The numbers of jays recorded giving repeated scolds, and therefore amenable to MANOVA analyses, are as follows: in the lab, 12 jays scolding hawks, 11 scolding cats, six scolding crows, four scolding jays, three scolding squirrels, and seven control scolds before the stimulus was presented; in the field, seven jays scolding hawks and three scolding crows. All calls that were elicited by a known predator, regardless of whether we knew what jay did the scolding, were used in discriminant function analyses. For field experiments, this resulted in a sample of 59 calls in reaction to hawks, 30 to crows, and 17 to squirrels. In the lab, this resulted in 43 calls in reaction to hawks, 50 to cats, 18 to crows, 16 to squirrels, 11 to jays, and 61 control scolds when no predators were in the test arena. In order to discriminate among calls of different individuals we randomly selected 10 calls from each of our 39 captive jays. These 390 calls are analyzed together in the "all bird" and "age, sex, flock" factors and then analyzed within their respective age (adult and juvenile) and flock (TF = town flock, CCC = country club flock) categories. NS refers to probabilities that are greater than 0.10. Figures 18 and 20 give visual representations of some of these results.*

| | | Percent correct: classification by | | Improvement in classification | Significance of MANOVA | |
|---|---|---|---|---|---|---|
| Grouping factor | N | Chance | DFA | | Time, shape | Frequency |
| Lab Predators | 6 | 16.7 | 26.6 | 9.9 | 0.05 | NS |
| Field Predators | 3 | 33.3 | 48.8 | 15.5 | NS | NS |
| All Birds | 39 | 2.6 | 70.5 | 67.9 | <0.001 | <0.001 |
| TF Adults | 18 | 5.6 | 86.7 | 81.1 | <0.001 | <0.001 |
| TF Juveniles | 8 | 12.5 | 96.3 | 83.8 | <0.001 | <0.001 |
| CCC Adults | 7 | 14.3 | 82.8 | 68.5 | <0.001 | <0.001 |
| CCC Juveniles | 6 | 16.7 | 85.0 | 68.3 | <0.001 | <0.001 |
| Age | | | | | 0.05 | NS |
| Sex | 8 | 12.5 | 49.2 | 36.7 | 0.05 | NS |
| Flock | | | | | 0.01 | 0.05 |

cantly higher as the experiment proceeded. Only one bird did not shift frequencies as it became more motivated (it was mobbing a crow). The others increased the pitch of their peak frequencies by 80 Hz to 480 Hz throughout the five-minute mobbing sequence (seven mobbed hawks, two mobbed crows, and one mobbed a squirrel). Pinyon Jays' voices increase in pitch just like ours do when we become greatly alarmed and scream!

Why do jays not explicitly tell their flock members what type of predator has alarmed them? Perhaps they do, but our experiment was unable to detect the difference. Other birds give similar alarm calls to a variety of predators when they are very close, like in our lab experiments (Klump and Shalter 1984). This argument is weakened in our case because we placed predators at natural distances from nests in our field experiments and also failed to detect any predator-specific calls (Fig. 18A). A likely reason for the lack of predator identity in Pinyon Jay alarm calls is that jays respond to all predators in a similar way; they immediately fly to the upper portions of trees and then approach and mob the intruder be it a snake, hawk, or coyote. When one response like this is sufficient for many predators there is no reason for natural selection to increase the variety of alarm calls. Species that tell each other specifically about predators do so because they need to gauge their response to the type of predator. Vervet monkeys in east Africa offer a striking contrast to Pinyon Jays in this respect. Vervets give different alarm calls in reaction to eagles, snakes, and leopards (Seyfarth *et al.* 1980). Eagles pose their greatest threat from above and snakes threaten from the ground. Accordingly, vervets look up when they hear eagle alarm calls but look down when they hear snake alarms. Leopards are the most serious predator because they often crouch in bushes waiting to spring on an uninformed vervet. Vervets cannot waste any time looking up or down for leopards and accordingly they run straight up trees when they hear a leopard alarm call! The need for vervets to give exact reports about predators is contingent upon their need to react differently to leopards, eagles, and snakes. Pinyon Jays do not have such a need and this may explain why we failed to find differences in their alarm calls.

### Other information in Pinyon Jay alarm calls

Although lacking in an explicit description of the predator, Pinyon Jay alarm calls are certainly not devoid of information. The most conspicuous information (besides the general alert signal) is an individual signature of the jay crying danger. Time, frequency, and shape aspects of Multiple Racks vary between individuals giving each a unique look and sound (Fig. 19). When we plot calls in DFA space the ones produced by the same individual cluster into distinct groups (Fig. 20). What a contrast to the wide overlap between calls when reacting to particular predators (Fig. 18). Whether we analyze all 39 birds together, or break them down into age and flock groups, the classification function devised by DFA is very accurate at predicting which individual gave a particular call (Table 3). Knowing call microstructure allows us to improve our ability to assign calls to individuals by 70–80% (Table 3).

Age, sex, and flock membership may also be encoded by Multiple Racks (Table 3). We created eight age, sex, and flock groups from the calls of the 39 jays analyzed above for individuality. These eight groups resulted from categorizing jays into adult *versus* yearling age groups, male *versus* female sex groups, and Town *versus* Country Club Flock groups. We discovered that Town Flock birds gave longer, lower frequency and flatter (lower slope) Racks than did Country Club Flock birds. Flock differences were more significant than age and sex differences and accounted for most of the moderate improvement in classification ability attributed to age, sex, and flock in Table 3.

Each flock appears to have alarm calls with unique properties that are recognized in playback experiments, but we are not quite ready to claim that flocks have unique dialects. The Town Flock (TF) and Country Club Flock

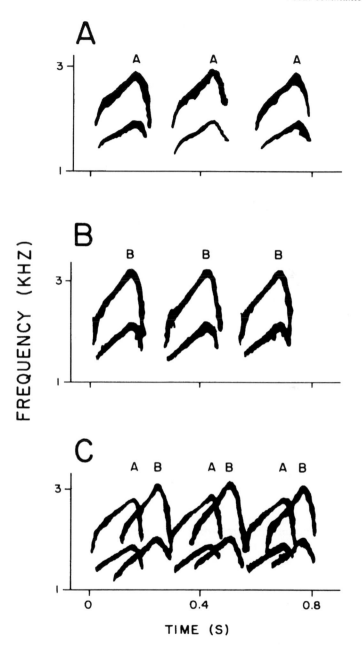

Figure 19 *Multiple Racks given by two jays in the field. Bird A's calls are shown alone in (A), bird B's calls are shown alone in (B), and both are shown together (in natural time) in (C). Differences in shape and frequency range of calls gives each jay a unique signature.*

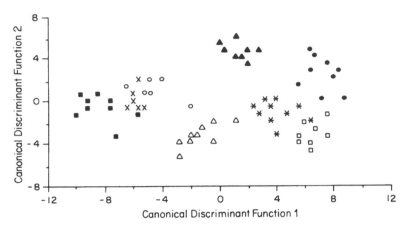

Figure 20    *Alarm calls of Town Flock juveniles plotted in discriminant function space. Each individual produces unique calls which cluster together in this plot. Clustering enables us to tell accurately which juvenile gave a particular call just by knowing the call's DFA co-ordinates.*

(CCC) have experienced different associations with humans over the last decade that may account for the flock-specific alarm calls we recorded. We have continuously captured, handled (weighed, measured, and banded), and observed the TF and only rarely intervened in the life of the CCC. The major differences in these flocks' alarm calls are consistent with greater alarm or fear by the CCC members; CCC calls are steeper and reach higher peak frequencies than TF calls. These are the same differences we just discussed within individuals as they become more alarmed. Perhaps CCC birds just react differently to captivity than TF birds. This is an intriguing suggestion because it implies dramatic behavioral changes in flocks that may have occurred in only 15–20 years. More flocks need to be studied.

If flocks do not have dialects then how do we account for the recognition in our playback tests? We suggest that individual recognition is responsible. Alarm calls are highly individualistic and jays could simply choose to support individuals with whom they are familiar. Recognition of individuals in the flock allows flock members to be distinguished from birds in other flocks and, more importantly, it allows for differentiation between birds within a flock.

## MOTIVATION OF ALARMED PINYON JAYS

The likelihood of an animal carrying out some action can be thought of as its motivation (Zahavi 1982). Motivation is an important message to include in vocal signals because it allows listeners to gauge their response to the urgency of a situation. Urgent situations, for example, intense scolding, can be reacted to quickly, while less important situations can be slowly acknowledged or ignored. Pinyon Jay alarm calls provide excellent indications of their senders' motivations. Moderately motivated jays use Racks that are relatively flat (low slope) and low in pitch. As an individual's alarm increases its Racks become steeper, higher in pitch, and harsher due to inclusion of upper frequency harmonics. Additionally, highly alarmed jays may include Rackas, Nears, and Chirrs in their sequence of Racks.

Gene Morton (1977) pointed out that graded systems of alarm calls are especially adaptive in species where kin live together or where co-ordinated group responses are effective at thwarting predators. Pinyon Jays fit Morton's criteria exactly. Flocks are conglomerates of many families so that relatives are in close association. Further, jays quickly form a tight flock when under siege by attacking *Accipiter* hawks and falcons. Such group responses should lower the chances of a predator successfully capturing a jay. We have not recorded alarm calls preceding group flights, but they seem to be of very high pitch and include Rackas or Chirrs. Flock life has favored individuality in alarm calls in addition to graded calls indicative of motivation. In a large group it is important to know who sees a dangerous situation and how dangerous they perceive it to be. Urgent danger signals from a mate or kin suggest that the listener's fitness is in immediate danger and a rapid response could be advantageous.

### Fearful or fearless mobbers?

If you watch a Pinyon Jay scold a dangerous predator you cannot help but feel admiration for the seemingly fearless jay. Pinyon Jays approach predators ten times their size, often coming within half a meter of certain death, in order to scream their protests. When we tethered a Great-horned Owl on the ground, mobbing jays walked up to within 1 m of this dangerous predator. Listening closely to a scolding jay's voice, however, suggests that it really is not fearless. Instead it sounds fearful. Higher pitched calls that characterize jays' voices as they mob dangerous predators also characterize the calls of a wide variety of animals motivated by fear (Morton 1977, Zahavi 1982).

Other properties of Pinyon Jay alarm calls suggest that scolding jays are hostile and indecisive. Chevron-shaped calls such as Racks, chips used for scolding by small birds, and barks and grunts used by mammals are indicative of animals motivated by indecision (Morton 1977). Pinyon Jays are presumably indecisive about whether they should approach or retreat from a potential predator. Chevron-shaped calls also indicate something of interest has been found. The harsh, wide frequency tone of Racks probably indicate hostility. This harsh tone is easy to locate, allowing others to find the predator and join in cooperative mobbing.

Bear with us as we take a look at the cognitive implications suggested by the voice of a scolding jay. Ultimately the jay is motivated to expunge a potential threat to its fitness, but it seems that the jays' psychologies do not agree with their genes' demands for self-preservation. The scolding jay's voice does not reveal pleasure. On the contrary, it reveals a psyche torn by fear, hostility, and indecision. The jay is in a mental state similar to that of an animal physically cornered by a predator, or a human in the face of adversity. These are the times that we seem to rise above our normal abilities and move mountains or slay giants. Is it too outlandish to think that a Pinyon Jay might not respond in a similar fashion? We think not. A scolding jay may be physically outmatched by a predator, but in terms of mental attitude the jay has the upper hand. Pinyon Jays scold predators like David attacked Goliath; they put mind above matter and usually win. Most certainly, "strength in numbers" also applies to mobbing Pinyon Jays.

CHAPTER 5

# Flock composition

*"During the winter (Pinyon Jays) collected in immense flocks, and in
one instance Dr Coues estimates their number at a thousand or more."*
(Baird *et al.* 1874)

Early investigators, such as the eminent Dr. Coues referred to in the opening
quote, were awed by the size of Pinyon Jay flocks. The large "flocks" that
impressed early naturalists were likely to have been several flocks momentarily aggregated at locally abundant foods. However, they stimulated our
interest in Pinyon Jays because these flocks were among the largest known
avian social groups. Naive to the ways of jays, we asked two basic questions.
How many individuals comprise a flock of Pinyon Jays? Further, how many of
each age and sex are found in a flock? Little did we know when we began
studying Pinyon Jays that it would require ten years of daily observations to
answer such simple questions!

   Why were these basic questions so difficult to answer? Two reasons stand
out. First, male and female Pinyon Jays look alike. Second, attempts to count
the individuals in a flock as they mill about in front of you (without missing or
recounting some) produces eyestrain, headaches, and confusion. To solve
these problems we began capturing all the jays in a flock living in and around

Flagstaff, Arizona (Photo below). We adorned each jay with a unique set of colored leg bands (rings) so that we might be able to keep track of each individual's activities. After two years most jays in the "Town Flock" were marked and two permanent feeding and trapping stations were established within the flock's home range.

*Ponderosa pine habitat on the flanks of the San Francisco mountains, near Flagstaff, Arizona. This picture is typical of the home range inhabited by the Town Flock. From Marzluff and Balda (1990).*

The efforts of two colleagues, Gene Foster and Jane Balda, were crucial to our success during these years. Gene lived at one of the feeding stations and centered her activities around the jays. Every day for 12 years, when the jays visited her house all human activity paused to observe, count, capture, and mark jays. Jane lived at another feeding station and kept tabs on birds there as well as those dispersing to a neighboring flock. Additionally she spent hundreds of hours keeping records of jay sightings, banding, and weighing. The efforts of Gene, Jane and numerous students Gene employed enabled us to keep over 95% of the jays in our study flock marked from 1972 to 1983. We were then able to follow the breeding activities of recognizable jays in order to determine their sex (only females incubate) and the identity of mated pairs. We could conduct a "roll call" every time we saw the flock at the feeding stations in order to determine how many jays resided in our flock and when particular birds entered or disappeared. Marking newly fledged young each year enabled us to determine the age composition of the flock with finer and finer detail each year as known-aged jays grew older. Many graduate students from Northern Arizona University, most notably Diana Gabaldon and Pat McArthur, participated in this activity. Knowing who parented which young allowed us to glimpse the genetic structure of the flock.

We learned early on how to band nestlings. Parents are not keen on having shiny objects in their nest so we had to boil our aluminum bands with eggs to dull them. Parents also have color vision, as we learned the hard way! Red (the color of blood) and white (the color of fecal sacs) prompted parents to remove bands of these colors, often with the nestlings attached! After finding a few color-banded nestlings evicted from nests we changed our system and no longer color-banded young until they visited our feeders as a creche.

*A mother jay evicts one of her nestlings in an attempt to remove a white band, mistaken as fecal matter, from the nest.*

In this chapter we provide a detailed backdrop for the activities within the Town Flock by describing its constituency. We will examine the age, sex, and genetic structure of the flock. Rather than focus on average structure, our aim is to give you a feel for the dynamic nature of a flock's structure. In particular, age and sex structure change throughout a year and between years, whereas the constituency of relatives that an individual has within the flock changes as the individual ages.

## FLOCK SIZE AND AGE COMPOSITION

### Seasonal changes

The annual production of juveniles in late spring and early summer dramatically alters a flock's size (Fig. 21) and age composition (Fig. 22). In early spring, before most juveniles have integrated into the flock, the flock is at its smallest size and composed of 75% adults (birds two or more years old) and 25% yearlings (birds born the previous year). Young produced by flock members make up one third of the flock during the summer and their addition results in a 52% increase in summer flock size over spring flock size. Many juveniles born in late spring and summer die before fall. However, flock size continues to increase through the fall as jays dispersing from neighboring flocks briefly join our study flock. Flock composition shifts further towards juveniles in the fall as most dispersers are young of the year. At this time over 40% of the flock's members are juveniles. Flock size declines through the winter as many dispersers leave or die and some flock members also die. Age composition throughout the winter is similar to that throughout the summer; approximately 30% of the flock members are juvenile, 20% are yearlings, and half are adults.

### Annual changes

Year-to-year changes in the age structure of a flock reflect the constancy of the environment. In constant environments populations will attain stable age distributions. That is, regardless of whether the population is increasing, decreasing, or remaining the same size, it will always include a constant proportion of individuals in each age class (e.g. 60% juveniles, 20% yearlings, 10% two-year-olds, and 10% three-year-olds). If we observe changing age structures we may suspect that environmental conditions are also changing and attempt to determine which conditions are responsible for the observed age changes. Here we examine annual changes in age composition within each of the four seasons and begin to link environmental fluctuation to changes in the flock's make-up.

*Spring*    Flock size and age composition are most consistent from year to year during the spring (Fig. 23). Annual variability is reduced during the spring

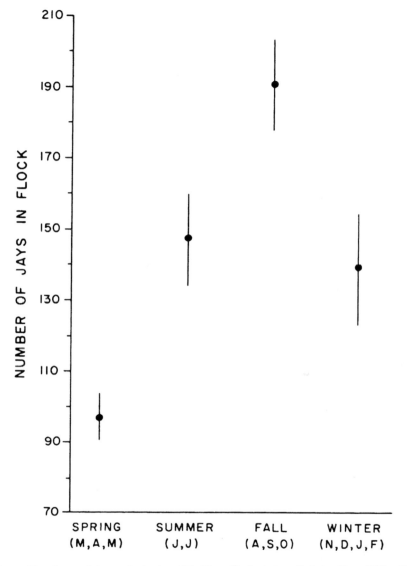

Figure 21 *Seasonal changes in the size of the Town Flock. Average flock size (from 1973 to 1982) is represented by the point and, as an indication of variation in size, the vertical lines represent one standard deviation above and below the average. From Marzluff and Balda (1990).*

because the volatile juvenile cohort has been reduced by mortality and emigration during the fall and winter months. One conspicuous annual change in spring composition is the occasional incorporation of juveniles in May. These juveniles come from successful early nesting attempts which occur when spring snows are light, nest predators are few, and pine seeds are abundant (see Fig. 56).

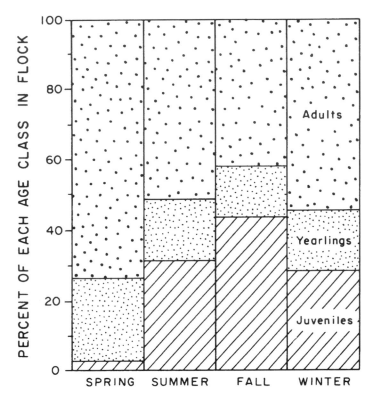

Figure 22   *Seasonal change in the age composition of the Town Flock. Percentages are average values over the time span 1973–1982. From Marzluff and Balda (1990).*

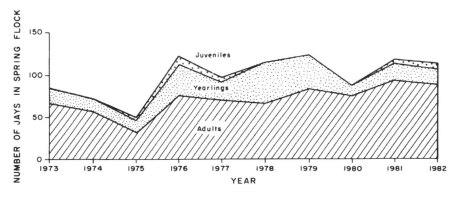

Figure 23   *Annual variation in the age composition of the Town Flock during the spring (March, April, May). Adults represent all birds two or more years old. Total flock size, which is averaged in Figure 21, is here represented by the uppermost point each year. The number of jays of each age is indicated by the vertical distance between the line above and the line below each age.*

A substantial reduction in spring flock size, especially in the number of adults, occurred in 1975. However, this decline was not associated with any unusual climatic events and may represent an artifact of our sampling regime. This spring followed the bumper pine crop of 1974 (Fig. 4) during which time we rarely counted the individuals in the flock because it relied on naturally abundant pine seeds and rarely utilized our feeding stations.

*Summer*    Variation in summer flock size and age composition is due to variable nest productivity and the resultant number of juveniles in the flock. Flock size was highest in 1974, 1977, and 1978 when many juveniles were produced and lowest in 1973, 1975, and 1982 when few were produced (Fig. 24). As we

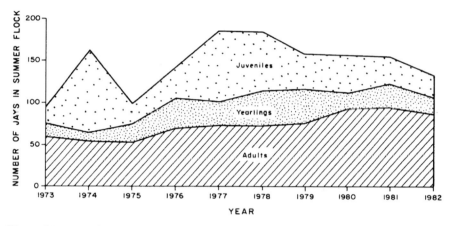

Figure 24    *Annual variation in the age composition of the Town Flock during the summer (June, July). See legend in Figure 23.*

discuss in more detail in Chapter 8, three factors are important determinants of nest productivity: the size of the previous fall's pine crop, the severity of spring snowfall, and the abundance of nest predators. These factors, and others including unpredictable chance events (such as kids with pellet guns!), interact to influence the number of juveniles in the summer flock. Spring snowfall appears to exert the major influence: when snows are light (1974, 1977) juveniles are abundant and when snows are heavy (1973) juveniles are rare (Fig. 25). Abundant nest predators (especially Common Ravens) may account for some low productivity in relatively lean snow years. For example, snowfall in 1979 and 1982 was similar to that recorded in 1977, but during winter surveys only five ravens were spotted in 1977 compared to 32 in 1979 and 42 in 1982. This eightfold increase in ravens and associated predation on jay nests may account for lower productivity in 1979 and 1982 when compared to 1977. The colored rings we placed on jays' legs became a prized possession of a child living near the nesting jays in 1982. His marksmanship further reduced productivity in that year. Pine seed abundance has less effect on the

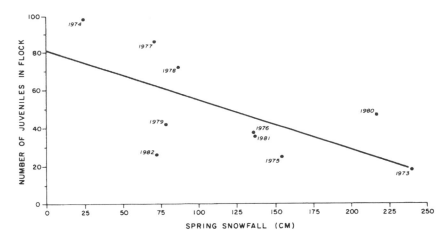

Figure 25    *Annual variation in the number of juveniles in the Town Flock during the summer in relation to spring snowfall. Years are given next to each point. Spring snowfall is the total accumulating during March, April, and May. Line represents the best-fitting regression (r = −0.65, P < 0.05) of juvenile abundance with snowfall.*

production of juveniles. Seasons following the bumper crops of 1975, 1977, and 1981 were not consistently productive. This may be peculiar to our study area where the extra eggs laid after bumper crops are commonly destroyed by spring snowstorms.

These are interesting correlations between factors important to nesting success and the subsequent composition of the flock, however, much is unaccounted for. Take for example 1978 and 1980. Few seeds were produced, predators were abundant, and snowfall was heavy (at least in 1980) yet productivity was high. The reasons why remain a mystery known only to the jays!

*Fall*    Annual variability in flock size and age composition is most pronounced during the fall (Fig. 26). Variability in juvenile numbers is especially pronounced and results from separate processes: productivity within the flock and dispersal from other flocks. Large numbers of juveniles in 1974 and 1977 were the result of successful breeding by members of the flock. In contrast, abundant juveniles in 1975, 1978, 1980, and 1982 resulted when wanderers from other flocks associated with our flock for periods of several days to months. At the extreme, our flock tripled in size in the fall of 1978 as hundreds of juveniles, born elsewhere, flooded into Flagstaff.

Summer moisture (rainfall in June, July, and August) appears to be an important correlate of juvenile abundance. In moist summers insects, an important food for juveniles, are abundant. Juveniles should therefore find it easier to fend for themselves in moist summers than in dry ones and their survival until the fall should be correspondingly greater. This hypothesis is

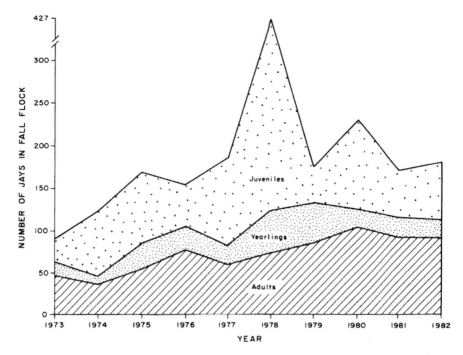

Figure 26　*Annual variation in the age composition of the Town Flock during the fall (August, September, October). See legend in Figure 23.*

supported by the records: following wet summers, more juveniles born in the Town Flock reside there in the fall (Fig. 27).

What about juveniles dispersing from elsewhere who are most abundant following dry summers (Fig. 27)? We do not know what the weather conditions in their natal areas were, but it is safe to say that they were drier than in Flagstaff, which because of its high elevation, gets more summer rain than lower elevation pinyon-juniper woodlands. In dry years Flagstaff may be especially attractive to young jays in search of insects because it represents the only oasis within miles. Watered lawns and gardens in town add to the oasis atmosphere of Flagstaff. Sometimes this oasis can become a cruel mirage. We once saw a group of young foraging on a recently fertilized lawn. These young were inexperienced foragers and ate many fertilizer pellets. Within hours two were dead.

*Winter*　Few individuals immigrated successfully (remaining as a breeding flock member) into our study flock, therefore annual fluctuations, so pronounced in the fall, begin to dampen by winter (Fig. 28). An unusually large number of foreign juveniles remained into the winter in our flock in 1978, perhaps because of the unusual drought conditions that year. Other than this

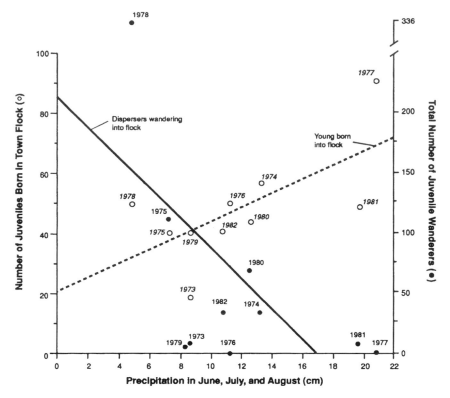

Figure 27    *Yearly variation in the number of juveniles in the Town Flock during the fall in relation to summer precipitation (rainfall in June, July, and August). Juveniles born in the flock (open points, dotted line) are more abundant after moist summers. Juveniles wandering into the flock (solid points, solid line) are more abundant after dry summers. Years are given next to points. Lines represent best-fitting regressions (natives: r = +0.65, P < 0.05; wanderers: r = −0.58, P < 0.05) of juvenile abundance with summer moisture.*

blip, flock size and composition mimic those characteristic of summer. Especially prominent is the abundance of juveniles following successful breeding by flock members in 1974 and 1977 when springs were warm and snowfall minimal.

*Detailed age structure*

Our long-term marking program allowed us to view more and more of the age structure of the flock each year (Fig. 29). The most striking result of our efforts is the realization that many jays in the Town Flock are quite old. Roughly 20% (average in 1975–1982 = 22%) of the jays in the flock are five or more years old and 5% (average in 1980–1982 = 5%) are ten or more years

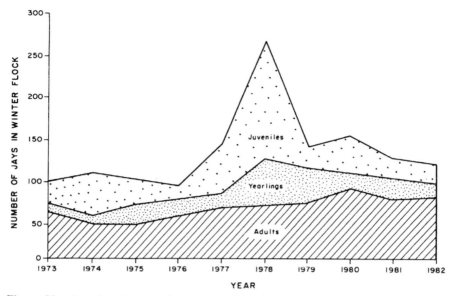

Figure 28 *Annual variation in the age composition of the Town Flock in winter (November, December, January, February). See legend in Figure 23.*

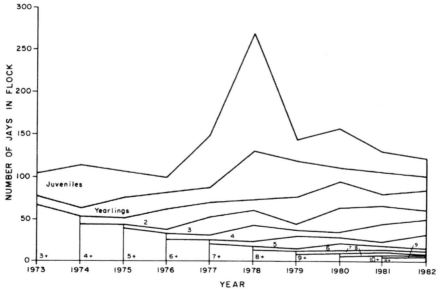

Figure 29 *Detailed age composition of the Town Flock in winter. Plot is same as in Figure 28 except that adult age classes are delimited. The number of jays per age class is indicated by the vertical distance between the line above and the line below the age class. Ages followed by a + indicate jays that were at least two years old at the start of our study. This unknown-aged cohort comprised 37.5% of the flock in 1973 and declined to comprise only 4% of the flock in 1982. From Marzluff and Balda (1990).*

old. This stable core of older jays is in stark contrast to the extremely variable, but large, number of juveniles we have just discussed. Two other recent studies of social jays report similar age structures (Mexican (Gray-breasted) Jay, Brown 1986; Florida Scrub Jay, Woolfenden and Fitzpatrick 1984).

Juveniles produced in the few years when breeders were extremely success-ful continue to profoundly influence the age structure of the flock. This is most readily seen if we partition the flock into year classes, or cohorts, and follow them through the period of study as in Figure 30. Juveniles produced in 1974 and 1977 continue to dominate the flock as yearlings, two-year-olds, three-year-olds, four-year-olds, and even five-year-olds when they alone comprised 10% of the flock in 1979 and 13% in 1982. Ecologists refer to single age groups that predominate for years such as the 1974 and 1977 cohorts as "dominant age classes". Juvenile cohorts greatly inflated by dispersers (1975, 1978, 1980) had a less lasting influence on flock composition, again reflecting the emphemeral association of most dispersers with our study flock.

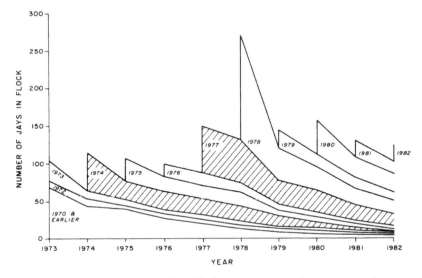

Figure 30    *Year class structure of the Town Flock during the winter. Data points are the same as in Fig. 29 but here we have emphasized the impact each year class has on flock composition by following a cohort of juveniles through our study. Age classes begin as juveniles produced in the year labeling each class. Vertical width of the class indicates the number of individuals of that class still alive in a given year. Shaded classes represent extremely productive breeding seasons of 1974 and 1977 when many juveniles were born.*

*Why are simple questions so hard to answer?*

We began this chapter being surprised at how long it takes to answer simple questions such as: What is the composition of a flock of Pinyon Jays? Let us take a more detailed look at the number of juveniles found in a fall flock of jays as a way to illustrate why such simple questions are so frustrating. This frustration is not unique to the study of Pinyon Jays; it is a common attribute of ecological investigations because many inter-related factors simultaneously influence processes.

Summer rainfall is only one of many factors that in concert influence the number of young in a flock during the fall. Summer rainfall accounts for only 42% and 34% of the year-to-year variation in the abundance of juvenile natives and juvenile wanderers, respectively (Fig. 27). Nearly two-thirds of the annual changes in juvenile abundance must be accounted for by other factors. As we have argued, spring snowfall, predator abundance, and possibly availability of pine seeds influence the number of young born in the spring and summer (Fig. 25). Therefore, the number of young varies wildly from year to year before summer droughts begin to whittle away juveniles. As a result there may be many juveniles in the fall flock following breeding seasons like 1978 even if later conditions are extremely dry. On the other hand, few juveniles will be found in fall flocks following poor breeding seasons, such as 1981, despite extremely wet summer conditions. At least one other factor, availability of mating opportunities in the natal flock, also influences the fall movements of juveniles. Juveniles are likely to wander from their natal flock into neighboring flocks when their chances of obtaining a mate at home are slim. When there are few birds of the opposite sex in their natal flock dispersal should be great. The number of juveniles in our fall flock therefore is likely to depend on our flock's sex ratio and the sex ratios of neighboring flocks in addition to the extent of summer rains and success during the breeding season (see also Chapter 11).

We are left with an incomplete answer to the simple question: Why were there more juveniles in our fall flock in 1978 than in 1973? As is often the case in ecology, multiple processes (e.g. reproduction, mortality and dispersal), each of which are influenced by multiple factors (e.g. climate, food, predation, and mate availability), together determine a seemingly simple property such as the abundance of juveniles. Ecologists rarely understand, let alone have data necessary to quantify, all the factors and processes at work, so we can in many cases offer only incomplete answers to simple questions.

Long-lived organisms, living in variable environments, like Pinyon Jays, complicate things further. Each year of study often produces only one data point relevant to a portion of a question. Many more points than the few we present in Figure 25 are needed to rigorously assess a relationship. The outlying points such as 1978 and 1980, which we discussed may not be outliers at all after 30 or 40 years' of data are collected. Our initial surprise raised in the introduction thus is not really that startling. We are lucky to have answered a few simple questions after only ten years of daily observations!

### Sex composition of a flock

The sex ratio, usually expressed as the number of males per female, is an important attribute of social groups. Rarity makes individuals of the least abundant sex more valuable to members of the opposite sex than *vice versa*. Accordingly, members of the rare sex often initiate breeding at an earlier age and with older mates than members of the opposite sex (Hunt 1981). An ability to breed at a young age also can influence other options available to young birds. In Florida Scrub Jays, for example, females disperse and breed at younger ages than males which often remain in their natal territory and help their parents breed rather than compete for breeding opportunities elsewhere (Woolfenden and Fitzpatrick 1984).

Members of the abundant sex often compete intensely for access to members of the rarer sex (Emlen and Oring 1977). This intrasexual competition can be a powerfully creative force that may even counteract the forces of natural selection. Darwin recognized this potential in his 1871 book on the subject. He included the process by which intrasexual competition influences behavior and morphology under the heading of *sexual selection*. The peacock's tail, the stag's antlers, and the human male's repertoire of "come-ons" are monuments to the power of sexual selection.

In their competition for mates, male birds often defend areas containing resources needed by females. Males are then forced to disperse widely in search of the highest quality territories (Greenwood 1980). Thus a single property of a social group, sex ratio, can influence decisions concerning when to breed, where to breed, and with whom to breed. Here we document the seasonal and year-to-year changes in the sex ratio of the Town Flock. Later we discuss how the sex ratio influences breeding behavior (Chapter 7) and the decisions of yearlings, particularly whether they should breed, disperse, or help their parents breed (Chapter 11).

We used breeding behavior and body size to estimate the sex ratios of various age classes. Breeders can be quickly sexed because only females incubate. This enabled us to sex nearly every bird in the Town Flock that was four or more years old. Males average 13 g (12% of the average male's weight) heavier and have bills 2–3 mm (7% of the average male's bill length) longer than females. Although males and females overlap slightly in size, we were able to confidently call very large birds males and very small birds females, even if we never observed them breeding. These criteria allowed us to sex 35% of the yearlings and 68% of the two- and three-year-old jays in the flock.

Female Pinyon Jays are the rare sex. This is striking whether we look at the Town Flock (Figs. 31 and 32) or an unprovisioned flock in New Mexico (Ligon and White 1974). The usual male bias in sex ratio is due to higher female mortality, perhaps because incubating and brooding females are vulnerable to predators and/or because energetic costs to them are high (see Chapter 12). In addition, females disperse more frequently than males which may expose them to dangerous situations in unfamiliar surroundings (Chapter 11). Given that females and males play different roles in the flock (Chapter 9), it is not surprising that they show different patterns of mortality.

There is considerable year-to-year variability in the sex ratio of adults. On average, 57% of the adults in a flock in January are males (Fig. 31A). This male-bias is equivalent to 1.3 adult males per adult female. In two of our study years (1974, 1979) adult females slightly outnumbered adult males in January. In three other years (1980, 1981, 1985) males were over 50% more abundant than females. Greater female mortality through the season is apparent as the adult cohort becomes more male-biased at the end of the year (Fig. 31B). In December, an average of 59% of the adult flock members are males. In other

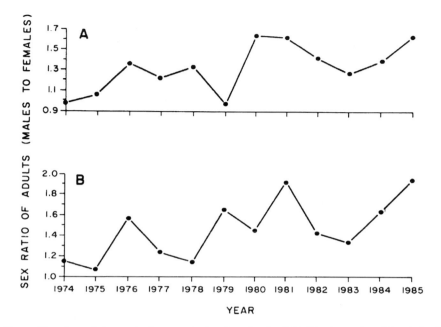

**Figure 31** *Annual changes in the sex ratio (number of males divided by the number of females) of adults in the Town Flock. Sex ratio in January (A) is contrasted with the sex ratio of the survivors in December (B). A ratio of one represents an equal number of males and females. Ratios greater than one indicate that males outnumber females and ratios less than one indicate that females outnumber males.*

words, each male is in the company of 1.5 females. Adult sex ratios were always male-biased at this time with some years (1981, 1985) having nearly two adult males to each adult female.

Yearling sex ratios change more within a year and are more variable between years than are adult sex ratios (Fig. 32). In late summer and early fall, 74% of the yearlings in our flock were males (Fig. 32A). Yearling sex ratios are exceptionally volatile at this time of the year. Take for example three sequential years, 1975–1977. Males outnumbered females 6:1 in 1975 and 8:1 in 1977, but the intervening year saw females slightly outnumbering males. As we discuss in Chapter 11, most yearling immigrants are females and more immigrate in years when yearling males are abundant. This adjustment of yearling immigration to a flock's sex ratio results in seasonal amelioration of the usual male bias. For example, the strong male bias in late summer was reduced after fall immigration when 65% of the yearling cohort were males (Fig. 32B) and further reduced after spring immigration when 59% of the yearlings were males (Fig. 32C).

Wild annual fluctuations in yearling and adult sex ratios are a further indication of the variable environment inhabited by Pinyon Jays. Climatic severity, pine seed abundance, and predator abundance all vary from year to year and all can influence female survivorship. Female survivorship in turn

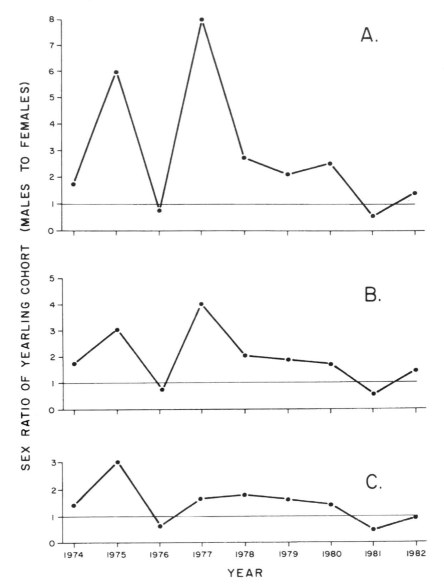

Figure 32    *Annual changes in the sex ratio of yearlings in the Town Flock. Sex ratio in the late summer prior to immigration during the fall (A) is contrasted with sex ratios after fall immigration (B) and after fall and spring immigration (C). From Marzluff and Balda (1989).*

determines the degree of male-bias in the sex ratio. As we will show in future chapters, this translation of a variable environment into a variable social structure has important consequences on the options pursued by jays, especially young jays.

*Immigrants and natives in the flock*

Many jays in the Town Flock are not hatched there but have immigrated from other flocks. On average, 36% of yearlings in the Town Flock are immigrants. There is great year-to-year variation in the abundance of immigrants comprising the yearling cohort which we discuss in Chapter 11. Immigrants may bring new genes, cultures, and traditions into their adopted flock which may significantly alter a flock's habits, especially given the observation that in some years over half of the yearlings in the Town Flock were immigrants (Fig. 33).

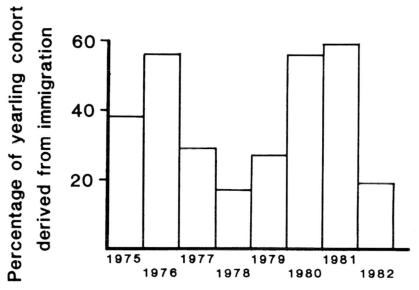

Figure 33   *Percentage of the Town Flock's yearlings that were immigrants. Immigrants could have joined the flock as juveniles or as yearlings. All other yearlings were born in the Town Flock.*

Sexual differences in the likelihood of immigration introduce sexual asymmetries into the social composition of the yearling cohort. As in most birds, female Pinyon Jays are more likely to disperse than are males. As a result, a slim majority of female yearlings (52% of those in the Town Flock) do not reside in their natal flock, but most yearling males (89% of those in the Town Flock) do. This means that young males and females live in very different social settings. In particular, young females are rarely in the company of their family members and relatives, but most young males spend their entire lives with their extended families. Let's investigate these extended families to determine differences in male and female jay's social spheres.

*Kin structure of the flock*

Remember from Chapter 1 that natural selection may favor co-operation amongst relatives because they hold some genes in common. Helping relatives may enhance an individual's evolutionary fitness which is measured by its genetic representation in future generations. Quantification of the kin structure of a social group allows us to appraise the potential importance of indirect kin selection to the evolution (or at least the current maintenance) of cooperation among group members. Indirect kin selection can only be relevant in social groups that include relatives and its potential increases as the degree of relatedness of group members increases in comparison to the degree of relatedness between members of different social groups.

Avian social groups typically are comprised of nuclear families (parents and their offspring) or occasionally a few extended families (parents, offspring, and more distant relatives). New World jays exhibit both types of kinship groups. Florida Scrub Jay groups consist of parents and up to six offspring (Woolfenden and Fitzpatrick 1984). Mexican (Gray-breasted) Jay groups have a more complex kin structure, often including two extended families within one group (Brown and Brown 1981). Pinyon Jays have the most complex kin structure known for group-living birds. The kin structure of their large flocks is more reminiscent of baboons than birds!

Pinyon Jay flocks are best described as troops comprised of numerous clans. Clans are true extended families consisting of two or three generations of parents, their sons, and their sons' mates. Occasionally daughters remain in their natal flock but, as we have just seen, many emigrate to neighboring flocks. Most sons remain in their natal flock and often mate with immigrant females or females from other clans in the flock. Clans are apparent when we draw the family genealogies of the flock (Appendix 5). However, they are not apparent when we watch jays in nature. Pairs associate in nature, but once their young become independent (about two months after fledging) it is rare to see clearly defined clans in which members associate preferentially with one another. Unlike many primates, we have never seen troops splinter along clan lines nor have entire clans even briefly disassociated from the troop.

Pair-bonds form between members of different clans and may act to cement unrelated birds into a cohesive, permanent group. It is likely that most, if not all, clans are linked through pair-bonds. In the Town Flock, we have determined how 339 individuals at least one year old are organized into a flock. Seventy three per cent of these jays are linked through pair-bonds between clans (Appendix 5). If we knew the complete family histories of all flock members we would probably find this percentage to be even higher.

From our point of view, then, the flock consists of many small clans of related jays. Since we are interested in determining how an individual's cooperative tendencies may be influenced by kinship let us take a jay's view at the kin structure of its flock. An average Town Flock jay in its first year of life in its natal flock is in the company of 3.45 relatives (Fig. 34). These relatives are primarily parents and full siblings, although an assortment of half siblings, aunts, uncles, nieces, nephews, grandparents, cousins, half cousins, half aunts, half uncles, half nieces, and half nephews are also in the flock. A young female immigrating into a flock would not be in the company of this diverse array of relatives. The relatives that co-exist with young Pinyon Jays (especially males)

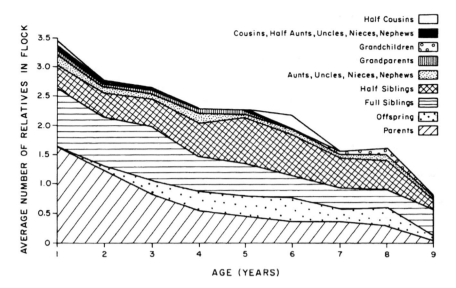

Figure 34    *The average number of relatives an individual in the Town Flock has living with it in its natal flock. All relatives alive during a year in a jay's life were counted. Averages per age category are based on sample sizes of 108, 78, 48, 33, 25, 19, 16, 10, and 9 individuals in their first, second, third, fourth, fifth, six, seventh, eighth, and ninth years of life, respectively.*

may be especially important in creating a genetic environment that helps to ease a young bird into the complex social environment of the flock.

As a jay ages the number and variety of relatives living in its flock changes. The number of relatives in its natal flock gradually declines so that an average nine-year-old jay, having outlived most of its relatives, is in the company only of 0.77 relatives (Fig. 34). The types of kin in the flock also change as an individual ages. Individuals begin to accumulate offspring in their second year and grandoffspring in their seventh and eighth years. Individuals do not have living grandparents after five years in the flock and after eight years parents are also dead. Nine-year-olds are primarily in the company of their siblings. Few offspring accompany these old jays in our study flock because recent generations have not been as successful as past generations (Chapter 9). In productive flocks this is unlikely to be the case and older jays might have a few more relatives, especially offspring and grandoffspring. If this is the usual case, then the absolute number of relatives that a jay has in its flock would be independent of age. As a result, older immigrants and natives would reside in similar genetic environments.

Anyone who has attended a large family reunion knows that all relatives are not equally related. Parents and offspring, as well as full siblings, are related by 50% (that is, they are expected to have 50% of their genes in common). In contrast, an individual is expected to have only 25% of its genes in common with its half siblings, aunts, uncles, nieces, nephews, grandparents, and grandchildren. More distant relatives share even fewer genes (e.g., half cousins only average 6.25% in common). Armed with this information, let us take a different look at the kin structure of a flock. We will produce a picture of an individual's genetic representation in the flock by multiplying the number of relatives an individual has in the flock by their degree of relatedness (0.5 for siblings, 0.0625 for half cousins, etc.). The sum of these products for all the relatives a given-aged jay has in the flock equals the number of "full genetic equivalents" an individual has in its flock. For example, an individual with two full siblings in the flock would have one genetic equivalent as half of its genes should be represented in each sibling. Genetic equivalents tell how many "identical twins" an individual has in the flock (of course these "twins" are split up into many, more distantly related jays)! The number of genetic equivalents decreases as Town Flock jays age (Fig. 35) because close relatives (parents, siblings, and offspring) decline in frequency with age (Fig. 34). This may again be peculiar to the Town Flock where productivity has declined in recent years.

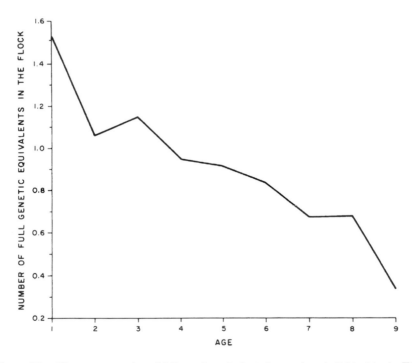

Figure 35   *The average number of full genetic equivalents (see text) an individual in the Town Flock has living with it in its natal flock. Sample sizes are the same as in Figure 34.*

We have emphasized the number of kin that coexist in a flock and have shown how this is a function of age. We must point out, however, that most jays in a flock are distantly related at best. A one-year-old that does not disperse has 3.45 relatives in its flock, but the other 100 or so jays in the flock are no more closely related than half cousins. Pinyon Jay flocks consist of a few close relatives embedded in a sea of genetic strangers and for this reason the average coefficient of relatedness between flock members is very low. In a flock of 150 jays, with a winter age structure and kin structure such as we have described, the average coefficient of relatedness between two flock members is less than 0.03 (Table 4). In other words, two randomly selected jays in a flock will, on average, share only 3% of their genes in common. This is approximately four times lower than the lowest average coefficients of relatedness within groups of co-operatively breeding birds (Brown 1987, pp. 200–204).

**Table 4.** *Average coefficient of relatedness among members of a Pinyon Jay flock. We have assumed a flock size of 150 with typical winter age structure. Further, we assume "unrelated" jays include some distant relatives not yet identified because of incomplete genealogies. Therefore we conservatively assigned a coefficient of relatedness of 2% to unrelated jays. The average coefficient of relatedness for each age class equals their total genetic equivalents divided by flock size. Weighted average coefficient of relatedness equals the sum over all ages of percentage of the flock in an age class multiplied by average coefficient of relatedness for members of an age class.*

| Age class | Percentage of flock in age class | Number of relatives | Number of unrelated jays | Genetic equivalents from relatives | Genetic equivalents from unrelated jays | Average coefficient of relatedness |
|---|---|---|---|---|---|---|
| 1 | 29 | 3.45 | 146.55 | 1.504 | 2.93 | 0.030 |
| 2 | 17 | 2.72 | 147.28 | 1.069 | 2.95 | 0.027 |
| 3 | 14 | 2.62 | 147.38 | 1.153 | 2.95 | 0.027 |
| 4 | 11 | 2.29 | 147.71 | 0.943 | 2.95 | 0.026 |
| 5 | 8 | 2.28 | 147.72 | 0.905 | 2.95 | 0.026 |
| 6 | 7 | 2.15 | 147.85 | 0.826 | 2.96 | 0.025 |
| 7 | 4 | 1.56 | 148.44 | 0.670 | 2.97 | 0.024 |
| 8 | 3 | 1.60 | 148.40 | 0.680 | 2.97 | 0.024 |
| 9+ | 7 | 0.77 | 149.23 | 0.330 | 2.99 | 0.022 |
| | | | | | weighted average = | 0.027 |

*Is indirect kin selection likely to be important to cooperating Pinyon Jays?*

We have seen that a jay shares very few genes with an average flock member. The intriguing question is: do they share enough genes for indirect kin selection to favor behaviors dangerous to an individual that, nonetheless, benefit the group (ie., altruistic behaviors)? It is possible for young jays (those three years old or younger) to recoup individual losses by aiding the flock. In fact, they should give up their lives for the flock. Look at it this way; young jays have more than one "identical twin" spread out over the flock (Fig. 35), so an altruist that dies to save these relatives keeps more than one genetic equivalent in the flock. A selfish jay that saves its own skin and lets its relatives die retains only one genetic equivalent (itself), whereas more than this die. Young jays should be altruists because this behavior may result in greater genetic fitness than purely selfish behavior. This assumes, however, that a young jay could save all its relatives from certain doom by performing some duty, hardly a possibility. Typical co-operative actions such as sentinel duty are unlikely to save all the sentinel's relatives at once. However, if done repeatedly, it may save many relatives; indirect kin selection would favor such actions even if they eventually cost the sentinel its life. This is an extreme example in support of indirect kin selection and one only applicable to a few age classes of jays. What about the general significance of indirect kin selection?

Biologists have argued over the importance of indirect kin selection since Hamilton first introduced the concept 25 years ago. An important point brought out by D. S. Wilson (1975, 1977) and experimentally confirmed by M. Wade (1980) is that some population structures offer especially fertile grounds for indirect kin selection. Indirect kin selection should be especially prevalent in populations that are subdivided into social groups whose members are more similar genetically with each other than they are with members of other groups. Theoretical investigations suggest that under these conditions groups of co-operators will outcompete groups of selfish individuals thereby enabling altruistic behaviors rapidly to evolve. Altruists benefit when group members share many genes because co-operation perpetuates the bene-factor's genes held in common with other group members. However, when group members share few genes co-operation does little for an altruist's genes and instead perpetuates many competing genes. Wade (1980) showed that populations of flour beetles structured into kin groups quickly evolved away from normal cannibalistic tendencies. However, when he disrupted this population structure by allowing frequent matings between members of different kin groups cannibalism persisted. If beetles the size of pinheads have social behaviors tuned to their population structure, surely complex verte-brates, like Pinyon Jays, do as well!

Gray-crowned Babblers studied by Johnson and Brown (1980) may provide a vertebrate example where population structuring has promoted co-operative behaviors. These birds live in small groups whose members co-operate in many ways, for example by helping at the nest and by warning each other about predators. Group members are closely related to each other but distinct from members of other groups. Therefore, helping flock members aids an indi-vidual in perpetuating its unique, flock-specific genes.

Pinyon Jays live in populations subdivided into social groups, however, their structure is unlikely to provide a climate conducive to indirect kin selection. First, group members share few genes. Second, successful families produce sons that stay in their natal flock and daughters that emigrate to neighboring flocks. This increases the degree of relatedness between members of different social groups. These two observations suggest that flock members are not sufficiently similar to each other nor are they sufficiently different from members of other flocks to favor co-operation within the group as a means of spreading an individual's genes. Such co-operation would spread a few of an individual's genes, but it would also spread many competing genes found in unrelated flock members. Indirect kin selection would, however, boost the evolution of cooperation and may have been especially important early in the evolution of sociality if primeval jay flocks, as we suggested in Chapter 1, included only a few extended families as in modern Gray-crowned Babblers and Mexican (Gray-breasted) Jays. At present, co-operation within a flock is likely to be maintained because it directly enhances co-operators' survival and reproduction.

In large societies like Pinyon Jay flocks indirect kin selection may be important, regardless of population structure, if relatives are recognized and preferentially aided (Brown 1987). Kin recognition would make it more difficult for competing genes to benefit from co-operative actions. Any casual observer who witnesses an entire flock of jays fly up in response to a sentinel's warning cries knows that warnings of impending danger are given not only to relatives. However, another co-operative activity, helping at the nest, is only practised amongst close relatives and as such the role of indirect kin selection in its evolution must be investigated further. First, however, we will set the stage for this investigation by examining the dominance structure within a flock. We will then focus on the monogamous pair, the basic building block of a Pinyon Jay flock. The association of a pair represents an extreme form of co-operation between two unrelated jays.

CHAPTER 6

# Dominance relationships within the flock

Watching a flock of Pinyon Jays move through their home range reminds us of watching a parade. Everyone moves in the same direction and is approximately evenly spaced so that movement is orderly and no collisions occur. Coordinated movements are critical for efficient use of space when foraging as well as when marching. This same pattern is seen when the flock takes to the air to fly to another location. Even though the flock flies rapidly and swirls about, we have never observed a mid-air collision, an event that could have disastrous consequences for the participants. Marching bands and precision flying corps practice for many months to perform in such tightly organized formations. Through long hours of practice, each member of the group becomes thoroughly familiar with the movements of every other participant, and can anticipate and rely on each individual to perform the proper maneuver at the proper instant. In most such groups there is usually a leader who gives commands and instructions. Individuals must always be aware of the signals given by this drum major or squad commander. Foraging and flying

Pinyon Jays, however, manage to execute the same kind of complex movement patterns without obvious practice and in the absence of a conspicuous leader. How do they do it? After many years of study we still cannot answer this intriguing question. We do, however, have some ideas about the dominance organization of the flock and know how this organization influences the ability of some individuals to capitalize on resources.

## WAR AND PEACE

### Contested resources

Outwardly a flock of Pinyon Jays appears as a harmonious, serene group of birds that co-operate in locating food, defending their young, and driving off dangerous predators. Any student of behavior must be impressed with this seemingly peaceful collection of birds performing their complicated maneuvers. We have seen birds feed and drink shoulder to shoulder on many occasions with hardly a scuffle. One such incident occurred when the birds discovered a large salt block. They crowded around the block so tightly that we could not possibly have detected what the birds were doing. The block was completely covered with birds and some birds were actually standing on the backs of others, yet fighting was rare (see illustration at beginning of this chapter).

In the study of animal behavior, animals are often classified as being either "distance animals" or "contact animals" (Hediger 1950). Distance animals will not tolerate the intrusion of another animal into what is known as its "individual distance". This is broadly defined as the space around the animal that is protected and defended from members of its own species. A classic example is the spacing pattern of European Starlings on telephone lines. Here, hundreds of birds line up at regular intervals. Some swallows, swifts, and nuthatches provide good contrasting examples of contact animals when they form dense roosting clusters. These two categories of behavior do not seem to fit our jays very well as they show no obvious and consistent individual distance, nor do they regularly cluster into tight groups. Pinyon Jays distribute themselves in space in a variety of patterns that appear to be dictated by the environmental conditions at hand. When resources are clumped, so are Pinyon Jays, and when resources are spread out, Pinyon Jays do the same.

Life in a flock of Pinyon Jays is not always a bowl of cherries. In late winter and early spring the distribution of flock members changes as pairs of birds begin to spend much time together. During this time, birds become aggressive towards other flock members. Mated females seem especially testy. Their hormones surge as the breeding season approaches giving them the avian equivalent of PMS which we call PBS (pre-breeding syndrome)! These females have a well-defined individual distance and any male other than their mate that enters this space will be repulsed. First, the female vocally assails the intruder with a Chirr (Chapter 4). If the approaching male is not deterred by this warning, the female may peck at him. We have seen such attacks escalate and birds become locked in combat with talons and bills brought into play.

During nest building we have also seen altercations over a choice morsel of nest material. Nest-building often begins while the ground is still covered with snow. The preferred sticks and grasses used in nest construction may be exposed only where differential warming has melted the snow or where wind has blown the ground bare. Birds crowd to these exposed sites and fill their bills rapidly with vegetation. We have seen numerous "tugs-of-war" between two birds that have the same piece of nest material in their bills. These types of competitive interactions usually end peaceably. However, things heat up when one jay robs construction materials from another's nest. Nest owners first attempt to drive the robbers away with vocalizations, but if this does not work, vicious fights ensue. Fighting birds lock feet and flap vigorously as they fall to the ground. Birds peck at each other with forceful stabs during these encounters. This combat at the nest is the most aggressive behavior observed during the year.

Aggression is also evident during seed caching and during cache recovery. If one bird observes another cache a seed it may fly directly over to the cache location and attempt to dig up the seed. It is the year's young that appear to specialize in such thievery. Adults attempt to reclaim their pillaged seeds and will chase off intruders.

### Assessment of dominance

In order to study dominance we attracted the Town Flock to feeding stations where we fed them sunflower seeds, mealworms, greased popcorn, and an occasional pinyon pine seed. One such feeder, which was visited regularly, was 1.3 m long by 0.66 m wide. It was not uncommon to observe more than 15 birds on this rostrum at one time. Here, as at the block of salt, the concept of individual distance became meaningless. Birds would cram onto this platform with some actually being forced off, or muscled to the edges where they were in danger of falling to the ground. When watching birds at these feeders one thing was obvious: Pinyon Jays have no table manners whatsoever. They eat with a rapid frenzy, scattering food in their wake, and displacing one another in rapid succession.

We were able to make detailed observations of aggressive interactions at feeders because we had accentuated the clumping of a needed resource, in this case food. In addition to reading color-band combinations to confirm that a certain bird was still alive, we kept track of as many aggressive interactions as we could record. Because of the relatively large number of displacements going on at the same time only a small proportion were ever recorded. When possible we noted the "winner" and the "loser" of such encounters. The winner was defined as the bird that maintained its position at the feeder after the confrontation. On the other hand, a loser was defined as the bird that lost its position at the feeder and either had to walk, hop or fly away.

Losers occasionally gave appeasement postures that also interfered with their ability to feed. Appeasement behaviors are easily identified signals of submission. Two signals of appeasement are Begging Behavior and Chin-up Posturing. Begging Behavior in an aggressive context is rather rare, but conspicuous when given. The subordinate crouches slightly with its head extended horizontally and its wings flapping or fluttering (see Chapter 7, Fig. 46). Courtship Feeding Chirrs often accompany this act. On occasion, a subordinate performs this act to a group of closely clumped birds as if to signal its status to the entire assemblage. Begging Behavior is typically used by females, and offspring still dependent upon their parents, to entice their mates or parents to feed them (Chapters 4, 7). Thus, begging is an interesting example of a behavioral act that sends more than one signal depending upon the context in which it is given.

The second appeasement behavior, the Chin-up Posture is also given as a response to a challenge by a dominant jay. During this act, the subordinate stands absolutely motionless with its bill pointed skyward. Not a single body feather is ruffled or quivered. Wings are held tightly against the body and legs are extended, pushing the jay's body above its neighbors'. The white throat patch is very conspicuous and stands in striking contrast to the blue body (see Chapter 7, Fig. 42). This posture may be maintained for up to 30 seconds.

*A subordinate jay points its bill skyward in the "chin-up" display to signal appeasement to a domi-nant jay at a feeder.*

At the feeder, a subordinate that performs an appeasement behavior is spared a peck from the dominant bird and is allowed to remain on the feeding platform. After a time, the subordinate may again resume feeding if it does not interfere with the feeding activities of the dominant. Thus, one must wonder why these behaviors were rare compared to the more common forms of submission: moving away from a dominant or flying off.

### Ethograms of aggression

In addition to scoring who won and who lost an encounter, we also recorded what the fighters did during the round. To this end we filmed feeding sessions with a high-speed movie camera that exposed 150 frames per second and then analyzed the movements of winners and losers in slow motion. We exposed many rolls of film to make complete records for 24 confrontations. A method used by pioneers in the study of animal behavior, the ethogram, provided an interesting summary of the behaviors of the winners and losers. An ethogram is a diagram showing the sequence of behavioral events and the transitions between these events.

The ethogram for winning was fairly simple (Fig. 36). An altercation was

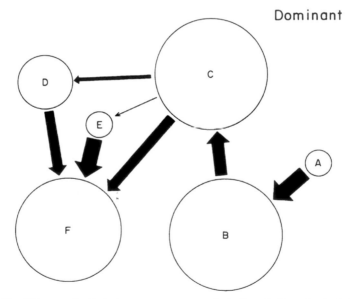

Figure 36   *Ethogram for birds that won encounters at the feeding platform. Circles indicate behavioral acts and arrows indicate transitions between acts. The size of the circles and the width of the arrows indicate the relative proportions of each act and transition. Acts of winners are abbreviated as follows: A = flies at subordinate (occurred only one); B = turns body and/or head to stare at subordinate (occurred 88% of the time); C = approaches the subordinate (occurred 92% of the time); D = pecks subordinate (occurred 23% of the time); E = turns away from the subordinate (occurred only once); F = occupies feeding position of the displaced bird (occurred 75% of the time).*

most often initiated when the dominant bird turned its head and/or its body toward the subordinate and looked directly at it (Fig. 36B). Next, the dominant leaned toward, or stepped toward, or lifted its leg as if to step toward the subordinate bird (Fig. 36C). The subordinate bird reacted to this approach by either giving a submissive posture, jumping away, or flying off. The dominant jay then moved into the feeding motion at this now unoccupied spot (Fig. 36F). Ninety per cent of the encounters in Figure 36 proceeded directly from B to C to F. Only 23% of the encounters included the dominant actually pecking at the subordinate (36D). On one occasion, the dominant bird turned away from the subordinate (36E), but the subordinate still left the platform.

In contrast to the ethogram for winning, the ethogram for losing was complex and variable (Fig. 37). Six different acts were involved in the process of losing and five of them were commonly executed. Ultimately, the losing jay left the feeding place it had occupied prior to the encounter (Fig. 37F). However, the steps preceding this retreat formed a tangled web. Seventy per cent of the time a subordinate responded to a dominant by either making a feeding motion towards it (Fig. 37A) or turning its head and/or body towards it

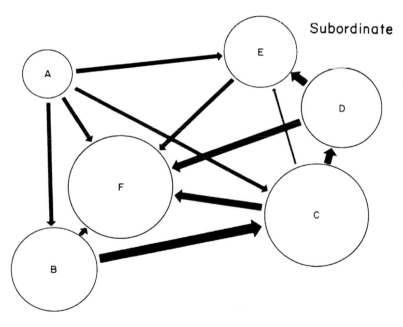

Figure 37    *Ethogram for birds that lost encounters at the feeding platform. Circles indicate behavioral acts and arrows indicate transitions between acts. The size of the circles and the width of the arrows indicate the relative proportions of each act and transition. Acts of losers are abbreviated as follows: A = makes a feeding motion towards dominant bird (occurred 17% of the time); B = turns head and/or body towards dominant (occurred 50% of the time); C = leans away or crouches in front of dominant bird (occurred 42% of the time); E = steps away from dominant (occurred 38% of the time); F = leaves the feeder (occurred 75% of the time).*

*A dominant jay flies at a subordinate on the feeder who crouches and moves away.*

(Fig. 37B). The submissive bird then either leaned away from or crouched before the dominant (Fig. 37C), turned its head away from the dominant (Fig. 37D), and/or raised a foot as if to step away from the dominant (Fig. 37E). Losers proceeded from each of these acts to retreat (Fig. 37F).

It is interesting to note that very few aggressive encounters included one bird inflicting harmful blows upon the other. Aggressive encounters usually ended with the winner simply glancing at the loser and apparently causing little physical harm. But, is this really true? The dominant bird maintains its position at the feeding platform and may even gain access to choice feeding spots. On the other hand, the subordinate is prevented from feeding, if only for a few seconds. Additionally, the subordinate usually gyrated through many more behaviors (each exacting some amount of time and energy) than did the dominant (compare the ethograms of winning and losing).

Why do subordinates perform such complicated maneuvers when they are challenged by dominants? The answer may be that they have much to lose by being displaced from a rich feeding location. Remember that the flock stays together as a unit most of the time, so when it leaves a feeding spot, *all or most* birds leave together. There is little doubt that some birds eat more than others. Thus, it behoves each bird to obtain as much food as is possible while the flock is eating. Better fed birds should be healthier, have greater survival, and produce more offspring than poorly fed birds.

The complicated moves of the subordinates may be their way of escaping notice from the dominant birds without actually leaving the feeder. All of these acts cost the loser time and energy, but if the loser can stall long enough, the dominant bird may be distracted or actually displaced by a more dominant bird, thus allowing the subordinate more time at the feeder. Subordinate jays can be thought of as underdogs in a heavy weight boxing match. Their "bob and weave" tactics may allow them to remain in the ring long enough to collect a respectable payoff.

### Controlled aggression

Why do dominant birds use such subtle signals to displace another bird? Why not use very overt, striking signals in confrontation? The answers to these questions are probably rooted in the social lifestyle of Pinyon Jays. Such restrained or controlled aggression is a characteristic of many animal societies (Lorenz 1966).

As we argued in Chapter 1, natural selection must be maintaining the complex social system of Pinyon Jays. All behaviors performed by members of the flock can be thought of as being either disruptive or beneficial to the social stability of the flock. Power struggles between individuals can be both, depending upon the situation and the results of these conflicts. For example, if aggressive behavior results in a stable social hierarchy, this may be beneficial because each member of the society will learn its place and can operate accordingly without having to deal with constant challenges and the ensuing turmoil. Clearly, as the number of aggressive acts increases, the cohesion of the flock decreases. The same can be said for the *intensity* of aggressive acts. What if a dominant bird signaled its status and intentions to all or most members of the flock with a single, bold act? Now, the receivers of this signal would have a number of options, including: submitting without further challenge, leaving the challenged site, ganging up on the dominant, shunning the dominant at a later time, or ignoring the signal. Flock stability would be enhanced if the receivers used the first option, but stability would be disrupted if any of the other options were pursued. Moreover, stability would be likely to degrade because bold advertisers may cause injury or draw the attention of predators. Furthermore, rare advertisements may be misinterpreted or simply missed by some flock members. Subtle advertisements can be repeated with little cost until each flock member knows their position in the social fabric of the flock. In this way subtle displays enhance social cohesion.

We have seen that life in a flock of Pinyon Jays is not as harmonious as it first appears. When resources are contested, fighting among flock members occurs, but it is almost always in a subdued fashion. Flock integrity is therefore maintained. Now we will turn our attention to the flock's social hierarchy.

Seasonal changes in aggression and social status

*Methodology*

We recorded aggressive acts when we could determine the identity of both participants as the flock visited our feeder in 1972 and 1973. From these observations we calculated the probability of the different cohorts (a group of birds of the same sex and age) engaging in an aggressive encounter. This was done as follows. If all interactions between flock members occurred at random, then the chance of being involved in a fight should be predicted by how often any given cohort visited the feeder. This is the "expected" number of conflicts. To calculate this expectation per cohort, we first divided the total number of visits of all individuals belonging to all cohorts by the total number of encounters we observed. This number was the probability of any bird engaging in an aggressive act on any given visit. This probability was then multiplied by the number of conflicts for that cohort. This "expected" number of conflicts was then compared to the actual number of conflicts we observed for that cohort. Data on these social interactions were lumped into three periods corresponding to different stages in the annual cycle of the flock: June–September, October–December, and January–May.

*June–September*

During late summer and autumn the flock is fragmented into groups of birds that are in different stages of breeding (Chapter 1, Fig. 5). Those pairs that lost their initial nests are re-nesting in small colonies, and those that succeeded in their initial attempts are feeding young. Non-breeding yearlings have formed a wandering band. Adult birds are also molting and coming to the feeder on an irregular schedule. Fledglings are in creches which move around the home range and often merge at the feeders. These creches contain some, but not all, the adults. Thus, birds do not have equal opportunities to engage all other flock members in aggressive behavior.

Aggression at this time of the year is quite high. We observed 2904 birds visiting our feeder and they engaged in 960 aggressive acts. Thus 33% (960 out of 2904) of visits end up in aggression. Put another way, on average any given bird, on any given visit would have a 33% chance of engaging in an aggressive conflict. However, there is no average bird or average cohort in the flock. Each interacts at a different level of aggression (Table 5). Adult males, for example, fought on 43% of their visits, whereas adult females fought only 22% of the time. Yearling females were the least aggressive, fighting only 11% of the time. It is interesting to note that juveniles were not the meek and timid "newcomers" one might expect them to be. They initiated 640 fights (two-thirds of all those we recorded) and fought on 36% of their visits to the feeder. Adult males fought slightly more than expected with other adult males and females, and fought less than expected with the juveniles. Adult females also fought less often than expected with the juveniles. These young were very belligerent, and were constantly squabbling amongst themselves, having 149 more aggressive

**Table 5.**   *Total number of visits to feeder and aggressive acts for each cohort of jays in the Town Flock from June to September.*

| Cohort | Number of visits to feeder | Number of conflicts observed | Number of conflicts expected | Probability of initiating a conflict |
|---|---|---|---|---|
| Adult male | 342 | 148 | 113 | 0.43 |
| Juvenile | 1796 | 640 | 593 | 0.36 |
| Yearling male | 235 | 84 | 78 | 0.36 |
| Adult female | 258 | 58 | 85 | 0.22 |
| Yearling female | 273 | 30 | 90 | 0.11 |
| Total | 2904 | 960 | | |

encounters than predicted. However, they had fewer than expected squabbles with all other cohorts.

Adult males are the dominant cohort during this season (Fig. 38). They defeated adult females in 63% of their encounters, yearling males 59% of the time, and yearling females 86% of the time. However, these tyrants only dominated juveniles 59% of the time. Juveniles were on equal status with yearling males and dominated yearling females 61% of the time.

In summary, aggression during this four-month period is distinguished by fighting and squabbling among the juveniles, and the overall domination by the adult males.

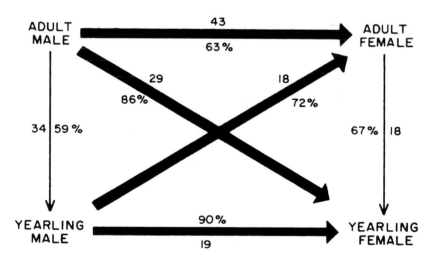

Figure 38   *Winners and losers of interactions between adult and yearling cohorts in the Town Flock from June to September. Arrows point from winners to losers. Numbers along arrows indicate the number of times members of winning cohorts defeated members of losing cohorts. Percentages are the percent of interactions between cohorts that were won by the dominant cohort. Thick arrows indicate that the proportion of wins was statistically significant (greater than a 50:50 ratio, P < 0.05, Binomial test).*

*October–December*

As fall gives way to winter the flock exists as a single unit that forages, caches seeds, and roosts together (Chapter 1). On rare occasions, small groups splinter from the flock. The young birds raised during the summer are now fully integrated into the flock. Thus, all birds have an equal opportunity to interact at the feeder.

The overall level of aggression at this time of the year was slightly less than during the previous months. During this four-month period, we recorded 5728 birds visiting our feeder and 1423 aggressive encounters (Table 6). On a single visit to the feeder, every bird had roughly a 25% chance of engaging in a confrontation. However, not every cohort interacted according to this expectation (Table 6). Adult males fought on over half (55%) of their visits to the feeder. This was the highest seen for any cohort at any time of the year. In contrast, the other cohorts fought on only 10–13% of their visits. Most of the aggressive acts by the adult males were directed at other adult males, but they also fought slightly more than expected with all other cohorts except the juveniles. In fact, all cohorts engaged these young birds less than expected. Adult females seemed particularly averse to fighting with juveniles as they had 130 fewer encounters with juveniles than expected according to the calculated probability. These young birds appeared to have equal social status with all other birds in the flock except adult males. Further, they interacted with each other at below chance levels, thus showing a decided decline from earlier in their lives.

**Table 6.**    *Total number of visits to feeder and aggressive acts for each cohort of jays in the Town Flock from October to December.*

| Cohort | Number of visits to feeder | Number of conflicts observed | Number of conflicts expected | Probability of initiating a conflict |
|---|---|---|---|---|
| Adult male | 1809 | 995 | 448 | 0.55 |
| Yearling male | 387 | 49 | 96 | 0.13 |
| Juvenile | 1706 | 202 | 423 | 0.12 |
| Adult female | 1300 | 127 | 322 | 0.10 |
| Yearling female | 526 | 50 | 130 | 0.10 |
| Total | 5728 | 1423 | | |

Adult males continued their domination of the flock. They won 76% of all interactions with adult females. They dominated yearling males in 80% of encounters and won 87% of encounters with yearling females (Fig. 39). During this time, adult males clearly dominated juveniles, winning 72% of these interactions. Adult females were dominant over yearlings of either sex. Yearling females were clearly subordinate to all other cohorts.

In summary, this period is typified by high levels of aggression by the adult males. They continue to dominate all cohorts in the flock. All other cohorts demonstrated similar, but relatively low, levels of aggression.

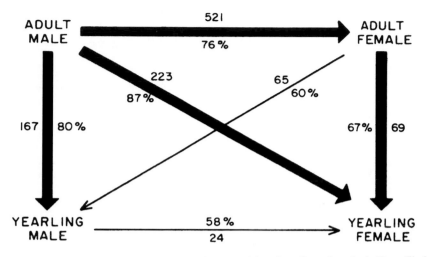

Figure 39    *Winners and losers of interactions between adult and yearling cohorts in the Town Flock from October to December. Legend is the same as in Figure 38.*

*January–May*

As spring approaches, courtship continues and nesting begins. The flock is still merged into a single unit but yearlings occasionally split off from the main body of the flock. During this period the flock regularly visited our feeder and we recorded 11 007 individual bird visits. Spring and summer represent an ebb in aggression as only 14% of these visits resulted in aggression. Again interactions were not distributed equally across all cohorts. Adult males were still the most aggressive cohort in the flock, fighting on about 25% of their visits to the feeder (Table 7). Still, this is only about half as aggressive as they were earlier in the fall. This is the lowest level of aggression these males demonstrated throughout the entire year. Second-year males (referred to as yearlings in the preceding sections), most of whom did not breed the year before, became more aggressive than the remaining cohorts which experienced a relatively sharp

**Table 7.**    *Total number of visits to feeder and aggressive acts for each cohort of jays in the Town Flock from January to May.*

| Cohort | Number of visits to feeder | Number of conflicts observed | Number of conflicts expected | Probability of initiating a conflict |
|---|---|---|---|---|
| Adult male | 4708 | 1157 | 671 | 0.25 |
| Second year male | 685 | 113 | 98 | 0.16 |
| Second year female | 878 | 57 | 125 | 0.06 |
| Adult female | 2812 | 141 | 400 | 0.05 |
| Yearling | 1924 | 100 | 274 | 0.05 |
| Total | 11007 | 1568 | | |

decline in aggression from the earlier period. Second-year females, adult females, and yearlings (referred to above as juveniles) were the least aggressive (Table 7). Their levels of aggression were the lowest for any cohort at any time of the year.

Although this was a time of relative calm in the flock as viewed from the overall levels of aggression, such was not the case for adult males, which had 330 more confrontations with other males than expected by chance. However, just the opposite was occurring between adult males and adult females. Males initiated 78 fewer fights with females than were predicted. This is probably a result of the preparation for breeding that is occurring at this time. All cohorts, especially adults, were still interacting less than expected with last year's young. These young also fought less than expected among themselves, but they were clearly dominated by adults and second-year males.

During the winter and early spring, adult males were again the dominant cohort and subdued all other cohorts (Fig. 40). These birds dominated adult females in 80% of their fights, yearlings in 79% of fights, second-year males in 66% of fights, and second-year females in 78% of fights. Second-year females were the most subordinate cohort in the flock.

In summary, this period is characterized by relatively low levels of aggression between cohorts, and a high level of aggression between adult males. Females and young birds rarely fight.

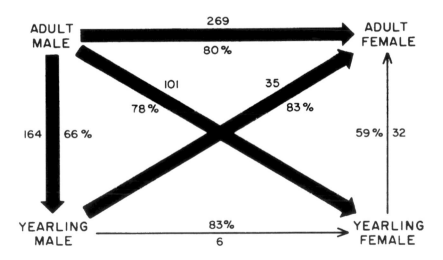

Figure 40    *Winners and losers of interactions between adult and yearling (now second-year birds) cohorts in the Town Flock from January to May. Legend is the same as in Figure 38.*

## WHY ARE JUVENILES AFFORDED SPECIAL STATUS?

An interesting question that arises concerns the apparent aggressive nature of juveniles and their integration into the flock. In most passerines, young are not tolerated after they become independent. In fact, in many cases they are actively driven off the territory. But, in Pinyon Jays these young are not only allowed to remain in the flock, their high levels of aggression during their first few months of life are tolerated. Juveniles confronted each other 29% more often than they would if encounters were randomly distributed among all flock members. A disproportionate number of these conflicts were initiated by young males. Adult males are the most dominant cohort in the flock, and it appears that male aggression is common from the onset of fledging. At this time of intense conflict among the juveniles, the weather is mild and natural foods are at peak abundance. Thus, minimal energy demands are placed on the squabbling juveniles. It is interesting that aggression drops off dramatically once the flock has reformed in the autumn. Without this reduction in fighting the flock could probably not function as a unified social unit. The mechanisms causing this decline in aggression are not known, but may involve the dominant actions of the adult males. The high level of aggression among these young birds may determine the social status a bird will later assume in the flock. Thus, social status may be determined very early in life and, as we argue in Chapter 11, will profoundly alter a jay's course in life.

Not only are young birds aggressive toward each other but they also are bold and fearless in their interactions with adults. We have seen juveniles push adults off the feeder, land on the backs of adults in order to obtain a feeding spot, and actually take food out of the mouths of adults. The adults seldom appeared to be threatened by these pushy behaviors and actually deferred to the young at concentrations of food. Adult females were the most passive cohort and showed the greatest tolerance of the young throughout their first year of life.

Why do adult birds endure these behaviors from young birds? We have no satisfactory answer, but as mentioned earlier this boldness is most evident during the mildest time of the year, when the cost of deferring to the young is minimal: adults can defer without placing themselves in jeopardy.

A CONFUSING DOMINANCE HIERARCHY

*Dominance of male Pinyon Jays*

There is little doubt that adult males are in aggressive control of the feeding platform, usually dominating all other cohorts and showing relatively high levels of aggression among themselves. Simply put, they fight more than any other sex and age class of jays. To some, this fighting and squabbling may bring back memories of the actions of barnyard chickens. Here the chickens, young and old, are constantly in a power struggle for control of food and mates. In fact, much of our original knowledge and conceptual thinking about social organization stems from work done on this lowly bird by scientists at the University of Chicago in the early part of this century. There, Nick Collias and his colleagues made careful observations of flocks of chickens and determined that they developed a relatively stable social system (Collias 1943). After a period of adjustment, chickens form a dominance hierarchy that we call a "pecking order" (Brown 1975). In this system every chicken knows its place, and there is one completely dominant bird that literally rules the roost! This bird, known as the "alpha male", can peck and win fights with every rooster in the flock. All males in the flock are subordinate to this male. Next in line is the "beta male". This bird is dominated by the alpha, but can himself dominate all the other males in the flock. This linear pecking order continues down to the last bird, who cannot defeat any bird and loses to every bird in the flock.

The intriguing question that prompted this analogy with chickens is: do adult male Pinyon Jays show a similar social hierarchy? To test for this we required marked birds and piles of data collected over a relatively long span of time. We have such data only for adult males because they frequently interact with one another.

Let's begin our search for the "alpha male" by examining interactions between September and December of 1972. This was a time of maximum interaction between males and we recorded 861 aggressive encounters between 47 different adult males. Many birds engage only infrequently in aggressive behavior and a few appeared to be obsessively aggressive. There did not appear to be any relationship between the number of encounters a bird engaged in and the proportion of those encounters that he won. For example, Light Blue-Yellow interacted 53 times and won only 36% of his skirmishes, and Mr Green fought on 52 occasions and won 58% of his battles. On the other hand, some birds fought seldom and won most of their engagements. For example, Light Green-Black fought only 12 times and won all 12, and White/White-Yellow fought 13 battles and won ten. Thus, the first point that became evident was that the biggest "bullies" were not necessarily the biggest winners.

If we examine all fights among males, then one clear-cut champion emerges, Light Green-Black. But, because he fought so seldom (only 12 times), we never suspected his high social position in the flock until the entire data set was analyzed. Light Green-Black can be characterized as the "strong, silent type". No other birds came close to achieving this commanding degree of dominance, although several males won roughly 70–80% of their fights (Table 8). Thus, as a first approximation, it appears that a few select males (seven or eight) are in a

**Table 8.**   *Percentage of aggressive encounters won by eight dominant males fighting among themselves, with other males, and with other cohorts in the Town Flock from September to December, 1972. Table entries are percentages followed by the number of encounters in parentheses.*

| Bird | Fights with all males | Fights with dominant males | Fights with other cohorts |
|------|-----------------------|----------------------------|---------------------------|
| Light Green-Black | 100 (12) | 100  (6) | 67 |
| Dark Green-Red | 82 (17) | 67  (3) | 89 |
| Dark Blue-Purple | 80 (44) | 40 (10) | 84 |
| Purple-Orange | 78 (51) | 63 (19) | 94 |
| Light Green-Yellow | 79 (24) | 57  (7) | 84 |
| White/White-Yellow | 77 (13) | 67  (3) | 84 |
| Yellow-Purple | 76 (49) | 33 (12) | 97 |
| Light Blue-Black | 64 (36) | 10 (10) | 92 |

dominance class above all other males. These birds, overall, won 83% of all their clashes with other adult males (Table 8), whereas the remaining 39 males won only 41% of all of their confrontations with males. Earlier we mentioned that adult males were the dominant cohort overall, but now it looks as if only a small subset of males may in fact be the dominant group.

Let's take a closer look at these dominant males. Collectively, they lost 56 battles with other males and 35 (63%) of these were to other dominant males. However, considering only the fights among males, we could not determine a linear dominance hierarchy among the dominant males. Relative ranks become very murky when we remove Light Green-Black (the obvious alpha) and Light Blue-Black (the obvious loser, Table 8). Two major problems arose in our attempts to arrange the dominant males. First, all birds in this class did not interact with all the other birds and some interacted far more frequently than expected. Second, there were some reversals that were difficult to interpret. Competition among these dominant males was stiff as suggested by the fact that, except for Light Green-Black, each male lost more fights against the dominant males than against the rest of the male cohort (Table 8).

Even within the most dominant cohort in the flock (the adult males) there are some real losers! Many males with nine or more fights lost 80% or more of these contests. The lowest ranking adult males included White-Orange (17% wins), Light Pink-Yellow (16% wins), Dark Pink-Light Pink (11% wins), Purple/White-White (15% wins), Purple/White-Orange (12% wins), and Dark Green/Light Green-Light Blue (15% wins). This last male appeared to be the lowest ranking adult male. None of the males in this lowest ranking group was among the most frequent fighters. These subordinate males did not defeat the members of the other cohorts in the flock any more frequently than they defeated other adult males.

The remaining 33 males were difficult to categorize because they either fought so little or won only 30–50% of their encounters.

## A further look at the male dominance hierarchy

Even after we collected over 850 observations of aggressive interactions we were still not certain of the type of social hierarchy exhibited by the males in our flock of Pinyon Jays. We were confused because some dominant birds were never observed interacting during this first year of study. Also, during the fall there was a major influx of vagrant birds into the flock, most of which either perished or moved on during the winter. These visitors clouded our view of the social hierarchy by interacting with a few resident males then leaving before we could document their status. In order to distill a clearer picture of the male social hierachy we studied interactions between males for another year (1973) and lumped our observations across the seasons together. This provided 1706 observations of aggressive encounters between 41 adult males.

We did not have complete records for all 41 males. Nine of them died before the end of the year so we were unable to determine their status and they were removed from the analysis. These included three of the subordinate males from the previous year (Purple/White-White, Purple/White-Orange, and

Light Green/Dark Green-Light Blue). Only one of the previously high-ranking males died (Yellow-Purple).

We had excellent views of fighting ability for 13 males who interacted on 100 or more occasions. Mr Green was again the most pugnacious bird. He fought 169 times and won 57%. In contrast, White-Orange was the second most combative bird fighting 151 times, but he won only 10% of his fights. This reinforces our earlier conclusion that the most aggressive birds are not necessarily the most dominant birds in the flock.

The relative ranks of males that we sketched out during the first fall of study appeared to be quite stable. Light Green-Black continued to rule the flock. During 1973 he fought 111 times with other males and lost only three times. He was the king of the flock for the entire 16 months that we quantified dominance. On the other side of the fence, White-Orange was never able to elevate himself from his last place ranking. Given the large number of birds in the flock this seems rather remarkable, but it serves to underscore the stability of the flock through time.

In our continuing quest to determine the social hierarchy for the adult males, we selected the males that had the best win/loss records with all males and then attempted to arrange them into a linear order based on how they fared in contests with members of this select group. In order to qualify for this analysis a bird must have won over 50% of his battles and have been involved in at least 30 fights with other dominant males. Fourteen males fit this bill (Table 9). We followed the methods in Brown (1975) in order to construct a win/loss matrix (Table 10). Here the number of wins each bird has with every other bird is listed horizontally beginning with the alpha male and progressing to the least dominant male. Losses to each bird are arranged in vertical columns. Wins can then be summed horizontally and losses can be summed vertically. If a diagonal line is drawn through the matrix where each bird

**Table 9.**    *Win-loss records for the 14 most dominant males in the Town Flock fighting with each other in 1973. The identity of birds an individual defeated can be found by examining the rows in Table 10. Likewise, the identity of the birds an individual lost to can be found by examining the columns in Table 10.*

| Bird | Number of fights | Number of wins (%) | Number of losses |
|---|---|---|---|
| Light Green-Black | 57 | 56 (98) | 1 |
| Purple-Orange | 78 | 63 (81) | 15 |
| Dark Blue-Purple | 89 | 59 (66) | 30 |
| Red-Red | 41 | 31 (76) | 10 |
| Light Green-Yellow | 53 | 28 (53) | 25 |
| Light Blue-Black | 39 | 23 (59) | 16 |
| Dark Green-Red | 61 | 34 (56) | 27 |
| Orange-Orange | 30 | 11 (37) | 19 |
| Dark Pink-Light Blue | 56 | 19 (34) | 37 |
| White-Light Blue | 66 | 23 (35) | 43 |
| Light Pink-Light Green | 55 | 15 (27) | 41 |
| Mr Green | 68 | 12 (18) | 56 |
| White/White-Yellow | 50 | 13 (26) | 37 |
| Light Blue-Dark Blue | 38 | 4 (11) | 34 |

**Table 10.** *Win-loss matrix for the 14 most dominant males in the Town Flock fighting with each other during 1973. Entries are the number of times that the row bird defeated the column bird. This results in the row bird's wins appearing above the diagonal and losses appearing below the diagonal. Dashes indicate that no conflicts occurred between these two birds. Letters are used to code each bird's identity and are assigned to the 14 birds in the order that they are listed in Table 9 (A = Light Green-Black, N = Light Blue-Dark Blue, etc.).*

| | Column Bird | | | | | | | | | | | | | |
|---|---|---|---|---|---|---|---|---|---|---|---|---|---|---|
| Row Bird | A | B | C | D | E | F | G | H | I | J | K | L | M | N |
| A | | 4 | 3 | 4 | 7 | 3 | 11 | 1 | 6 | 5 | 4 | 2 | 3 | 3 |
| B | 0 | | 11 | – | 8 | 0 | 2 | 4 | 6 | 5 | 5 | 13 | 6 | 2 |
| C | 0 | 1 | | 2 | 5 | 7 | 1 | 1 | 1 | 5 | 10 | 14 | 7 | 5 |
| D | 0 | – | 2 | | 0 | 3 | 3 | 1 | 8 | 5 | 3 | 5 | 0 | 2 |
| E | 0 | 5 | 3 | 2 | | 3 | 2 | 1 | 5 | 1 | 1 | 1 | 4 | – |
| F | 0 | 1 | 3 | 0 | 1 | | 1 | – | 2 | 2 | 3 | 6 | 2 | 2 |
| G | 0 | 4 | 1 | 0 | 0 | 0 | | 2 | 4 | 7 | 1 | 2 | 6 | 7 |
| H | 0 | 0 | 2 | 0 | 0 | – | 0 | | 1 | 0 | 2 | 2 | 4 | – |
| I | 0 | 0 | 2 | 0 | 0 | 0 | 4 | 2 | | 2 | 4 | 1 | 2 | 2 |
| J | 0 | 0 | 1 | 0 | 2 | 0 | 0 | 5 | 0 | | 4 | 2 | 2 | 7 |
| K | 0 | 0 | 0 | 0 | 2 | 0 | 2 | 0 | 0 | 2 | | 7 | 0 | 2 |
| L | 1 | 0 | 0 | 0 | 0 | 0 | 0 | 1 | 4 | 5 | 0 | | 1 | 0 |
| M | 0 | 0 | 2 | 2 | 0 | 0 | 0 | 1 | 0 | 1 | 4 | 1 | | 2 |
| N | 0 | 0 | 0 | 0 | – | 0 | 1 | – | 0 | 3 | 0 | 0 | 0 | |

would encounter himself (see Table 10), then a perfectly linear hierarchy occurs when all of the wins fall above this diagonal. As can be seen in Table 10 this is not the case for our male Pinyon Jays. There are numerous reversals indicated below the diagonal line where subordinates defeat their superiors. These reversals are especially common for two high ranking males, Purple-Orange and Dark Blue-Purple. A recent quantitative technique developed by Appleby (1983) allows us to test the linearity of this hierarchy. This test indicates that, despite these reversals, male Pinyon Jays are organized into a significantly linear hierarchy ($\chi^2 = 52.04$, DF = 22, P < 0.001).

Our win/loss matrix clearly identifies Light Green-Black as the alpha male. He is followed by a group of six birds, Purple-Orange, Dark Blue-Purple, Red-Red, Light Green-Yellow, Light Blue-Black, and Dark Green-Red. These six birds, who we will refer to as "dominants", also won more than 50% of their encounters with other males in this group. The remaining seven males, who we will refer to as "subdominants", lost over half of their encounters within the dominant male group and were no match for the top seven birds.

The fact that reversals in the social hierarchy of these 14 males occasionally occur should come as no surprise. In a group as large as a Pinyon Jay flock reversals can come about in a number of ways. In a large, ever-moving flock one bird can easily surprise another. They may also be due to mistaken identity during feeding frenzies at the platform. Some of the reversals may also be the result of an individual's family background. It is possible that fathers, sons, and brothers may defer to one another in some fashion unknown to us. Unfortunately, these data were gathered before we knew the kin structure of our study flock and therefore we cannot assess its influence on dominance. Age may also explain some of the reversals we noticed. In many species dominance is gained with increasing age or experience. All we knew about our top-ranking males was that they were at least two years old. It would be interesting to know which of these birds were in the prime of their lives, which were up-and-coming, and which were in their twilight years.

### Changing of the guard

In 1976 the alpha male (Light Green-Black) disappeared. Presumably he died. We have no quantitative data on the process leading to the crowning of a new king, but qualitative observations suggest the following scenario. For about three weeks after his disappearance, no overt differences were observed when the flock visited our feeders. Shortly thereafter, a subdominant male (Orange-Orange) became very aggressive and often initiated encounters with many members of the dominant male group including the former beta, Purple-Orange. Orange-Orange was mildly successful in these challenges and it appeared that he might succeed in his "take-over bid". Suddenly, Orange-Orange became a very subdued member of the subdominant group and another subdominant, Dark Green-Red took over as the alpha male. Purple-Orange remained as the beta male until he suffered an injury and dropped in status. Interestingly, Dark Green-Red remated with the former alpha male's widow at the time of his ascension to alpha status. Apparently this female either preferred dominant males or was of very high quality and was thus selected by dominant males. Alternatively, a male's dominance may be influenced by his mate. Given the subordinate nature of female Pinyon Jays, we doubt this, but cannot rule it out.

### CONSEQUENCES OF STATUS

#### Does status influence visitation rate to the feeder?

We kept track of the total number of visits each male made to the feeder in order to determine if this might be influenced by their status. A slight trend is evident (Table 11). Mr Green, a subdominant, was the most frequent visitor for the year, followed by the two top-ranking birds in the flock. The lowest ranking subdominant male (Light Blue-Dark Blue) made the fewest visits to the feeder. This suggests that dominants visited the feeder more regularly than subdominant males. However, the relationship among the top 14 males is not significant.

Subdominants may prefer to sit in the trees surrounding the feeder rather than enter the fracas where they will be beaten by a more dominant bird. However, in a single, carefully timed visit to the feeder, a subordinate might be able to collect dozens of seeds and make up for time lost while waiting for the feeder to clear. Although we do not have data on the number of seeds taken by jays during these visits, laboratory studies suggest that subordinate birds do make up for lost time and seeds when caching.

**Table 11.** *Total visits to the feeder by the 14 most dominant males in the Town Flock and the frequency with which they fought other flock members. Total numbers of fights are listed in the table and frequency of fighting is listed in parentheses. For example, Light Green-Black visited the feeder 274 times and fought with males on 104, or 38%, of these visits. On 57, or 21% of his visits he fought with one of the other top 14 males, and on 65, or 24%, of his visits he fought with a jay in one of the other cohorts.*

| Bird | Number of visits | Fights with all males | Fights with dominant males | Fights with other cohorts |
|---|---|---|---|---|
| Light Green-Black | 274 | 104 (38) | 57 (21) | 65 (24) |
| Purple-Orange | 311 | 202 (65) | 78 (25) | 75 (24) |
| Dark Blue-Purple | 232 | 192 (83) | 89 (38) | 95 (41) |
| Red-Red | 253 | 126 (50) | 41 (16) | 78 (31) |
| Light Green-Yellow | 272 | 100 (37) | 53 (19) | 40 (15) |
| Light Blue-Black | 246 | 78 (32) | 39 (16) | 41 (17) |
| Dark Green-Red | 246 | 158 (64) | 61 (25) | 79 (32) |
| Orange-Orange | 252 | 60 (24) | 30 (12) | 22 (9) |
| Dark Pink-Light Blue | 213 | 190 (89) | 56 (26) | 88 (41) |
| White-Light Blue | 267 | 168 (63) | 66 (25) | 78 (29) |
| Light Pink-Light Green | 205 | 170 (82) | 55 (26) | 48 (23) |
| Mr Green | 326 | 196 (60) | 68 (21) | 88 (27) |
| White/White-Yellow | 238 | 104 (44) | 50 (21) | 26 (11) |
| Light Blue-Dark Blue | 136 | 134 (98) | 38 (28) | 82 (60) |

Nancy Stotz (1991) tested Pinyon Jays of known social status in cache and recovery sessions in which they interacted with jays of other social status. Subordinate birds, which often lost seed caches to dominant, thieving jays, tended to cache at higher rates than dominant birds. Dominant birds made relatively few caches. These dominant jays also probed many more empty sites when searching for seeds than did subordinates. Since, in this laboratory setting, dominant birds had few of their own seeds to search for, they may have been looking for the seed caches of other jays. Such subtle effects may also have been occurring at our feeding station, however detecting them when dozens of wild birds were present simultaneously would have been difficult.

*Does status influence aggression?*

We used the visit frequency data discussed above to calculate the probability of a male fighting when he visited the feeder (Table 11). We then looked for correlations between status and aggression as indexed by this fighting probability. No clear picture emerged. The bird which fought most frequently per visit was Light Blue-Dark Blue, the lowest ranking subdominant male. This bird rarely came down to the feeder, but when he did, he fought often with the other birds there, especially the adult males. He fought an average of 1.86 battles per visit and 98% of these were with other males. Dark Blue-Purple, a high ranking dominant, was the second most frequently engaged bird. On average, he fought 1.62 times per visit and most (83%) of these skirmishes were with other males. Dark Pink-Light Blue, a subordinate, was the third most aggressive bird. He fought an average of 1.56 times per visit, primarily

(89%) with other males. Light Pink-Light Green ranked fourth in aggressiveness. He fought an average of 1.33 times per visit, primarily (82%) with other males. The alpha male ranked tenth in aggressiveness. Apparently, the dominant male is not the one that is most often engaged in confrontations with his peers or other flock members. It appears that some birds were more inclined to fight than others, but this inclination does not strongly reflect a bird's current social status. We cannot help but wonder if the fighters are in fact birds that are trying to improve their social position in the male cohort. Accordingly, the alpha male rarely fights, perhaps because he has little to prove and therefore does not need to waste his time beating the other males.

One pattern was fairly obvious. The birds that were the most aggressive primarily fought other males. A highly significant correlation exists between the total number of fights and the proportion of those fights that were with other males (Spearman's $r = 0.87$, $P < 0.01$). Thus it appears that the birds that are prone to be fighters do most of their scrapping with other males.

### Interactions of dominant males with others in the flock

Some interesting trends are apparent when we look at the first fall's data on the interactions of the dominant males with the other cohorts in the flock. As mentioned earlier, as a group the adult males dominate all other cohorts. Surprisingly, however, the ranking for the dominant males within their cohort did not predict how they fared with the other cohorts (Table 8). Take Light Green-Black, for instance, who dominated the dominant adult males, but only won 67% of his battles with all other cohorts. In contrast, lowly Light Blue-Black rarely defeated dominant males, but won 92% of his fights with other cohorts. The second lowest ranking male in the dominant group, Yellow-Purple, also won the vast majority of his fights with other cohorts. Subdominant males were not the only males successful against other cohorts. Purple-Orange, a high ranking male, also consistently defeated members of other cohorts.

The 14 dominant males all won a high percentage of their fights with other members of the flock (Table 12). A couple of interesting points deserve mention. First, the dominant seven males lost a total of only 30 encounters with the other cohorts during the course of an entire year. Of these, 19 (63%) were to juvenile birds. The subdominants lost a total of only 48 fights to members of other cohorts and again many of these (23 or 48%) were to juveniles. Thus, even the highest ranking males defer to the young of the year at feeding stations. Second, subdominant males lost encounters to yearling females at a greater rate than the dominant males did. Were some of these males attempting to establish bonds with these females? Another question that will have to remain unresolved!

### Morphological correlates of dominance

Were there any morphological differences that might act to bestow dominance on some birds and subordination on others? Intuitively, one might think

**Table 12.**   *Percentage of fights won by the 14 most dominant males in the Town Flock when they engaged flock members in other cohorts.*

| | | Fights with: | | |
|---|---|---|---|---|
| Bird | Juveniles | Yearling males | Adult females | Yearling females |
| Light Green-Black | 89 | 95 | 96 | 100 |
| Purple-Orange | 64 | 95 | 98 | 100 |
| Dark Blue-Purple | 67 | 95 | 98 | 100 |
| Red-Red | 80 | 96 | 97 | 100 |
| Light Green-Yellow | 100 | 100 | 100 | 75 |
| Light Blue-Black | 72 | 100 | 94 | 100 |
| Dark Green-Red | 86 | 93 | 100 | 100 |
| Orange-Orange | 20 | 75 | 67 | 100 |
| Dark Pink-Light Blue | 92 | 96 | 93 | 100 |
| White-Light Blue | 50 | 92 | 97 | 60 |
| Light Pink-Light Green | 86 | 100 | 96 | 86 |
| Mr Green | 57 | 95 | 93 | 67 |
| White/White-Yellow | 100 | 80 | 100 | 100 |
| Light Blue-Dark Blue | 64 | 94 | 96 | 80 |

of physical size acting in such a fashion. Larger birds might be able to "throw their weight around" so to speak, or show off their large weapons (bills) to other birds. This appears only to be the case with respect to bill length for male jays. The 14 most dominant males had bills averaging 35.6 mm long (SD = 1.4) and weighed an average of 111.1 g (SD = 4.6). Our sample of subordinate males (N = 17) had bills averaging 34.7 mm long (SD = 1.2) and weighed an average of 110.1 g (SD = 4.1). Dominant males had significantly longer bills than subordinates (t-tests comparing means for bill size: t = 1.9, DF = 29, $P_{(one-tailed)} < 0.05$). However, body mass did not differ significantly between dominants and subordinates (t = 0.63, DF = 29, P > 0.50). Among the most dominant males, larger size was not linearly correlated with status (Table 13).

**Table 13.**   *Bill lengths and body masses for the 14 most dominant males in the Town Flock. Measurements are averages of multiple measures made on each bird during its lifetime. Males are listed from the most to the least dominant.*

| Bird | Average bill length (mm) | Average mass (g) |
|---|---|---|
| Light Green-Black | 38 | 119 |
| Purple-Orange | 34 | 102 |
| Dark Blue-Purple | 36 | 113 |
| Red-Red | 37 | 112 |
| Light Green-Yellow | 36 | 112 |
| Light Blue-Black | 36 | 104 |
| Dark Green-Red | 36 | 114 |
| Orange-Orange | 34 | 107 |
| Dark Pink-Light Blue | 38 | 109 |
| White-Light Blue | 35 | 113 |
| Light Pink-Light Green | 36 | 114 |
| Mr Green | 34 | 113 |
| White/White-Yellow | 34 | 116 |
| Light Blue-Dark Blue | 35 | 108 |

However, the alpha male had the longest bill and weighed more than any bird we have ever measured in the flock. Thus, even though stature may not be a sure ticket to dominance (all dominants are not larger than all subordinates), it may be an important prerequisite to ruling the flock.

## *The losing behavior of dominants and subordinates*

Earlier in this chapter we discussed different ways that birds reacted when they lost a confrontation. Birds either moved away, flew away, or gave one of two appeasement postures. In our first fall's data, we examined 56 losses by the most dominant males and 805 losses by the most subordinate males in order to see if birds in these groups lost skirmishes differently. In the vast majority of cases, the losers either flew off or moved away from the dominating bird. Therefore, there were no overall significant differences in the behavior of dominants and subordinates. However, the use of appeasement behaviors showed an interesting difference. Dominant males gave the Chin-up Appeasement posture during only 5% of their losses and never gave the Begging Appeasement posture. In contrast, subordinate males gave the Chin-up Posture in 9% of their losses and gave the Begging Appeasement in 3% of their losses. A similar relationship held when we analyzed our second year of data. The seven dominants gave appeasement gestures during only 4% of their losses. The seven subdominants used appeasement gestures in 11% of their encounters. All the other lower ranking males gave appeasement gestures in 17% of their losses. It appears that dominant birds do not often stoop so low as to appease their superiors! Instead, they simply move to a new location where there is less likelihood that they will be challenged.

## *The influence of status on reproductive success*

Dominant jays have greater access to food items (and presumably other items in short supply) than subordinate jays. Does this translate into a reduced ability of subordinates to rear offspring? In order to answer this question we compiled the number of yearlings produced by the 14 top ranking males and ten of the lowest ranking subordinate males. We restricted our analysis to the breeding seasons of 1973, 1974, 1975, and 1976 because we assumed status changed little during these years. The top ranking males produced significantly more yearlings from 1973 to 1976 than did the subordinates (average for dominants = 1.86 yearlings, average for subordinates = 0.5 yearlings, t = 2.21, DF = 16.7, P = 0.04). The alpha male, Dark Green-Black, was not extraordinarly fecund during the four years we observed him. He produced only one yearling. However, his successor, Dark Green-Red, was the most successful (10 yearlings) and longest-lived (16+ years) male we have observed. The long-time beta male, Purple-Orange, also left a substantial genetic legacy. He produced five yearlings from 1973 to 1976.

In summary, dominance is clearly beneficial. Dominant males have high fitness (they produce many yearlings). The proximate mechanism that allows these males to experience high fecundity is unknown, but probably is related to their ability to procure limited resources. In particular, as we will argue in Chapter 7, they may be able to obtain high quality mates as well as high quality foods.

### Potential biases in our observations

We have documented patterns of social interactions in the hope that they might clarify our view of the social organization within Pinyon Jay flocks. The wide ranging nature of these birds necessitated that we install feeding stations within their home range in order to observe the interactions between flock members. One question that is often raised about these types of data is: is the dominance of a bird identified at a feeding station specific only to that location or is it applicable to other areas and conflicts? In many species of birds an individual may be dominant at one site and subordinate at another site, depending upon the importance of that site to the survival and reproduction of the individual. This is referred to as site-specific dominance (Brown 1975). Here we presented data from only one site, but observations were also made at another site and those data were qualitatively consistent with the data we presented. Subordinate birds at one feeder were also subordinate at the other feeder which was over 3 km away. Similarly, the same dominant birds were in charge at both feeders. We conclude that the dominance relationships we have documented are site-independent. This is not to imply that the dominance relationships we have documented hold in all situations. They probably do not. Subordinates, for example, may become very aggressive and dominate superiors in order to guard their mates, nest sites, and caches. Indeed, when pairs of birds were allowed to cache and recover their seeds together in a laboratory some interesting role reversals occurred (Stotz 1991). A subordinate bird at its cache site was able to maintain control of the site when challenged by a dominant bird. If a dominant bird visited a subordinate's cache site, the latter bird could supplant the dominant. Although these reversals did not occur between all pairs or on all occasions when a dominant approached the cache site of a subordinate, they were relatively common events.

Another potential criticism of these data is that our color bands somehow influenced a bird's social status. Each bird was randomly assigned a color band combination, but it is possible that red bands, for example, lower a bird's status while black bands increase status. Although this seems pretty far-fetched, Nancy Burley (1985, 1986) has presented convincing evidence that color bands influence the behavior of Zebra Finches. To rule out this possibility in our study, we performed a simple experiment. After we determined that Light Green-Black was the alpha male and that White-Orange was the least dominant male, we exchanged their bands. Now, White-Orange wore the dominant Light Green-Black badge of superiority. We continued to make observations as we had done before the switch and noted no changes in behavior or status. Thus, we conclude that band color bestowed no special status on any particular bird. Birds probably recognized each other using the same clues they used before the plastic shackles were attached to their legs.

One final caveat must be made about the whole system we used to obtain these data. Here we are addressing an age-old question about the ability of humans to observe nature in a natural setting. The problem, simply stated, is that once humans enter the scene the behaviors of animals change in response to their presence. What we are able to observe is determined by the animals' responses to our physical intrusion into their lives. As a general statement this is rather egocentric. However, the establishment of a feeder, visited on a regular basis by our flock of Pinyon Jays may have changed their lives in an unnatural way. For example, without a consistent feeding stage Light Green-Black might have been able to exert his "dominance at the feeder" type of behavior only rarely. His flock mates would have been unable to observe his tyranny on a regular basis and would not have been able to learn that he could and would beat them in such a situation. The ability of dominants continually to advertise their status for choice morsels of food at our feeders may have bestowed a unique social hierarchy upon our study flock that is lacking in other, more natural flocks. We do not feel that this is likely because the foods of Pinyon Jays often come in highly concentrated locations. Cones clumped on a pinyon pine tree closely approximate feeders and force Pinyon Jays to crowd together and occasionally fight. This should produce a social hierarchy in unsupplemented flocks that is similar to the one we documented in the Town Flock.

CHAPTER 7

# The dating game

The pine nut harvest dominates the daily activities of each Pinyon Jay during the fall, however, once the flock's larders are stocked, jays begin to prepare for the upcoming breeding season. They display, chase, and sing to each other in an avian version of the dating game. Young jays begin to compete with their peers in order to prove their desirability as partners to jays of the opposite sex. A few widowed older jays also compete for new mates, but most older jays do not compete. Instead, they court their lifelong mates in the avian rendition of foreplay that functions to stimulate and synchronize their reproductive conditions.

   The pair-bond is the most stable association between unrelated jays. This behavioral bond appears even stronger than bonds between relatives. The enduring quality of Pinyon Jay pair-bonds casts pair-bond formation in another light; as the active process of choosing a lifelong mate. Viewed in this way courtship and pair-bond formation may have profound influences on an individual's lifetime fitness. The jay that courts and successfully pairs with a high quality mate can expect to benefit by this choice for the remainder of its life. Conversely, jays unable to secure high quality mates may be doomed to a life of unsuccessful reproduction.

In this chapter we discuss the static and dynamic qualities of pair-bonding. We describe the behaviors that make up Pinyon Jay courtship, emphasizing the temporal progression of displays. Next, we investigate the cues that jays use in their assessment of mate quality. Interestingly, there are no obvious physical traits that identify high quality mates. For example, bigger is not always better. Instead, mate quality often depends upon the compatibility of the partners' characteristics so, as in beauty, mate quality is in the eyes of the beholder! With this introduction to what courting jays are doing we attempt to understand why pair-bonds are monogamous and typically last for life. After all, if fitness is dependent upon attaining a high quality mate, why not try again if at first you don't succeed?

## COURTSHIP BEHAVIOR

Unlike some birds (e.g. many grouse, Ruffs, and birds-of-paradise) that congregate on traditional courting grounds, known as leks, solely to conspicuously display their breeding ability and/or availability, courtship displays of Pinyon Jays are subtle and easily overlooked. Courting jays usually break off from the foraging flock for short periods and then re-enter the flock as if nothing unusual has happened.

The first courtship behavior of the season is termed Silent Food Transfer. It is seen regularly from mid-November through to February and involves the transfer of a morsel of food between males and females. Many different food items are passed including insects and shelled pine seeds. This inconspicuous behavior is often performed rapidly with one bird holding the food item at the tip of the bill and the second bird taking the item in its bill. Occasionally the recipient gives a slight quivering of the wings during food transfer. As the breeding season approaches, Silent Food Transfer becomes more frequent and occurs at greater distances from the nucleus of the feeding flock.

Silent Food Transfer is performed by yearling birds and may be important in initial pair formation by birds of this age group. We have seen first-year birds offering tidbits of food to three or four birds in succession. Such offerings may indicate a male's ability as a provisioner and aid females in their choice of a mate, as suggested by Nisbet (1973) for Common Terns.

Beginning in late December females become more excited and/or demanding when males approach them with food. Females crouch before their males with head and bill pointed upward at a slight angle and bill opened wide (Fig. 41). Their wings are flapped and fluttered vigorously, as the female vocalizes loudly with a variety of loud Kaws, Chirrs, and soft musical notes. This behavior, known as Courtship Begging, becomes more intense as courtship proceeds. In January females often follow males on the ground or through the trees performing Courtship Begging. This act is very similar to the begging by juvenile Pinyon Jays as their parents approach with food. First-year birds also perform Courtship Begging.

We use the term Courtship Begging to be consistent with our previous reports. However, "begging" may be a poor choice of words to describe the motivation of females at this time of the year. Begging implies that the caller is submissive but Smith (1980) has argued convincingly that in many species the

Figure 41 *Courtship Begging postures of female Pinyon Jays. Wings are flapped vigorously as females utter Courtship Feeding Chirrs. From Balda and Bateman (1972).*

opposite may be true. Begging birds may be dominant individuals expressing their status and *demanding* to be fed. Begging results because the dominant's high level of aggression is inhibited by its drive for food and sex. Female Pinyon Jays do not dominate their mates during the breeding season (Chapter 4) so it seems unlikely that Smith's hypothesis accounts for courtship begging. Begging Pinyon Jays may indeed be subordinates honestly advertising their need for food.

Courtship Begging is an intense, ritualized display that females give during courtship, nest-building, egg-laying, and incubation. The commotion associated with this behavior has always perplexed us, for why should females attract attention to themselves and their nest by begging so loudly? We routinely locate jay nests by listening for begging birds. Presumably many predators can do likewise, making this behavior costly. It is hard to imagine that crafty predators such as ravens and crows do not use begging as a clue to help them locate jay nests. Ravens are able to follow wolf howls to locate carrion (Harrington 1978) so why not jay cries?

Begging calls may, however, contain sounds that are difficult to locate. A study by Redondo and Arias de Reyna (1988) investigated the sound structure of many corvid begging calls (unfortunately not including any Pinyon Jay begs). They discovered that calls of species that are most susceptible to predation (those nesting in open nests) degraded over shorter distances than calls of species that are less susceptible to predation (those nesting in cavities). Greater sound degradation may limit predators' abilities to locate the exact location of begging calls. This may thwart some predators hunting for Pinyon Jay nests but an entire colony of begging jays would still attract many predators to the general vicinity of the colony.

Pinyon Jays may further foil predators' attempts to locate their nests by begging only when they perceive the coast to be clear. When parents approach nestlings and detect nest predators they scold the predator with harsh Racka and Multiple Rack calls that cause the young to crouch motionless and silently in the nest. In addition, males may not approach nests and/or females may leave their nest during incubation to beg and be fed some distance away if predators are near.

We conducted an experiment in 1987 in order to determine if adults adjust their approach to the nest depending upon the proximity of predators. We took on the role of predator, which was easily accepted by the jays as they routinely scold, dive, and mob at us when we climb nest trees to monitor nesting productivity. (In fact, we were often convinced that some pairs recognized our cars and began scolding when we drove near their nests, especially if our ladder was on top of the car.) In the experiment we stood close to (within 10 m) and far (greater than 30 m) from 27 incubating or brooding females. As males approached we recorded the pairs' responses. These jays adjusted their behavior to the situation at hand. When we were close, 12 of the males did not feed their mates, but all 27 fed when we were far from their nest. Three of the non-feeding males entered the nest tree, apparently before they detected us, and then became motionless. At one low nest, we approached to within 1 m of the motionless jay and he would not move, apparently convinced we could not see him! When males did feed their females, the location of feeding depended upon our proximity to the nest. When we were close, feeding was rarely at the nest (33.3% of 15 feedings). However, when we were far, feeding was typically at the nest (80% of 15 feedings).

Perhaps there are also advantages to begging that allows it to persist despite occasional disadvantages. Such an advantage must apply to many birds and all corvids, as courtship begging is observed in most monogamous birds and begging by nestlings is virtually universal among all birds (Skutch 1976, Smith 1980). As discussed above, begging may sometimes be a display of status, but from a functional standpoint it simply stimulates a bird with food to "hand over the goods". As such, a parent with food is akin to a robot that will not feed its mate or its young until feeding is triggered by begging. Begging may, however, have an additional more cognitive effect. It may confer to the male that the beggar *needs* to be fed. Female jays obtain over 80% of their daily food intake from their mates and, of course, nestlings are totally dependent upon their parents for food. By signaling a need for food, parents would be stimulated not just to feed, but to keep feeding as long as the need persists. In addition, as begging is graded in intensity it could indicate the recipient's relative need for food. Consistent with this idea is the increased vigor of begging as courtship gives way to incubation and females become more reliant on their mates to provide them food. Similarly, nestlings may adjust their intensity of begging to their need for nourishment. Parents can then adjust their feeding rate to meet the needs of their nestlings rather than allow feeding rate to be determined solely by food availability (Hussell 1988). The idea that begging is not simply an on/off signal but instead is a graded signal that promotes continued feeding and feeding in proportion to dietary need is the Pinyon Jays' version of the human adage: "the squeaky wheel gets the grease"!

Begging louder in order to get more food would seem to be vulnerable to cheating. That is, young could increase their growth rates by continuing to beg even when satiated. However, fooling parental jays is not so easy. On several occasions we have tethered all the nestlings from a single nest at a site where we could easily observe parents feeding them. Invariably, the largest nestling continued to beg even when its mouth was stuffed with food, whereupon the parent actually removed food from the nestling's mouth and fed it to a brother or sister. In this case, too much squeak does not result in more grease!

The most subtle courtship display, Silent Sitting, becomes obvious as the pre-nesting period proceeds and the above two behaviors become more frequent. The pairs participating in courtship feeding leave the feeding flock and perch silently next to each other in the foliage. During these bouts birds appear to sleep, perform Silent Food Transfer, perform Chin-up Appeasement postures (Fig. 42), sing soft musical songs, and allopreen (Fig. 43). The song, the most musical ever heard from this species, is a combination of soft whistles, pure notes, harmonic coos and chirps, and lasts for 10–15 seconds. One bird will often sidle up to its mate after the Chin-up posture so physical contact is made. Mates will then preen each other about the sides of the head and nape of the neck. Mates sometimes act like teenagers on their first date by repeatedly sidling toward and away from each other while Silent Sitting.

The subtlety of Pinyon Jay courtship is exemplified by a display known as the Swagger Walk. Courting birds sometimes walk side by side in open spaces

Figure 42   *Chin-up Appeasement posture performed during Silent Sitting and aggressive interactions between jays. The white throat patch is conspicuously exposed during this display of submission or reduced aggression. From Balda and Bateman (1972).*

on the ground in this strutting display that is more reminiscent of drunken sailors than of birds. Both jays are upright and their tails are horizontal. Their tails and wings are vibrated intermittently as they strut and move their heads back and forth in exaggerated movements. Feathers on the lower belly and head are erect but contour feathers, malar feathers, and white feathers of the chin are compressed. Swaggering jays periodically stop, touch bills, and allopreen. We had observed jays for nearly ten years without noticing this display until Berger and Ligon (1977) reported its occurrence in captive jays. Once alerted to Swagger Walking we routinely observed it in wild jays.

Swagger Walks often end with Stick Manipulation, where males pick up a twig, pine needle, or tuft of grass in their bill and carry it toward their mate. Males then fly to nearby trees and sit with material in their bill. If the female does not follow, the vegetation is dropped and the male returns to his female. As courtship advances, females respond to Stick Manipulation by following males up into the trees. Now males lead females into the foliage and place sticks on a crotch between them and silently feed the females. This activity, focused at a specific spot in the tree, is termed Crotch Sitting. Occasionally, the stick is carried to a second or third location where food transfer again occurs.

Figure 43  *An allopreening Pinyon Jay scratches its mate's neck. During this equivalent of pri-mates' mutual grooming jays scratch, caress, and refurbish each others' feathers, especially around the head and neck where single birds cannot reach (C.B.).*

The sites selected for Crotch Sitting are not used exclusively by a single pair. Sometimes six to eight birds will visit the same site with no apparent conflicts, and at other times a single male will aggressively chase other birds off. It was not uncommon for Crotch-sitting birds to visit other pairs' sites during this period. Sites resemble adequate nest sites but nests rarely are built there. Some particular sites, however, may be used year after year for Crotch Sitting.

At about the time Stick Manipulation becomes conspicuous another behavior becomes prominent. From two to 12 birds fly rapidly over the foraging and courting birds, Kawing loudly while performing steep dives and sharp turns above and below tree level. At times as many as four discrete groups participate in these strenuous flights. It appeared to us that one bird was always either a leader or a pursuer in these dramatic flights, which consisted of both adult and yearling birds. Upon landing in a tree the birds chased the leader up and down through the branches in a spiral pattern before launching off on a continued flight.

Originally Balda and Bateman (1972) believed these flights served to strengthen pair-bonds as in European Jays (Goodwin 1951) and Rooks (Marshal and Coombs 1957). However, pair-bonds are maintained throughout the year whereas flights are only a spring ritual. Presently, we can be certain only that flights are performed by non-nesting birds in the midst of nesting birds. They may aid in mate assessment by displaying vigor and flight ability of several potential mates simultaneously.

The Precopulatory Display is the jays' most striking visual signal. We have seen it only once in nearly 20 years of watching free-ranging jays and its initial documentation was again dependent upon observations of captive jays (Berger and Ligon 1977). The displaying male assumes a hunched position in a tree or on the ground with his wings spread in a caped fashion and his tail fanned and held upright (Fig. 44). The male stays close to the female, circling her by jumping from branch to branch or walking around her. His head feathers are erect and he may fan his wings and turn slowly to show his back and wings to his mate. The female remains sleek with only her head feathers erect and appears to watch the male by turning her head toward him. She may crouch and beg as she does during Courtship Feeding. Finally, the male steps up on the female's back and copulates. Both jays quiver their wings during copulation and the female utters a Copulation Chirr.

When courtship is at its peak, pairs of birds are continually leaving and re-entering the foraging flock. As with most monogamous birds, courtship must progress through a set series of behaviors that are sequentially organized. Interruptions and distractions need to be minimized during these acts. Thus,

Figure 44    *Precopulatory display of male Pinyon Jays. Males (left bird) bow, cape their wings, and strut around their mates.*

courting birds remove themselves from the hustle and bustle of the foraging birds in order to court in more secluded surroundings, an interesting compromise for such a social creature.

The duration of courtship varies greatly from one year to the next but was noticeably shorter following a bumper pinyon pine crop, thus allowing rapid breeding when food is abundant. Once Courtship Begging is common, nesting is usually initiated within six weeks.

Above we have described the most obvious acts of courtship as they occur temporally progressing toward the actual building of the nest. Below we also describe the spatial aspects of courtship for this wide-ranging species. As mentioned, the home range of a Pinyon Jay flock is exceptionally large and well traversed. Starting in late December or early January flock movements become more restrictive as more time is spent on and near areas previously used for nesting and caching.

By the time Courtship Begging is common and thereafter, flocks move little for a span of one or two hours. The flock forages quietly at this time as it slowly moves through an area. Individual pairs separate themselves at the periphery of the flock, some performing Silent Sitting, others performing Stick Manipulation or Courtship Begging. Flights may be common. Dull blue-gray yearlings often form the nucleus of the foraging flock at this time.

At various times and for undetermined reasons long distance flights are performed. A small number of birds start to Kraw or Kaw softly and this is joined by other foragers. Soon most birds are loudly calling in a resounding din that can be heard for more than a kilometer. During this commotion all courtship ceases as the flock reassembles and flies above the trees to a new foraging and courting area up to 2 km away.

Throughout the course of a day a flock may visit up to eight different areas where courting and feeding occur. Our study flock outside Flagstaff maintained a traditional breeding area within its home range for at least 14 years and often courted on and near this area. The Town Flock nested in a number of different areas over the years and courted in many of these (Fig. 45). No strong relationship existed between the use of courting areas and eventual use of the areas for breeding, perhaps because jays often courted on or adjacent to caching areas which were near but not coincident with nesting areas. Areas that have the proper habitat characteristics to make them optimal caching grounds may not contain the proper characteristics to be good nesting areas. Thus, the two areas may be separated by some distance.

The best description of flock movements during courtship is that they are erratic in time and unpredictable as to distance and direction. This may be an important anti-predator behavior which reduces predators' abilities to locate nesting jays. It is certainly effective at frustrating human observers!

THE PAIR-BOND

Pinyon Jays appear to follow the avian norm and form monogamous pair bonds. In his important summary of bird mating systems, David Lack (1968) calculated that over 90% of all birds were monogamous. Our evidence for

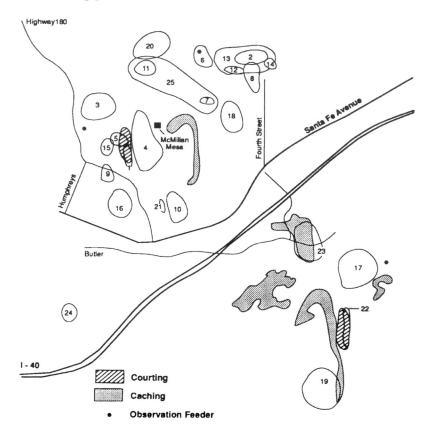

Figure 45    *Courting, breeding, and caching locations within the home range of the Town Flock.*

monogamy, as was Lack's, is circumstantial. We have never seen more than two adults tending a nest and pairs remain in close contact throughout the courtship and breeding season. It is possible that jays occasionally copulate with birds other than their mates. Nancy Stotz once observed one male mount a female and a second male rush over and push the first male off the female. Extra-pair matings seem unlikely, but they remain possible because in a colonial species many birds are reproductively active at the same place and time, and in forest species actual copulations are difficult to witness. We must hedge our claim that Pinyon Jays are strictly monogamous until genetic tests indicate that all young in a nest are parented by the male and female in attendance. Recent experiments have shown the necessity of such caution in assigning mating systems even to well studied species. In Red-winged Blackbirds, for example, Bray *et al.* (1975) showed that females were not loyal to a single male as had been assumed. The investigators vasectomized some territorial males and found that 69% of the eggs laid by females living on their territories were fertile! Neighboring males, not bumbled vasectomies, were responsible. Extra-pair copulations have also been observed in "monogamous" Cattle Egrets, European Starlings, Acorn Woodpeckers, Bobolinks, Mallards, and Lesser Snow Geese (reviewed in Westneat *et al.* 1990).

Pair-bonds appear to be formed throughout the year. Breeding jays that lose their mates during nesting often obtain new mates within a few weeks and breed with them the same year. Courting occurs throughout the year and juveniles exhibit some courtship feeding and chasing in their first autumn. Thus young jays probably begin establishing pair-bonds when they are only three or four months old. Most, however, do not finalize pairing and breed until they are two years old (Fig. 46). Females, being the rarer sex do not have to wait long for a mate and breed earlier in life than males (average age of first reproduction for females is 1.56 years *versus* 2.0 years for males, Fig. 46).

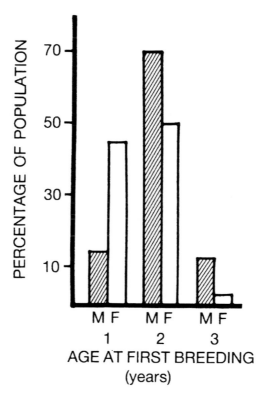

Figure 46    *Age of first breeding in Pinyon Jays. Most birds breed by age two, but a few do not breed until they are three years old. Percentages are based on 63 males and 66 females. From Marzluff and Balda (1988c).*

Bonds are probably maintained throughout the year. Pairs do not form conspicuous duos during the non-breeding season, but they do respond to each other in captivity during this time. We captured four known pairs from the Town Flock and tested their reactions to their partners *versus* reactions to opposite sex members of the other pairs (Table 14). We placed two birds in an experimental chamber for 15 minutes and measured three responses: the number of contact (Near-er) calls they gave, the number of times they hopped or flew between perches, and the number of times they vocalized together. We call the latter vocalizations "paired calls" because they are given by one bird in apparent response to a call given by the other bird. These two calls had to occur within five seconds of each other to count as being "paired". Although only four pairs were tested the results are clear. Birds were more active with their mate than with a non-mate of their flock. We will always remember the reaction of one female after we had separated her from her mate for three weeks. In contrast to her previous moribund reactions in experiments, when we reunited her with her mate she flew wildly about the chamber, calling, and sidling up to her mate. If any reaction by an animal can objectively be called "happy", this was it! Mated jays called more, appeared to stimulate each other by calling, thus inducing paired calls, and moved more than did unmated jays (Table 14). In addition, when we placed groups of jays in an outside aviary and recorded distances between them we found that pairs perched closer to each other than they did to their fellow flock-members of the opposite sex (Table 14). This evidence suggests that mates are recognized in the fall and winter. Studies on captive and free-living jays demonstrate that mates recognize each other during the breeding season as well (see Chapter 8). Mate recognition throughout the year suggests that pair-bonds are maintained throughout the year by subtle vocal, and perhaps visual, signals.

**Table 14.**    *Evidence for recognition of mates outside of the breeding season. Tests are described in text and were conducted from September to January in 1984–85 and 1985–86. The mean, standard deviation, and number of pairs tested are given for responses to mates versus responses by the mated birds to other opposite sex adults from the flock. Statistical tests (Mann-Whitney U-tests) indicate that even though only four pairs were tested, mates were responded to more vigorously than other flock members (the significance levels for all responses was less than 5%).*

| Response | In presence of mate | | | In presence of opposite sex flock member | | |
|---|---|---|---|---|---|---|
| | Mean | SD | N | Mean | SD | N |
| Contact calls | 49.0 | 15.7 | 4 | 12.2 | 5.4 | 6 |
| Hops or flights | 38.8 | 6.8 | 4 | 19.5 | 17.1 | 6 |
| Paired calls | 25.0 | 34.1 | 4 | 0.7 | 0.8 | 6 |
| Distance apart (cm) | 20.1 | 20.8 | 3 | 144.3 | 59.9 | 6 |

Jays maintain pair-bonds with the same partner for life. We have documented only three cases of divorce, or dissolution, in over 100 pair-bonds (Fig. 47). All other pairs remained mated until death did them part! On average, pairs were together for 2.5 years (N = 107, SD = 1.8) before one partner died, but 10% (N = 11 of 107) of all pairs remained together for over five years and one pair remained mated for ten years. When a jay's mate dies it quickly re-pairs with another unmated individual. This allows some individuals to have several mates throughout their lifetime, although most jays

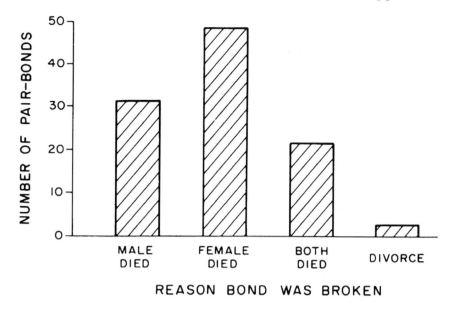

Figure 47    *Reasons for the dissolution of 104 pair-bonds. We define "divorce" as the dissolution of a pair-bond when both partners survive the year and mate with other jays rather than re-pairing.*

have only one mate (Fig. 48). Our oldest male (he was at least 16) holds the record for having the most mates. He had six, and showed a definite preference for young females in his last years of life!

Our documentation of permanently monogamous pair bonds raises three interesting questions: 1) How do jays select their lifelong mates? 2) Why are jays monogamous? and 3) Why do they remain with the same mate year after year? These questions will lead us into three very active areas of behavioral research. We will attempt to review appropriate theories in each area and show how Pinyon Jays can be used to test and modify many of these theories. Entire books have been written on each topic alone so our discussions will be incomplete and are intended to whet the readers' appetites and highlight the uniqueness of Pinyon Jays. All of these questions are difficult to answer completely and necessitate investigating the possible fitness consequences of, and ecological constraints on, permanent monogamy.

MATE CHOICE

Ethologists, ecologists, and evolutionary biologists have long been interested in the study of mate choice. Halliday (1983, p. 4) defines mate choice as: "any pattern of behavior, shown by members of one sex, that leads to their being more likely to mate with certain members of the opposite sex than with others". The key point is that "certain" individuals are chosen as mates, thus mating is not random. The long standing interest in mate choice revolves around this non-randomness. Darwin (1871) wrote an entire book on an important evolutionary consequence of mate choice that he called sexual selection (see Chapter 5). Sir Ronald Fisher (1930), in one of his many important contributions to biology, enhanced Darwin's theories with mathematical proofs showing how many ornaments and armaments of animals could have evolved through sexual selection. Behavioral ecologists have focused on the mechanisms of choice. We refer the interested reader to the book edited by Patrick Bateson (1983) for a summary and extension of many theoretical and empirical studies of mate choice. We will briefly review some of these ideas and questions before suggesting how and why Pinyon Jays choose mates.

## Theories of mate choice

Let's first deal with the mechanisms of mate choice. Darwin proposed that sexual selection had two basic components: 1) competition among members of one sex for the opportunity to breed with members of the other sex, and 2) choice, or preference, for certain members of one sex by members of the other sex. Current researchers attempt to determine the relative importance of these two processes in producing observed mating patterns. In Elephant Seals, for example, enormous bull seals fight amongst themselves and the winners do the majority of the mating (LeBoeuf and Peterson 1969). This mating by dominant males is a common pattern in many birds and mammals. Bull seals weigh hundreds of kilograms more than the females so one would be tempted to conclude that only *competition* within one sex (males) enables dominant males to be the breeders, and that the small females could not (or would not dare!) exercise any *choice* in the matter. But they do! Females choose to mate with dominant males by screaming when subdominants try to mount them (Cox and LeBoeuf 1977). Dominant males respond to female screams by lumbering over and displacing subdominants before they copulate. In elephant seals, as in most well studied species, it is clear that both intrasexual competition and intersexual choice are active in mate choice.

Once the relative importance of competition and choice has been determined, more specific proximate questions can be asked. Most importantly, which sex competes and which sex chooses? Robert Trivers (1972) proposed that members of the sex that contribute little parental care compete for access to members of the sex that contribute the bulk of care. Additionally, members of the sex that invest heavily in care should be choosy about who they mate with. This elegantly simple theory, based on the relative amount of parental care each sex invests in its offspring, has been confirmed in many vertebrate and invertebrate animals. In most mammals, for example, males contribute little more than sperm to offspring, whereas females invest heavily in care. As predicted, males typically compete to mate with females and females exercise considerable choice. In monogamous birds, both sexes make substantial contributions to parental care (birds do not have mammary glands) and both sexes compete and are choosy. Males, however, often compete more strongly for females than *vice versa* because females are less abundant than males, owing to sex-biased mortality rates. Nancy Burley (1977) extended Trivers's theory of choosiness by examining variation in selectivity within a sex. She suggests that high quality mates can afford to be more choosy than low quality mates. Burley's (1977, 1986) empirical studies with Rock Doves and Zebra Finches substantiate many of the theory's claims and suggest that mate choice may be dependent upon the quality, as well as the quantity of parental care.

Investigations into the evolutionary consequences of mate choice have centered around the important question: why are mates chosen instead of randomly accepted? Put another way, what is the advantage of being choosy? Several hypotheses attempt to answer this question. Mates are hypothesized to be chosen because they control important resources, have good genes, have good parenting skills, or are aesthetically pleasing. Choosiness can evolve if selective individuals live longer and produce more or better quality offspring than unselective individuals. Choosiness, even for aesthetic traits that do not themselves influence fitness, may evolve because attractive individuals may have greater access to high quality mates and/or be able to extract more parental investment from their mates than unattractive individuals (Burley 1986).

Proponents of the "good genes" theory suggest that mates are selected for their genetic quality. In his review, Searcy (1982) suggests that choice for genetic quality can mean three things: 1) The fittest males are chosen because their offspring will also be fit, 2) Mates are chosen because their genotype is compatible with that of the chooser, and/or 3) Attractive mates are chosen because their offspring will inherit this sex appeal and experience high mating success, as long as the tastes of the choosers don't change.

Mate choice for genetic complementarity leads to a novel look at inbreeding. Relatives, by definition, share similar genotypes and they may carry gene clusters that have co-evolved with current aspects of their environment. All genes in such clusters, or co-adapted complexes, are necessary for proper adaptation to the environment. Breeding with relatives may thus be beneficial because co-adapted complexes, held in common by relatives, are not disrupted (Bateson 1978, Shields 1982). Bateson (1978) demonstrated that Japanese Quail may select mates based on their genetic complementarity. His experimental quail were given a choice between siblings, cousins, and unrelated birds as potential mates. They selected cousins more than either other group and those mating with cousins attained higher fitness than those mating with members of the other groups. Such "optimal inbreeding" may allow for the preservation of co-adapted gene complexes without running too great a risk of inbreeding depression.

Halliday (1983) argued against the importance of choice for good genes because the payback to the choosers isn't made until their offspring mature. He claims that most evidence suggests that mates are chosen for their immediate paybacks, particularly their contribution as parents in the current reproductive effort. An individual's parenting ability may be assessed during mate choice through a variety of evidence, such as the quality of its resources, especially its territory, its age and/or experience as a breeder, its dominance, and its past breeding performance. Complementarity of parental behaviors may also influence choice. After all, most successful parents function as a single unit, not two independent, uncoordinated ones.

Many humans would readily believe that aesthetic traits may form the basis for deciding with whom to mate. It is difficult, however, to conceive that an animal, in which attaining fitness is paramount to the survival of its genes, would use such a frivolous criterion in such an important decision. Nancy Burley (1986), however, has empirically demonstrated that choice for aesthetic qualities by Zebra Finches may influence fitness. She applied red (attractive) and green (unattractive) color bands to finches' legs. Amazingly, finches (especially males) with "attractive" red bands had higher reproductive success than finches with "unattractive" green bands. Band color itself obviously cannot influence reproduction. Instead, attractive males increased their longevity and lifetime reproductive success because they obtained better mates and extracted more parental care from these mates than unattractive males. The extra care provided by mates of attractive finches even occasionally lightened males' work loads enough to allow them to court and breed with a second female. Sexy Zebra Finches may double their potential success by becoming bigamists. These results suggest that non-functional, aesthetic traits may be important in mate choice by animals as well as humans. However, some caution needs to be applied here because Zebra Finches have naturally red legs. Red bands may not simply be aesthetic traits; they may enhance a natural signal indicative of health and vigor.

Animals may use more than one indicator to guide their choice of a mate. In fact, if animals strive to obtain the best possible mate they should use all the information available to ensure they obtain a quality mate. A few studies suggest that multiple factors influence choice of mate. Rock Doves, for example, use age, experience, dominance, and color of plumage in mate choice (Burley 1981).

As you can see, mate choice is an extremely complex and important decision for animals. It should be very important to young Pinyon Jays for at least three reasons. First, as we have seen, most jays only have one mate for their entire life (Fig. 48); it better be a good one! Second, as we will discuss shortly, for those jays that outlive their first mate, the quality of the subsequent mate may depend on how well they did with their first mate. Lastly, as we will demonstrate in the next chapter, breeding performance is dependent to a large degree on parental behaviors such as nest-site selection and nest attentiveness. Choosing a poor mate can endanger current and future nesting attempts.

*Competition and choice in Pinyon Jays*

The processes involved in mate choice by Pinyon Jays have proven difficult to study in the field. However, Kris Johnson (1988a,b) investigated the relative importance of choice and competition in an aviary population of jays. She determined that both sexes compete for mates, but that males were slightly more competitive than females. Males with larger bills and heavier mass outcompeted smaller males. The winners of competition among females are also larger and perhaps older than the losers. Both sexes also choose among

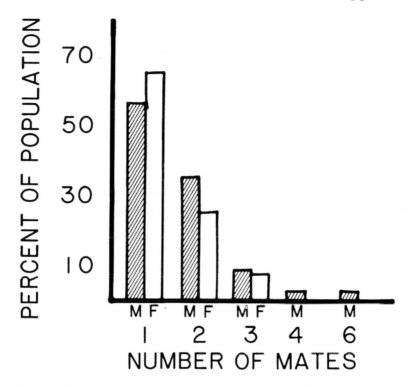

Figure 48  *Number of mates individual males and females had during their lifetimes. Percentages are based on sample sizes of 56 males and 60 females. From Marzluff and Balda (1988c).*

potential mates. Males prefer large, dominant females and may therefore reinforce female-female competition. Females, however, do not prefer large, dominant males. Instead, they appear to prefer brightly colored males with large testes. Bright color and testis size may indicate a healthy and vigorous male. It is hard to imagine how testis size is ascertained by females (sex organs of birds are internal) and this apparent choice may be an artifact of experimentation. Testes were measured after choice and it is plausible that testes increased in those males that were courted most vigorously by females. Thus large testes may be caused by female attentiveness, not *vice versa*.

The findings that both sexes are choosy about who their mates will be and that both compete for mating opportunities agree well with the theories we discussed. Both sexes should be choosy and compete for the best partner because they both invest heavily in parental care. Competition may be stronger among males than females because flock sex ratios are male biased (Chapter 5). A stronger relationship between size and competitive ability for males compared to females may produce the slight sexual dimorphism we reported in Chapter 1. However, in nature male size is not strongly related to dominance at food (Chapter 6) and therefore may be only weakly related to a male's ability to access females. Sexual selection may therefore only weakly favor larger size in males relative to females.

*Mate choice for size* Laboratory results suggest that big males and big females may often have conflicting ideas of what constitutes an ideal pair. Big males prefer big females, but big females do not prefer big males. (We use "big" here as shorthand for birds with longer than average bills and greater than average body masses.) Our field results suggest why this may be so. Large females may be preferred because male fitness is higher if a male's mate is large. Males mated to heavier than average females lived longer, produced more offspring, and raised fitter sons than males mated to lighter than average females (Table 15). Females, on the other hand, do not always benefit by mating with large males. In fact they live longer and produce sons that give them more grandchildren by mating with lighter than average males (Table 15). This evidence suggests that large females benefit from their association with small males, but this is only an *inference* that we think is likely. It is also possible that large, high quality females just have greater access to the small, high quality males.

**Table 15.** *The influence of mate size on fitness. We report the average longevity, fecundity, and fitness of sons produced by birds mated with heavier than average and lighter than average mates. The three numbers listed for each fitness component and type of pair are, from top to bottom, the average, standard deviation, and sample size.*

| Type of pair | Life span (years) | Yearlings while paired | Yearlings produced by sons | Life span of sons (years) |
|---|---|---|---|---|
| Males mated to heavy females | 7.62 | 1.50 | 1.43 | 5.2 |
| | 3.48 | 2.28 | 1.51 | 2.9 |
| | 13 | 18 | 7 | 11 |
| Males mated to light females | 5.91 | 1.11 | 0.86 | 4.2 |
| | 2.12 | 1.27 | 0.77 | 2.1 |
| | 11 | 9 | 14 | 21 |
| Females mated to heavy males | 6.38 | 1.69 | 0.67 | 4.1 |
| | 3.40 | 2.56 | 0.82 | 2.1 |
| | 13 | 13 | 6 | 8 |
| Females mated to light males | 7.08 | 1.07 | 1.20 | 4.7 |
| | 2.75 | 1.27 | 1.15 | 2.5 |
| | 13 | 14 | 15 | 24 |

In the field, large males do not always get their way. If they did, we should observe strong assortative mating for body size; that is, large males should pair with large females, leaving small males and small females to form the remaining pairs. This is not the case with respect to bill size (Fig. 49) or body mass (Fig. 50). Statistically, mating is random with respect to these indications of body size. This randomness may be another example of the mayhem that is sure to result when male and female interests conflict! Big males often

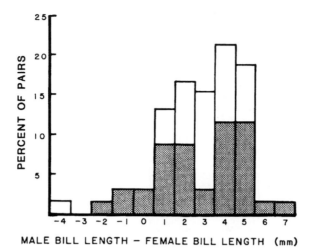

Figure 49   *Frequency distribution of differences in bill sizes of mates. Entire histogram gives distribution for all pair-bonds (N = 59). Shaded portion is initial bonds only (N = 33). Unshaded portion is subsequent bonds only (N = 26). From Marzluff and Balda (1988b).*

get the preferred large females, but females may often get their wishes as large females commonly pair with small males (Fig. 51).

Our evidence suggests that big females are good for males but that small males are good for females. Why? Big females may require less of an investment from their mates during the incubation and brooding stages of nesting. Their larger body mass gives them a greater cushion against harsh environmental conditions. This means that during trying times, such as during spring snow storms, big females can survive longer on the food their mates bring them

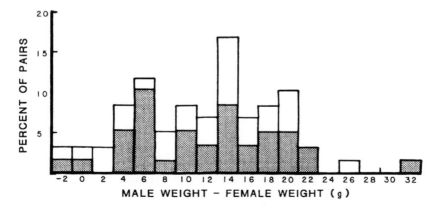

Figure 50   *Frequency distribution of differences in body weights of mates. Entire histogram gives distribution for all pair-bonds (N = 57). Shaded portion is initial bonds only (N = 32). Unshaded portion is subsequent bonds only (N = 25). From Marzluff and Balda (1988b).*

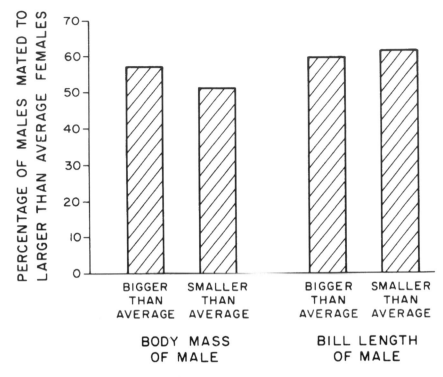

Figure 51    *Who mates with large females? The percentages of larger and smaller than average males that mated with larger than average females. Percentages are based on 59 pairs.*

than can small females. Male survivorship may benefit during these times because males can forage for longer periods before returning to their mates. Male fecundity would also benefit because males would have more time to find food for their mates and thus reduce a female's need to leave the nest and forage on her own. The less females have to forage for themselves the lower the chances that unattended eggs or young will freeze or be eaten by predators. When females become severely energy-stressed during storms, we have often observed them eating their eggs and young before abandoning the nesting attempt (Balda and Bateman 1976). Large females can endure energetic stress longer and would be less likely to make cannibalistic mates!

Females may not prefer large males because they are too domineering. Average females are more equal in size, and perhaps in dominance, to small males. Relatively equal dominance within the pair may ensure compatibility. Our only measure of compatibility is the rate of intrapair squabbling at feeders. Mated Pinyon Jays rarely fight; only 38 (2.9%) of 1309 fights between males and females involved mates. Pairs of light males and heavy females fought less at feeders ($\bar{x}$ = 0.6 fights per 2.5 years) in comparison to other pairs ($\bar{x}$ = 1.75 fights per 2.5 years). As we are already out on a limb, let's take the argument one step further. Greater compatibility and the lower rates of fighting that it confers may lead to longer life spans of males and females which lengthens pair-bond duration. Such pairs average longer pair-bond durations than other pairs which would allow both partners more opportunities for successful reproduction.

Alternatively, females may prefer small males for other reasons. Maybe small males invest more than large males in parental care because the latter are more concerned with maintaining their macho image. This may be the case in Dark-eyed Juncos as aggressive males with high testosterone levels make poor parents (Ketterson, pers. comm.). Weight-lifters on steroids and juncos pumped up on testosterone (also a steroid) may be all bark and no bite!

*Other indicators of mate quality*　Body size is not the only, nor the most important factor influencing Pinyon Jays in their choice of mates. A very strong mating pattern among wild jays is assortment for age (Fig. 52). Young jays nearly always mate with jays of their own age, and older jays often mate with the few available birds of their own generation. One banded female appeared especially fond of males her own age. As a yearling in 1973 she paired with another yearling. In 1976 she was four years old and paired with the only unmated four-year-old male. She obtained her third, and last, mate in 1978 when she was six years old. He was the only unmated six-year-old male known in the flock!

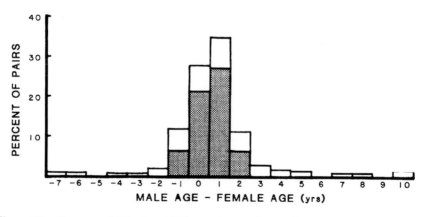

Figure 52　*Frequency distribution of differences in ages of mates. Entire histogram gives distribution for all pair-bonds (N = 141). Shaded portion in initial bonds only (N = 87). Unshaded portion is subsequent bonds only (N = 54). From Marzluff and Balda (1988b).*

We lack experimental evidence that jays actively select mates similar in age to themselves. However, our results suggest that they should. Pairs in which partners were within one year of each others' ages remained paired longer, produced more surviving offspring, and produced sons that lived longer than did pairs of more disparately-aged partners. Pair compatibility may again be important to the success of similar-aged partners. Partners of similar age have had similar experiences in the flock and with their environment. They also have similar expectations of future life and as a result may be selected to take similar risks (Chapter 12). (Young human males take more risks than older ones and, as all insurance dealers know, they often exhibit driving habits more compatible with their peers than with their parents as a result.)

Similar-aged mates do not benefit from equal division of labor during breeding as suggested for Herring Gulls (Burger 1987). Partners differing in age by one year or less divided the feeding of nestlings and cleaning of their nest in the same way as partners differing more in age (N = 15 pairs with age difference ≤ 1 year: male fed nestlings 71.5% of time, SD = 12.2, female cleaned nest 67.7% of time, SD = 17.4; N = 11 pairs with greater age difference: male fed nestlings 71.4% of time, SD = 12.3, female cleaned nest 62.8% of time, SD = 18.0; t-test for feeding: t = 0.03, P = 0.98; t-test for cleaning: t = 0.69, P = 0.50).

Before speculating further we must experimentally determine whether mates are chosen for their age and whether age of partners influences their fitness.

Living with the same, individually recognized jays for many years enables each Pinyon Jay to become familiar with flock members to whom it is not mated. When a jay's mate dies, it selects a new mate from a pool of these familiar jays that also currently happen to be without mates. Familiarity with many jays may enable older Pinyon Jays to choose mates on the basis of their past reproductive performance. Jays may monitor and keep track of the productivity of their flock members. Later, when they are in search of a new mate they could use this information to pick a previously successful, or proven, breeder. Assortative mating for previous success suggests that this may occur. We know the past reproductive history of 32 males and females that formed subsequent pair-bonds. Nearly three-quarters of previously successful males and over half of the previously successful females remated with other previously successful birds. In contrast, only one-third of the previously unsuccessful males and 18% of the previously unsuccessful females remated with previously successful birds. On the whole, reproductive stars mate with other stars and reproductive duds remate with other duds more often than expected by chance (Fig. 53)!

How can a bird weighing less than 150 g keep track of, and later recall, the past reproductive performance of its many flock members? Reproduction may be monitored as the breeding season draws to a close. We often observe groups of adults and yearlings peacefully visiting the nests of their flock members. Visitors approach the nest and conspicuously peer in at its contents. Parents do not harass these visitors. Productivity may also be monitored when fledglings gather into creches and are still being fed by their parents. Successful and unsuccessful parents are segregated at this time as only successful ones visit the creche. Unsuccessful parents re-nest or molt and rejoin the non-breeding yearling flock. Jays would be able to tell which adults met similar fates to themselves by noting their companions at this time of the year. Recalling the past success of a few hundred flock members does not seem like a monumental task for a bird that annually recalls the locations of thousands of buried pine seeds!

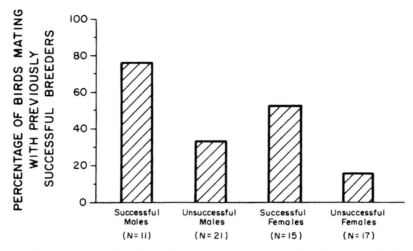

Figure 53   *Assortative pairing for previous success; studs mate with studs, duds with duds! Percentages are based on 11 successful males, 21 unsuccessful males, 15 successful females, and 17 unsuccessful females.*

Past performance is not a foolproof indicator of mate quality. A pair of previous duds averages as many young as does a pair of former stars. Two wrongs can make a right because, regardless of the past, many pairs fail to produce any young. Additionally, a previously unsuccessful jay is not entirely at fault for its past performance. Its mate may have been at fault. If there is no fitness advantage to using past performance as a factor in mate choice, why do jays appear to use it? It may be a valuable additional clue to the potential productivity they can expect from a mate. Occasionally this potential may be realized as it was for Dark Green-Red and Light Blue-Purple (see Appendix 5 1976 K for their family history). These jays were both previously successful and after their first mates died, they paired and produced more offspring than any other pair (17 crechlings and five yearlings during five years). No pair in which one partner was unsuccessful has been even half as successful as this. Potential windfalls of reproductive success like those Dark Green-Red and Light Blue-Purple experienced may select for the use of prior success as a cue in mate choice.

*Why do Pinyon Jays choose mates?*   Pinyon Jays appear to select mates using a variety of functional criteria. Prospective mates appear to be screened for their age, plumage brightness, body size, previous reproductive history, and possibly other factors, including aesthetic ones (plumage brightness may also be aesthetic), we have not investigated. Using many indicators enables jays to select mates that enhance their own reproduction and longevity. Multiple criteria may be used in mate choice, but one rule appears to govern choice for many criteria: choose a mate similar to yourself. Similarity in age, size, and past reproductive performance means that partners have had similar experiences, have similar demographic schedules, and may be on relatively equal footing with respect to dominance. What we have quantified as similarity may translate into compatability for jays and we emphasize that jays appear to choose mates with whom they are compatible.

Our evidence is most consistent with a "good parent" hypothesis of mate choice; compatible partners are selected because they make good parents. We certainly cannot rule out the "good genes" hypothesis and, in fact it may also be an important reason why large females are preferred as mates by males. We need to know if large mothers produce large, and hence attractive, daughters. Unfortunately, because most daughters do not breed in their natal flock we have data on body sizes of only four mothers and their daughters. This is not enough to test this aspect of the "good genes" theory. Body size is genetically controlled in male Pinyon Jays as in other birds (e.g. Darwin's finches; Boag and Grant 1978). That is, large males have large sons. We suspect that further data will show a similar pattern for female Pinyon Jays. This may, however, be a moot point as preliminary evidence suggests that in Pinyon Jays choice of a genetically superior mate does not guarantee success. Large individuals may be genetically superior and they may bestow large size upon their offspring, but large jays may be less successful than small ones if their mates are incompatible in size. For example, large females are very fit when mated with small males (the two partners are then similar in size) but much less fit when mated to large males (the partners are then more disparate in size, Fig. 54). Current evidence suggests that mates are selected for their compatible parental abilities and not for the hereditary aspects of their body size. It is still possible that all large females are not created equal and that the high quality large females get the best (small) males while the low quality large females get the poor (large) males. Therefore, the intrinsic quality of individuals, not compatibility with their mates, may determine success. Until we get a date with a jay there will always be questions concerning the process of mate choice, the indicators used to select mates, and the rationale behind the use of these indicators!

## THE EVOLUTION OF MONOGAMY

Picture yourself as a jay. Would you leave more offspring throughout the course of your life by channeling all your parental care to one mate and brood or would you be better off spreading your effort among two or more families? An important consideration is the effect of spreading out parental care. If care is spread too thinly, then regardless of how many young are parented, few may receive sufficient care to survive and the parent's fitness will suffer. In many birds, both sexes contribute substantially in the rearing of young. Therefore, as David Lack (1968) pointed out, any reduction in care of young by one parent, as would be required by a polygamous bird, would likely lower the number of young a pair could produce. (We follow Wittenberger's (1979) definition of polygamous to mean any non-monogamous mating system.) But, can the polygamous mate make up the lower productivity by rearing a few more young with another bird? If this is often possible, a polygamous mating system

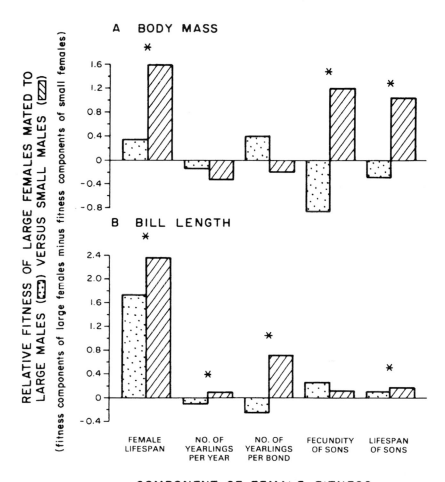

**Figure 54** *Fitness of large females relative to small females often depends on the size of their mate. We have plotted differences in fitness components of large females and small females mated to large and small males. For example, in our first comparison, life span of heavy females minus life span of light females equals 0.3 years for females mated to heavy males and 1.6 years for females mated to light males. Thus, relative to small females, large females live five times longer if they mate with small males than if they mate with large males. Positive differences indicate large females are more fit than small females. Negative differences indicate small females are more fit. Five components of fitness are compared for two measures of female size: body mass (A) and bill length (B). Dotted bars indicate the relative fitness of large females mated to large males. Shaded bars indicate the relative fitness of large females mated to small males. In seven of ten comparisons, large females were relatively more fit when mated to small males than when mated to large males.*

will result in higher fitness to those participating in it and a genetic or cultural propensity towards polygamy will spread through the population.

Parental care is most easily shared when the young are less than totally dependent on the parents, particularly when young are born nearly independent (nidifugous or precocial young) or are fed on abundant, easily obtained foods like seeds. However, most birds raise young that begin life naked, helpless, and entirely dependent upon their parents' care. In such cases, any reduction in parental care by one parent is likely to cause the entire nesting effort to fail; monogamous individuals will therefore out-reproduce polygamous ones and monogamy will evolve. This is especially likely when both parents care for their offspring. Corvids fit these categories and therefore the need for biparental care may reinforce monogamy. A recent experimental study using Black-billed Magpies clearly demonstrated the selective advantage of monogamy (Dunn and Hannon 1989). This study found that half of all pairs studied, normally fail to produce fledglings each year, but all females who had their mates removed (N = 20) failed to fledge young. The need for biparental care may be the single most important reason that most birds are monogamous (Mock and Fujioka 1990).

We learned in 1982 that Pinyon Jay nestlings are dependent on both of their parents' care. We were monitoring parental behavior at a nest tended by a ten-year-old male and a seven-year-old female. All was well until we approached the nest one day to find a dead nestling on the ground and three remaining young begging noisily in the nest. A few minutes later the female returned, fed her babies, and left again. During five hours of observation the female continued to feed her young but she rarely probed the nest to remove parasitic fly larvae and the male never visited the nest. Puzzled, we drove home and, as luck would have it, we located the male, dead in the middle of the road (see vignette for Chapter 12). Our oldest known-aged male had been hit by a car 100 m from his nest. His mate continued feeding the young, but each day one more young was dead below the nest, having fallen from the nest searching and stretching for food or avoiding the bites of parasites. The female, even though very experienced and previously successful at rearing young, could not bring any of the young to independence alone. Presumably the young of a two-timing male or female would suffer a similar fate, thus reinforcing the evolution of monogamy in Pinyon Jays.

An inability by one parent to successfully rear young is, however, unlikely to be the sole factor responsible for the evolution and/or maintenance of monogamy. We know that one parent alone cannot rear young, but perhaps one parent with a contribution from the other could rear a brood. If the needed contribution was relatively minimal then monogamy would not be essential. For example, if a female only needed half of a male's care to rear the brood, then each male could have two females and rear twice as many young as a monogamous male. As long as one sex requires 50% or less of the care the other sex can offer, then a non-monogamous mating system could evolve. The less care required of one sex, the greater is the chance that members of that sex can successfully breed with more than one member of the opposite sex. However, members of the rare sex may wreck this theory by demanding more care from their partners than is required to successfully reproduce. Breitwisch

(1989) argues that in male-biased, monogamous birds, females command high parental investment from their mates because they can desert poor providers and remate with good providers. As females demand more parental care from males the possibility of males tending two or more females diminishes. However, high quality females may go so far as to become bigamists.

The necessity of incubation may also foster monogamy. In Pinyon Jays, as in most birds, only females have brood patches allowing them to incubate eggs. Therefore, until a male is able to install an electric blanket in his nest, he must entice a female to incubate his eggs. Incubation is costly for females because they must expend energy incubating and they are vulnerable to predators while they remain on the nest. It is not uncommon to see one cost of incubation in Pinyon Jays: a nest empty except for broken eggs and a few female breast feathers stuck to them. Apparently a hawk or an owl snatched the female from the nest, crushing her into her clutch before hauling her away. Males reduce the cost of incubation, and indeed may make it possible in cold climates, by supplying much of the female's food. Male Pinyon Jays provide 80% of their mates' diets during incubation. Ford (1983) argued that a female's need for male assistance may favor monogamy more strongly than the nestlings' need. This is certainly reasonable in Pinyon Jays, because it is difficult to imagine that a male could support two incubating females and himself during the cold and often snowy breeding season.

One other factor may favor monogamy, at least by males. Females may enforce it! We played a dirty trick on a pair of jays in 1972. We removed a mated female from the flock for seven weeks. During this time her mate formed a new pair-bond with another female. They were raising fledglings when we released the original female back into the flock. The original female immediately began badgering the new pair (forming a "love triangle") until after one week the new pair-bond was broken and the original bond reestablished. The deserted female did not successfully rear her young. Female Pinyon Jays apparently fight for their rights and this aggression may inhibit any male's attempt toward polygamy.

WHY ARE PAIR-BONDS PERENNIAL?

We have argued that there are many reasons why jays breed with only one member of the opposite sex at a time, but why should this be the same individual year after year? Kittiwake Gulls, for example, are monogamous but when a pair is unsuccessful at producing young they often divorce and try breeding with another bird (Coulson 1966). Not so in Pinyon Jays. Thirty-eight pairs of jays failed to produce crechlings their first year yet remained paired the next breeding season. None divorced. Ten of these pairs failed in their first two years and two even failed in their first three years without divorcing. In fact, none of these 38 pairs ever divorced. Nearly every pair we have studied has failed one year and remained paired provided both partners lived. Five pairs have remained mated despite failing to produce crechlings for four or more consecutive seasons. Pinyon Jays remain paired through thick and thin!

Divorce would not pay if past performance was not indicative of future performance or if parental care did not influence performance. We have alluded above to the importance of care and discuss it further in Chapter 9; suffice to say that it matters a great deal! Future performance can be predicted from a pair's past performance. Over three-quarters of the time, a pair that is successful their first year will succeed again (21 of 27 pairs or 77.8%). In contrast, over half of the pairs that fail in their first year never succeed (11 of 20 pairs or 55%). Given this information, divorce certainly seems like a worthwhile option after a pair initially fails to reproduce.

When an expectation, such as divorce after failure, is not carried out, biologists typically re-examine their expectations. Most commonly, we search for adaptive reasons that make the counter-intuitive result beneficial to individual animals. In the present case, we had no idea why unsuccessful jays remained paired for life but we were confident that there must be some advantage to faithful mates.

There wasn't. There is a strong relationship between pair-bond duration and the total number of yearlings a pair produces (Fig. 55), but this is simply quantifying the obvious: the longer two jays survive, the more breeding attempts they have and the more chances they have at producing offspring. As long as a jay remains paired for three or more years it has a good chance of replacing itself in the population. This relationship does not address the benefit of remaining with one mate. As long as a jay remains paired with someone (or more than one) for three years, it has a good chance of succeeding to pass its genes on to the next generation. A jay's chances of success do not increase measurably by remaining with *one* mate (Table 16). The only benefit of mate

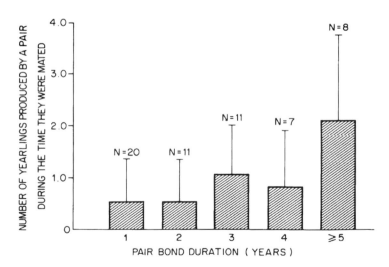

Figure 55 *The influence of pair-bond duration on the average number of yearlings that a pair of jays produced during the time they were paired. Height of hatched bars indicates average productivity. Thin line indicates one standard deviation above the average. The number of pairs averaged for each duration of pair-bond is given above each bar. Initial and subsequent bonds are combined.*

**Table 16.** *Influence of pair-bond duration on annual production of crechlings, yearlings, and the probability of nesting successfully. Sample sizes (N) are all pairs of a given bond duration for which reproductive data were known. Reproductive data were known for some pairs at only one bond duration and for others at several durations. Initial bonds are those formed between two jays that have never bred. Subsequent bonds are those formed between two jays that have bred before. Initial/subsequent bonds include one jay that had previously bred and one that had not.*

| Duration of pair bond (years) | Initial bonds | | | | Initial/Subsequent bonds | | | | Subsequent bonds | | | |
|---|---|---|---|---|---|---|---|---|---|---|---|---|
| | Crechlings | | | Proportion of pairs successful* | Crechlings | | | Proportion of pairs successful* | Crechlings | | | Proportion of pairs successful* |
| | N | Mean | SD | | N | Mean | SD | | N | Mean | SD | |
| 1 | 55 | 0.82 | 1.14 | 0.40 | 11 | 1.0 | 1.48 | 0.36 | 40 | 1.28 | 1.36 | 0.58 |
| 2 | 26 | 1.50 | 1.27 | 0.69 | 2 | 1.0 | 1.41 | 0.50 | 27 | 1.0 | 1.24 | 0.48 |
| 3 | 15 | 1.13 | 1.19 | 0.60 | 1 | 1.0 | – | 1.00 | 12 | 1.75 | 1.36 | 0.75 |
| 4 | 10 | 1.30 | 1.06 | 0.70 | 1 | 4.0 | – | 1.00 | 6 | 1.83 | 1.86 | 0.50 |
| 5–10 | 14 | 0.93 | 1.38 | 0.50 | 6 | 1.0 | 1.10 | 0.67 | 4 | 2.25 | 2.06 | 0.75 |

| Duration of pair bond (years) | Yearlings | | | Proportion of pairs successful† | Yearlings | | | Proportion of pairs successful† | Yearlings | | | Proportion of pairs successful† |
|---|---|---|---|---|---|---|---|---|---|---|---|---|
| | N | Mean | SD | | N | Mean | SD | | N | Mean | SD | |
| 1 | 48 | 0.33 | 0.63 | 0.25 | 13 | 0.54 | 0.88 | 0.31 | 39 | 0.85 | 1.11 | 0.46 |
| 2 | 25 | 0.64 | 0.86 | 0.44 | 5 | 0.20 | 0.45 | 0.20 | 21 | 0.76 | 1.26 | 0.43 |
| 3 | 15 | 0.33 | 0.62 | 0.27 | 2 | 0.50 | 0.71 | 0.50 | 11 | 0.91 | 0.94 | 0.64 |
| 4 | 7 | 0.57 | 0.79 | 0.43 | 1 | 0.00 | – | 0.00 | 6 | 0.00 | 0.00 | 0.00 |
| 5–10 | 13 | 0.54 | 0.78 | 0.38 | 6 | 0.33 | 0.52 | 0.33 | 4 | 0.75 | 0.96 | 0.50 |

* successful is defined as producing at least one crechling
† successful is defined as producing at least one yearling

fidelity appears to be slightly greater average reproduction in the second year of an initial bond (Table 16). We can take a closer look at this potential benefit by comparing the same pairs in their first and their second year of an initial pair bond. We have such data for 18 pairs. There is a slight, but statistically non-significant, benefit to remaining with the same mate for two years (in their first year pairs produced an average of 0.83 crechlings [SD = 1.15] and 0.42 yearlings [SD = 0.69] and in their second year an average of 1.44 crechlings [SD = 1.29] and 0.63 yearlings [SD = 0.90]). Remaining paired with the same mate for one, two, three, four, five or more years doesn't benefit a jay much, especially after the second year. The real significance of the data in Table 16 is the great variation in productivity among pairs mated for the same length of time (note the large standard deviations relative to the averages for each pair-bond duration). This variation indicates that factors other than pair-bond duration have a substantial influence on productivity. We examine these factors in Chapter 9.

It could be argued that the benefits of mate fidelity are more subtle than the production of offspring. Perhaps, for example, fidelity allows pairs to nest earlier in the year, lay more eggs, or expend less energy courting and breeding. However, these benefits do not translate into the tangible benefit of greater reproductive success. Therefore, they cannot be important to the evolution of mate fidelity; they would not facilitate the spread of faithful genes.

When we first analyzed these data we were dumbfounded. There was no advantage to mate fidelity, yet it persisted in our population. We could predict which pairs could improve their chances for success by divorcing, yet they didn't. Why not? Lesser Snow Geese, Ring-billed Gulls, and House Wrens do not divorce because divorcees would be unable to keep control of their territories and the resources therein (Cooke *et al.* 1981, Kovacs and Ryder 1981, Freed 1987). This potential cost of divorce is not applicable for Pinyon Jays because they are not territorial. Our best guess at why Pinyon Jays are faithful to their mates is that their extremely social life gives them no other option. Members of a pair are together in a flock throughout the year and we have seen that females (and presumably males) are capable of defending their mates from other flock members. An ability continually to monitor their mate's activities in the flock may allow jays to enforce fidelity on each other. Life in a society with stable membership enables jays to recognize each other and perhaps use past performance as a key to mate suitability. If, as our data suggest, this indicator is utilized in the process of mate choice then unsuccessful jays may not benefit by divorce because their previously successful flock mates won't have them as mates! Lastly, the annual turmoil that would be created by re-pairing among many jays each year would undermine the tranquil stability characterizing Pinyon Jay flocks. Perhaps divorcees would not co-operate with their former mates thereby dividing a co-operative flock into small competing factions. The extra time and energy demanded by competition for mates and establishment of new pair-bonds may erode the basic benefits of flock life. Some pine seeds may go unharvested and some predators may not be detected by jays bent upon securing new mates. Certainly the onset of reproduction would be delayed and less easily adjusted to the abundance of pine seeds. In short, social life may constrain divorce in Pinyon Jays because the benefits of sociality are only fully realized by populations of faithful partners.

Ideally, we would compare the sociality and individual fitness of jays in flocks comprised of faithful and unfaithful jays. This is not possible, but we can look at the characteristics of the few pairs in our flock that did divorce in order to improve our understanding of why divorce may occasionally occur. Mate incompatibility appears at the root of some divorces. One of the pairs that divorced was the most dissimilar-sized pair we have quantified. The male was 33 g heavier, and had a bill 7 mm longer, than the female. In another case the male had a physical impairment (a crippled wing) that may have reduced his ability to co-ordinate activities with his mate. Apparently, only when gross differences in mates preclude their abilities to complement each other's activities is divorce an option. This may reinforce choice of compatible partners.

## Pairs are more than males and females

The close, co-operative association of mated Pinyon Jays causes us to often view the pair as an entity in its own right; not a male and a female, but a pair. This view was typical of early ornithologists who saw pairs co-operating for the pair's benefit. More recently, ornithologists view the pair-bond as a "grudging truce" in which both partners behave only in their own best interests (Mock 1985). This view does not exclude mutualistic co-operation among partners, but it emphasizes their selfish tendencies. We are not completely comfortable with such an individualistic view of the Pinyon Jay pair-bond. Certainly we agree that animals act in their own self-interest so as to increase their genetic contribution to future generations. Therefore, the pair should properly be viewed as an association between two selfish jays. Males and females do what is best for themselves, not what is best for the pair. But in species like Pinyon Jays which are likely to breed with only one individual for their entire life, is there any difference in whether we view the pair as a pair or as a male and a female? If your lifetime reproductive success is determined by the joint success of you and your mate, then what is best for you is synonymous with what is best for the pair. In fact, selection should strongly favor individuals that behave so as to maximize the productivity of their lifelong mates because by so doing they maximize their own fitness.

Our evidence suggests that pairs have unique properties that are not simple products of the males and females which comprise them. Pair-bond duration may be one such property. Duration appears to be determined by more than the parents' survivorship probabilities. Paired males and females die the same fall or winter 20.8% of the time but if the pair members' deaths were independent, or random, events they should co-occur only 6.8% of the time. We suggest that mates affect each other's life spans. Compatible partners may enhance each other's survival, whereas incompatible ones may reduce their survival. The interaction of the pair influences each partner's life span so that pair-bond duration is not consistent with the average male's and average female's probability of survivorship. The pair's reproductive performance may also be an emergent property. A pair's productivity is predictable from year to year, but it is not dependent on the partner's productivities in previous pair-bonds.

Classically, we view natural selection as a process that works at the level of the individual. Individuals are successful or unsuccessful because of some of the traits they possess. Evolution occurs at the level of the population as the traits possessed by successful individuals become more prevalent, and those possessed by unsuccessful individuals become less prevalent in the population. If pairs have their own unique emergent traits that influence their fitness, then natural selection may also act at the level of the pair. Natural selection acts by judging the quality of a trait's expression; it cannot judge what is not expressed. Selection at the level of the individual cannot judge the quality of a pair's emergent traits because individuals do not express these traits. Pairs express emergent traits and for selection to judge their quality it must operate at the level of the pair. The result of this judgment is passed on to the individual, but it is the pair that is judged. In Pinyon Jays, for example, we have argued that partner compatibility is an important trait. Compatibility is not expressed by individuals, it is expressed by a pair. In order for compatibility to evolve in our population, natural selection must judge the quality of a pair's compatibility. Population composition could change (evolve) as individuals selecting compatible partners are favored over individuals selecting incompatible partners. Selection at the level of the pair, as we envision it, never acts against the selfish interests of individuals because an individual's interests are synonymous with the pair's interests. The important point is that the pair may jointly express some traits that are not expressed by individuals and natural selection is free to judge the traits.

Natural selection on traits expressed by a pair may counter selection on the same traits expressed by individuals. Take male body size, for example. At the individual level, large size may be selected for because it confers high competitive ability. However, at the level of the pair, small males may be favored because they are similar to females in size and therefore are more compatible as partners. Perhaps the reason sexual dimorphism in Pinyon Jays is not greater is because selection for compatible pairs restricts selection for competitive ability. The possible evolutionary significance of permanently monogamous pair-bonds serves to emphasize the functional significance of this behavioral bond between two animals.

CHAPTER 8

# When and where to nest

*"On March 3 [J. S. Ligon] returned to the colony and found nests in all the scrub oaks of sufficient size, but never more than one in a tree. As it had snowed many times since his first visit, the nests were damp from melted snow."* (Baily 1928)

Human expectant parents may pool their resources for years before they feel ready to undertake the responsibility of raising children. The decision of when and where to have children may depend on the mother's age, the couple's economic assets, the ages of their other children, the permanency of the couple's current employment, or the persistence of the couple's friends and relatives. In reality, however, many human births are not carefully planned; they are the result of a young couple's spontaneous romantic moods.

The decision of when and where Pinyon Jays should nest may be influenced by factors similar to those above (age, assets, permanency and duration of the pair-bond) with one major exception: Pinyon Jays are unlikely to allow activities influencing reproduction to be left to spontaneity. Natural selection would probably cleanse a population of its spontaneous members because spontaneous responses are often performed at inappropriate times or in inappropriate locations. Responses closely adjusted to current conditions are more likely to evolve as they may enable individuals to maximize their survival and/or reproduction. The timing of nesting and selection of a nest site are examples of behaviors that often appear to be spontaneous but in fact are usually closely in tune with environmental conditions. Anyone who follows birds closely is impressed when spring rolls around and, as if a switch has been thrown, individuals begin to build nests. But this is not a spontaneous reaction to a romantic moment. Instead, appropriate timing of nesting and placement of nests have been molded through generations by natural selection.

We begin this chapter by discussing nest construction and arguing that the onset of nesting is closely attuned to environmental conditions such as food abundance and spring weather. As a result jays rear their offspring when conditions are especially favorable. However, favorable conditions are difficult to predict in harsh and variable montane environments such as those inhabited by our study flock and the flock in New Mexico referred to by Baily in this chapter's opening quote. Therefore, jays often nest at clearly inappropriate times. We continue by investigating the appropriateness of jays' responses as they "decide" whether or not to nest colonially and where to place their nests in a colony and within a tree. These need not be conscious decisions. Natural selection acts like a Supreme Court by favoring individuals making wise choices and punishing those making poor choices. Through evolutionary time populations come to be composed primarily of individuals making appropriate responses not necessarily because individuals consciously weigh the costs and benefits of each response, but because individuals making appropriate responses produce more offspring and these offspring also have genetic and/or cultural tendencies to make appropriate responses.

We wish to emphasize three important themes throughout this chapter. First, typically maladaptive behaviors such as nesting early in the spring when snow storms are common may evolve if they occasionally repay their practitioners big fitness dividends. Second, we must consider all likely costs and benefits of behaviors such as colonial nesting in order to understand why it has evolved over alternatives. Third, many aspects of a complex environment that may influence behaviors do not favor the same response. In deciding where to nest, for example, climatic conditions may suggest one location while the type and location of predators favor another. As in human conflicts, compromised solutions often result, but old individuals use their experiences to make such compromises as beneficial as possible.

NEST CONSTRUCTION

Pinyon Jay nests take days to construct, and the parents can be proud of their efforts as these large, bulky, well insulated structures effectively protect eggs, nestlings, and brooding females from freezing night-time temperatures. During the nest building season, the flock normally feeds off the nesting ground for about 1.5 hours each morning after arousal. It then moves to the nesting grounds and individuals begin building. The first materials placed at the nest site are twigs varying in length from 6 cm to 35 cm. These sticks are carried cross-wise in the bill, most often by the male. With longer sticks in their bills, birds sometimes have trouble penetrating the dense pine foliage. We have often laughed at birds hovering at a pine bough as they attempt to fly through it with a long stick impeding their progress! The twig platform is secured to the crotch of branches and averages 162 twigs (Balda and Bateman 1972; N = 22 nests). Next, birds weave coarsely shredded grasses into the twigs. Both members of the pair bring these materials to the nest and look quite comical with their bills stuffed full of material. Females do most of the arranging of the nest lining and often arrange the nest for minutes after the male has departed in search of more material. Proceeding toward the center of the cup, grasses become progressively finer and more shredded. The inner bottom of the nest consists of a fine powdery material that appears to be woolly plant leaves, horse hairs, feathers, and fine rootlets. Jays are attracted to soft,

*A jay returns to its nest site with a large stick to weave into the outer portion of its nest.*

*Female Pinyon Jay with a bill full of lining material for her nest. (J. Marzluff).*

man-made material such as processed cotton, cloth, insulation, mattress material, etc. which they use for inner lining. This is often detrimental, however, as after snow or rain these materials do not dry as rapidly as natural materials. Wet nests are often abandoned as the moisture freezes during cold spring nights (Gabaldon 1978).

### TIMING OF BREEDING

Most temperate-zone, terrestrial, passerines nest in spring and early summer when young can be provisioned on the copious hatches of insects. Pinyon Jays, however, belong to a small select group of species that can, when conditions are favorable, breed at other times. In the Flagstaff area females may be incubating clutches of eggs by mid-February when night-time temperatures commonly drop below freezing and measurable amounts of snow often accumulate. In other years eggs may not be present until late April. Thus, timing of nesting may vary by 50 days within a single flock among years (Table 17). In southwestern New Mexico Pinyon Jays may even breed three times in a 12-month period; in the spring after a moderate pine crop, the following fall if a bumper pine crop is produced, and again early in the spring following the bumper crop (Table 17). Why this great variability in timing of nesting?

In most temperate-zone passerines the initiation of breeding is linked to increasing vernal photoperiods (spring day lengths) coupled with mild weather and adequate food (Baker 1938). Using field data and laboratory

**Table 17.** *Year-to-year variation in the onset of breeding by three flocks of Pinyon Jays. The year in the first column is for the fall of each seed crop. Breeding occurred the spring of the following year or occasionally (1969, 1974 in New Mexico) later in the fall of a bumper seed crop. Some breeding occurs each spring in Northern Arizona, but only once each year. In New Mexico breeding occasionally occurs twice a year and may not occur every year. Breeding after poor cone crops in New Mexico in April and May 1972, 1973 appears to have been stimulated by outbreaks of insects. Data for the Doney Park Flock are from Balda and Bateman (1972), and for the New Mexico Flock from Ligon (1978).*

| | Northern Arizona | | | New Mexico | |
| | Fall seed crop | Onset of next breeding | | Fall seed crop | Onset of next breeding |
| Year | | Doney Park | Town Flock | | |
|---|---|---|---|---|---|
| 1968 | Moderate | 31 March | | | |
| 1969 | Bumper | 9 March | | Bumper | early Sept. & early Feb. |
| 1970 | Light | 13 April | 15 March | None | none |
| 1971 | | | | None | mid April |
| 1972 | | | | Light | early May |
| 1973 | None | | 19 March | Light | early April |
| 1974 | Bumper | | 24 Feb. | Bumper | mid Aug. & late Feb. |
| 1975 | Light | | 1 March | None | information unavailable |
| 1976 | Moderate | | 20 Feb. | Light | mid Feb. |
| 1980 | Bumper | | 20 March | | |
| 1981 | Moderate | | 5 March | | |
| 1982 | Moderate | | 21 March | | |
| 1983 | Bumper | | 21 Feb. | | |
| 1984 | Bumper | | 12 April | | |
| 1985 | Light | | 5 March | | |
| 1986 | | | 14 March | | |

manipulations, Ligon (1971, 1974a, 1978) tested how these factors affected the timing of breeding in Pinyon Jays. He demonstrated that testicular growth was stimulated by increasing photoperiods (as is true for most passerine birds) but more importantly that testes showed accelerated growth in the presence of green cones and fresh pinyon pine seeds. Jays' testes also remained enlarged longer when cones and seeds were provided. Thus, pinyon pine seeds, when available in abundance, interact in a positive manner with increasing photoperiod to speed up gonadal growth and accelerate breeding. The maturation of green cones actually stimulates autumnal breeding (despite shortening photoperiods) by some Pinyon Jays in New Mexico (Table 17). Thus, the presence of green cones may override the effects of photoperiod.

Our field observations support Ligon's findings, as early breeding in northern Arizona is linked with the presence of large numbers of seeds cached the previous autumn. In the Doney Park Flock, breeding was nearly one month earlier in 1970, following a bumper seed crop, than in either 1969 or 1971 following lighter seed crops (Table 17). Abundant food may also allow the Town Flock to breed exceptionally early. This flock has access to numerous well stocked feeders in Flagstaff and accordingly breeds much earlier than the nearby Doney Park Flock (Table 17). Here is one case where well-meaning bird-watchers may be hindering the birds they feed. Often we feed birds to enhance their possibility of surviving long, cold and unproductive winters. These noble efforts, however, may actually be detrimental to Pinyon Jays because they stimulate early breeding when heavy snows can destroy nests.

Abundant pinyon pine seeds may also accelerate the Town Flock's onset of breeding. Breeding was earliest in 1975, 1977, and 1984 following moderate to bumper seed crops (Table 17). The availability of seeds cannot be the only factor influencing the onset of breeding in the Town Flock, however, as in 1981 and 1985 breeding was late despite large seed crops. In Flagstaff, heavy snowfall during the onset of breeding may negate the stimulatory effects of large seed crops. The inhibitory effect of spring snowfall may explain why breeding was nearly a month later in 1981 than in 1984 despite similar bumper seed crops in both years; the spring of 1981 was one of the snowiest ever recorded, with nearly 20 times more snow than 1984. In general, breeding in northern Arizona is earliest when pine crops are large and snowfall moderate (e.g. Fig. 56, 1975, 1977, 1984), and breeding is late when pine crops are small, regardless of snowfall. Breeding is especially early when pine crops are large and snowfall is minimal, as in 1984. One year (1985) stands in stark defiance of these claims; snowfall was moderate, pine crop was large, yet breeding was delayed. Several factors (not the least of which included the first author studying for his PhD qualifying exams) made following early nesting attempts by the flock difficult that year, thus early attempts that lasted only a few days

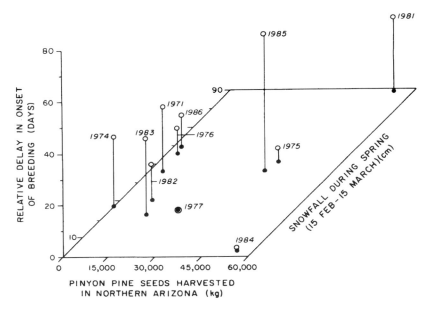

Figure 56    *The interaction of pinyon pine seed crop and spring snowfall on the initiation of breeding by the Town Flock of northern Arizona. Height above the snowfall/pine seed plane indicates the relative delay in breeding (onset in a year relative to 1977, the year of earliest breeding). The higher the point is above the surface, the greater is the delay in breeding.*

may have been overlooked. It is also possible that other factors such as disturbance of courting jays by predators and/or human activity may have delayed breeding.

In Pinyon Jays a selective premium seems to be placed on nesting as early in the year as possible. The timing of breeding in passerines is generally believed to occur when maximum reproductive success is possible (Lack 1968). Yet, incubating eggs and caring for helpless young in a snow storm hardly seems likely to increase this potential for success. In fact, early nesting by the Town Flock is rarely successful (Table 18). Nearly 40% of all nests in the initial colony are abandoned before eggs are laid because heavy snows fill nests. In contrast, only 16% of later nesting attempts are abandoned before egg-laying. This difference is statistically significant (37% of initial nests over seven years *versus* 16% of later nests over seven years, Wilcoxon matched pairs test, P = 0.016). Even if eggs are laid before spring snows thwart nesting they are often abandoned and freeze during snow storms. Females do not start incubating until the third egg is laid. Thus, if freezing temperatures occur when the first two eggs are exposed in the nest they may freeze. Twenty-nine per cent of nests with eggs in initial colonies failed during snow storms, but only 5% of nests in later colonies failed during snow storms (Wilcoxon test, P < 0.01). Given these significant negative impacts of early nesting, why does it persist?

**Table 18.** *Comparison of productivity in initial (1st) nesting attempts and subsequent (rest) attempts for the Town Flock. Nests started include any sites that pairs placed sticks in for at least two days. Question marks for survival of young indicate that adequate censusing of young was not possible.*

| | | | | | | | | | | | | | | Year |
| | 1981 | | 1982 | | 1983 | | 1984 | | 1985 | | 1986 | | 1987 | |
| Measure of productivity | 1st | rest | 1st | rest | 1st | rest | 1st | rest | 1st | rest | 1st | rest | 1st | rest |
|---|---|---|---|---|---|---|---|---|---|---|---|---|---|---|
| Number of nests started | 17 | 45 | 18 | 66 | 18 | 50 | 30 | 14 | 22 | 19 | 24 | 64 | 32 | 15 |
| Number of nests containing eggs | 4 | 34 | 1 | 51 | 10 | 40 | 29 | 14 | 21 | 18 | 15 | 57 | 31 | 15 |
| Percentage of nests started that were not completed | 76 | 24 | 94 | 23 | 44 | 20 | 3 | 0 | 5 | 6 | 37 | 11 | 3 | 0 |
| Number of nests with eggs failing from snow | 1 | 0 | 1 | 8 | 3 | 5 | 1 | 0 | 0 | 0 | 10 | 3 | 6 | 0 |
| Percentage of nests failing from snow | 25 | 0 | 100 | 16 | 33 | 13 | 3 | 0 | 0 | 0 | 67 | 5 | 19 | 0 |
| Number of nests fledging young | 2 | 8 | 0 | 16 | 6 | 4 | 12 | 2 | 2 | 11 | 0 | 14 | 9 | 2 |
| Percentage of completed nests fledging young | 50 | 24 | 0 | 31 | 60 | 10 | 41 | 14 | 10 | 61 | 0 | 25 | 29 | 13 |
| Number of young fledged | 5 | 27 | 0 | 59 | 20 | 17 | 46 | 5 | 6 | 43 | 0 | 48 | 37 | 9 |
| Number of young living one year | 1 | 11 | 0 | 9 | 1 | 5 | 11 | 0 | 0 | 7 | 0 | ? | ? | ? |
| Percentage of fledglings living one year | 20 | 41 | 0 | 15 | 5 | 29 | 24 | 0 | 0 | 16 | 0 | ? | ? | ? |

The critical requirement for early breeding to evolve is that the sum of all its costs be less than the sum of its benefits. The physiological drain of early reproduction appears minimal as pairs quickly re-nest following weather-induced abandonment. As early breeding is most common when food is abundant any energetic costs to breeders would be reduced in years of early breeding. The few young that a pair may produce early in the year may offset any costs if their survival is higher than survival of jays born later. Ligon (1978) proposed that early breeding by Pinyon Jays in New Mexico was adaptive because early-hatched young suffered less from parasitic fly larvae and were tended by better fed parents than were late-hatched young. Early-hatched young should therefore fledge in healthier condition and survive their first winter better than late-hatched young. Early-hatched young also have longer to mature physically, mentally, and socially before they harvest pine seeds, disperse, and/or obtain their lifelong mates. Greater maturity may therefore enhance their status and mating success in the flock.

The benefits of early nesting are rarely realized by fledglings in the harsh elevations of northern Arizona. Average fledging success of clutches is equal (27% of nests with eggs fledge) regardless of timing of nesting (Table 18). Moreover, early-hatched fledglings do not enjoy greater survivorship to one year of age (in fact it is lower) than later-hatched fledglings (annually 12% of early fledglings survive, N = 4 years, *versus* 20% of later hatched fledglings, N = 5 years).

The key to early breeding in northern Arizona may be that it costs little and occasionally pays off in a big way. Consider 1984: breeding was early following a bumper pine crop (Fig. 56), clutches were large (see Chapter 9), and few early nests failed because snowfall was minimal (Fig. 56, Table 18). Early nesting in 1984 paid big dividends in terms of higher fledging success and higher fledgling survival than late nesting which was completely unsuccessful because predation was extreme (Table 18). It is important to remember that early nesters that fail, simply re-nest later in the season, but once early breeding is forgone it cannot be repeated. Forgoing early breeding in the occasional year like 1984 would be disastrous to an individual's fitness. Occasional big pay-offs may therefore select for normally unproductive behaviors that entail little cost such as nesting in snow storms. Until Pinyon Jays in northern Arizona are able accurately to forecast spring weather (something humans have not yet mastered) they will continue to nest too early for their own good in most years!

## COLONIAL LIFE

Since the late 1800s Pinyon Jays have been known to nest colonially (Bendire 1895). Our colleagues who have visited jay colonies always respond with, 'you call *this* a colony?' They are skeptical of coloniality because nests are tens of meters apart, not tens of centimeters apart as in many typical colonial birds like gulls, terns, and auks. Any skepticism vanishes when all the parents of a colony return as a co-ordinated group to synchronously construct or tend their nests. Colonies of the Town Flock average 11 nests with a distance of

110 m between neighboring nests. However, they may contain from two to 30 or more nests with nearest neighbors spaced from 10 m to 600 m apart. Nests are spread uniformly throughout a colony. (See Appendix 1.)

Pinyon Jay colonies are situated on traditional breeding grounds within their home range. The Doney Park Flock had a traditional 100 hectare breeding area where it bred each of 14 years we observed them. This area was criss-crossed by a series of narrow dirt roads and adjacent to a large cinder cone that supported pinyon pine trees. We often observed jays caching seeds on the edges of this breeding area. Occasionally birds cached seeds directly on it. The Town Flock did not show the same consistent use of a single traditional area. Instead they initiated nesting on 24 different sites during 12 years of observation (Chapter 7, Fig. 45). Most of these sites have only been used for nesting one or two times over the years, but seven sites are repeatedly used and one site has been used nine times (Fig. 57). Most of the Town Flock's colonies are located at the periphery of the town of Flagstaff and within a few kilometers of well stocked feeders. Recently, the flock has nested in new colony locations south of town, perhaps because of increased human development around the formerly preferred northern areas. In some cases former nesting grounds have been destroyed and replaced with human dwellings.

All of the sites used by the Town Flock contain unimproved roads, and surprisingly most were within one kilometer of a school. A possible preference for school grounds was carried to an extreme in 1987. The main colony was situated in a previously unused area adjacent to a newly constructed school

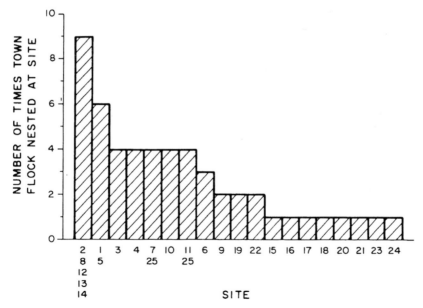

Figure 57 *Frequency with which the Town Flock nested on various sites within its home range (sites are indicated on Figure 45). Use of sites is from 1973 to 1977 and 1981 to 1987. Some sites considered distinct in Chapter 7 are lumped here because their geographic ranges overlap.*

and in the middle of a future school site! Nests are commonly lined up along the edges of unimproved roads and less commonly placed in forested expanses between roads. Both roads and school grounds are disturbed areas that may have high insect productivity. Human activity may stir up insects and frighten off predators; both may be advantageous to nesting jays (Gabaldon 1978). The Town Flock also caches seeds near their nesting grounds and occasionally we have observed them creating caches in and adjacent to their nesting grounds during the onset of nest building.

No two Pinyon Jay colonies are alike. They vary in the spacing between nests, the number of nests, and the synchrony of nesting. The time at which the colony was initiated may have a great deal to do with the particular characteristics of a colony. Colony size abruptly decreases as the season progresses (Fig. 58). Early colonies are usually large because all the breeders in the flock initiate the first colony. Previously unsuccessful pairs typically re-nest in smaller satellite colonies as the season progresses. In northern Arizona, snowfall commonly destroys early nesting attempts which may leave initial

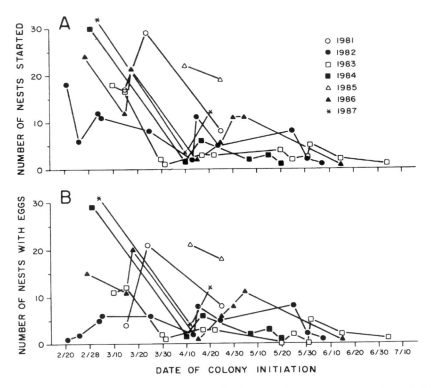

Figure 58   *Decline in breeding colony size for the Town Flock through the season. All nests initiated within a season (A) and nests in which eggs were laid (B) are drawn separately. Colonies initiated within a year have the same symbols and are connected by lines. Combining all years the seasonal decline in the number of nests initiated and number having eggs are both statistically significant (r = −0.64 and −0.46 respectively, N = 45 for both correlations, P < 0.01 for both).*

colonies with relatively few breeders laying eggs (e.g. 1981 and 1982 in Fig. 58B). Synchrony of breeders within the colony also is lower in late season colonies than in early season colonies. This is especially pronounced in larger colonies where all the nests in early colonies are typically initiated over a ten-day period compared to a 30-day period for colonies initiated later in the season (Fig. 59).

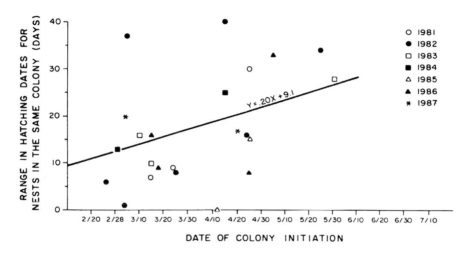

Figure 59   *Seasonal decline in synchrony of hatching within a colony of three or more nests. Synchrony is indicated by the maximum number days separating the hatching of any two nests in a colony. As this range in hatching date (y-axis) increases, synchrony can be viewed as decreasing. Colonies initiated within a year are given the same symbols as indicated in the legend. Hatching dates are significantly less synchronized in colonies of three or more nests initiated later in the year (r = 0.47, N = 23, P < 0.05).*

Figure 60 gives a picture of colony composition and nest initiation for the Town Flock. The seasonal trends discussed above are best illustrated within a year. Take 1984, for example, when nesting began on 21 February with an initial colony consisting of 15 nests with eggs laid over a ten-day period. In contrast, a later colony only contained four nests, each started on a different day over a protracted 24-day period. Similar trends occurred in 1981, 1983, 1986, and 1987. However, in 1982 a relatively large, synchronous colony was initiated late in the season.

Synchronization of activities within large colonies is striking. Within a five- to ten-day period all pairs commenced building stick platforms that would serve to anchor the nest in place. It was not unusual for us to observe between 20 and 30 pairs of birds building nests on the same day. Each day most birds were absorbed in the same stage of nest-building and few nests were started after the "season opener" (Fig. 61).

This severe synchrony brings up a series of interesting questions and speculations. Synchrony may be enhanced through a process known as social stimulation, or "The Frazer Darling Effect". Darling (1938) studied colonial-

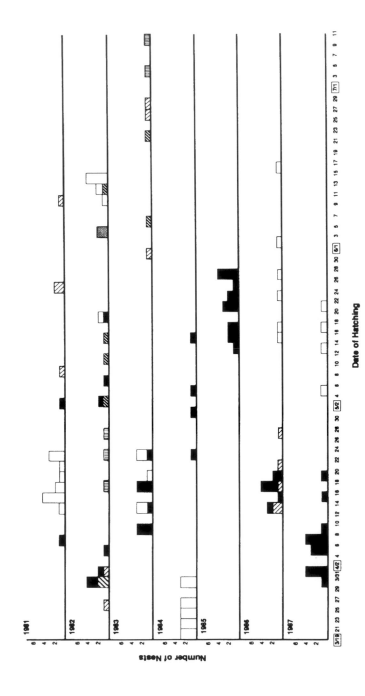

Figure 60   Colony sizes and hatching synchrony for the Town Flock. Different colonies within a year are shaded uniquely. Two colonies within a year that had nests active at the same time were spatially isolated.

# Chronology of Nesting

Figure 61    *Synchrony of nesting activities by members of the same colony. Each horizontal bar represents the stages of construction and development of young for one nest. Synchrony throughout nesting is indicated because most nests in a colony are at the same stage on the same date. 1981 data are from the Town Flock. 1970 data are from the Doney Park Flock.*

nesting gulls and hypothesized that vocal and visual signals of breeders stimulated other members of the colony so that all pairs became physiologically and psychologically ready to breed at about the same time. Furthermore, he hypothesized that social stimulation and synchronization would be greater as colony size increased. This has been confirmed in some, but not all subsequent investigations (MacRoberts and MacRoberts 1972). Our results suggest that social stimulation plays only a minor role in Pinyon Jay colonies. There is a slight tendency for early, large colonies to be more synchronized than late, small colonies. However, large early colonies are also more synchronized than large late colonies (Fig. 59). Therefore, the date of nesting, not the colony size (which determines the degree of social stimulation), appears more influential on synchrony within a colony.

Early colonies may be especially synchronized because resources are abundant, resources are equitably divided up among females, and/or some females delay breeding until the energy state of most females allows for breeding. The latter possibility is especially intriguing as it suggests that females who can quickly garner resources necessary for breeding may have to wait for their less efficient flock members. At the present time we have no data to support or deny any of the above. This synchrony of nesting, however, is important in that the flock maintains its integrity with all (or most) members performing similar tasks at the same time, which can lead to co-operation.

We are able to gain some insight into how jay colonies form by looking at nest placement through time. We typically locate colonies after a few birds have begun to construct their nests. We follow the formation of the colony until all nests are complete and plot nest locations on a map in the order in which they are initiated. We have done this for eight larger colonies (> 15 nests) initiated in 1981–1987. Many nests are initiated within a day of each other so we have grouped nests into three temporal sets. The first set is the largest and consists of nests started within a day or two of the season opener. A few days later a second set of nests is started and several days to a week later a few stragglers nest as a third set. Looking at the colony after each wave of nesting provides a "time-series photograph" of colony formation (Fig. 62).

Our glimpses of colonies through time suggests that they develop outwards from a central core. The first nests are predominantly central (48 of 81 or 59% of nests in the first set). The second set of nests are also predominantly central (30 of 56 or 53% of nests in the second set). The last set of nests, however, are rarely in the center of the colony (nine of 35 or 26% of nests in the third set). This tendency for later built nests to be placed on the periphery of the colony is statistically significant (chi square test comparing nests in center *versus* edge of the colony over three time periods, $\chi^2 = 11.31$, 2 DF, P < 0.01).

In many species, older birds often initiate nesting before younger, less experienced birds (Ryder 1981). There is a slight, but statistically nonsignificant tendency for this to occur in Pinyon Jays (Fig. 63A). Within a colony males attending the first set of nests averaged a year older than males attending the last set of nests. Given the inside-out pattern of colony formation we might expect younger birds, nesting later, to be found predominantly on the periphery of the colony. The slight tendency of older birds to nest before younger ones was not strong enough to produce consistent differences in the

*Two parents weave sticks into the coarse outer layer of their nest.*

ages of birds nesting in the center *versus* the edge of the colony (Fig. 63B). Gabaldon (1978) noted a similar lack of age-structuring within the colonies she studied.

CHOOSING A SPOT TO NEST WITHIN THE COLONY

Competition for nesting materials may set the inter-nest distances within a colony and produce the "inside-out" pattern of colony establishment we have

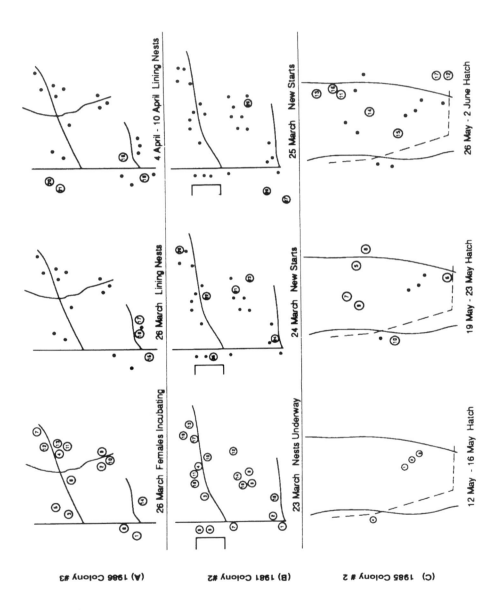

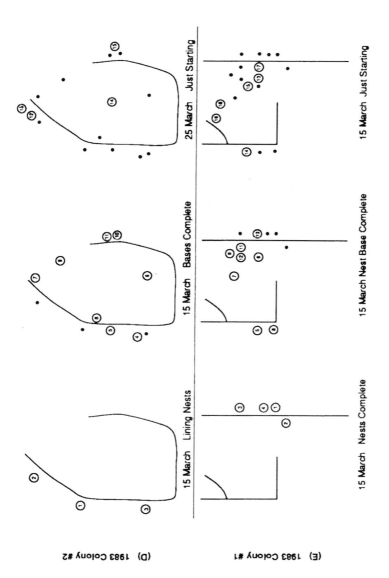

Figure 62 *Sequential maps illustrating the growth of five (A–E) Town Flock colonies. Each colony is depicted three times during the course of its development; earliest nests are shown in the left map, slightly later nests occupy the central map, and the latest nests are in the right map. Criteria for assigning nests to each map are listed below the map. Nests are represented by circled numbers when they first occur and as solid dots in later maps of the colony. Nests are numbered sequentially in order of their date of initiation to illustrate a colony's temporal development within each map. Colonies A (1986), B (1981), C (1985), and D (1983) show the common "inside-out" pattern of growth discussed in the text.*

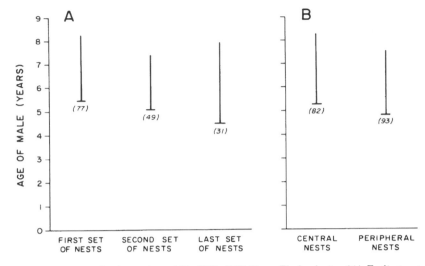

Figure 63    *Ages of males nesting within 1981–1987 Town Flock colonies. (A) Earliest nests are occupied by slightly older males than later nests (t = 1.34, P > 0.10). (B) Central and peripheral males differ little in age (t = 0.95, P > 0.10). Horizontal lines indicate average, vertical lines indicate standard deviations. Sample sizes are numbers in parentheses below horizontal lines.*

observed. Pairs often steal nest materials from their neighbors' nests rather than searching for unused materials. Theft of material may delay a pair's onset of nesting by several days or even weeks. Therefore, nests should be placed just far enough away from neighbors so that the neighbors' costs of stealing material are higher than their costs of getting new material. As more nests are built, the only way to stay far enough away from neighboring thieves is to nest on the colony periphery.

Alternatively, central nest sites may be actively preferred because they are of higher quality than peripheral sites. Predators are expected to encounter peripheral nests first and would have more difficulty penetrating into the colony center without being detected, mobbed, and driven from the colony. In Adelie Penguins, for example, Tenaza (1971) discovered that peripheral nesters had small clutch sizes and tended to flee quickly from intruders, which he attributed to increased predation on the colony edges. Peripheral nests in Pinyon Jay colonies do not appear to suffer extreme predation. Early reports by Balda and Bateman (1972) and by Gabaldon (1978) did not find consistent benefits to central nests, nor consistent use of central nests by individuals. In some years, central nests outproduced edge nests and in other years the reverse was true. Recently this lack of consistency has continued. Figure 64 indicates that in 1981–1987 central nests were actually slightly more vulnerable to predators than were edge nests. We conclude that natural selection should not favor choice of central nest locations. The "inside-out" pattern of colony establishment does not result because central locations in the colony are inherently of high quality. Rather, it is more likely to be a side-effect of spacing out to minimize competition and/or interference with neighboring pairs.

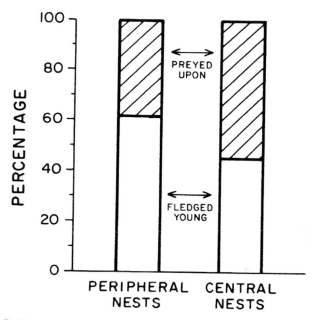

Figure 64   *Different fates of central and peripheral nests. All nests in this sample either fledged young or were preyed upon. Central nests were slightly more likely to be preyed upon and less likely to fledge young than were peripheral nests ($\chi^2 = 3.03$, DF = 1, $0.05 < P < 0.10$).*

Relatives in a flock appear to space their nests as far apart as possible (Gabaldon 1978). Diana Gabaldon first noted this tendency when she divided all nests for a year into two categories. Two nests were designed "near" if they were closer to their neighbor than the average of all inter-neighbor distances. In contrast, "far" nests were farther apart than the average inter-nest distance. Over a three-year period Diana located 19 pairs of nests of related birds. All relatives of both members of the pair were used in this survey. In 12 cases where relatives nested at the same time they nested in different, completely isolated, satellite colonies. In seven cases where relatives' nests were located within the same colony the distances between these nests was close to the maximum inter-neighbor distance known for the year. Only one pair nested "near" a relative. The same pattern held in 1981–1987. During this time we found 37 pairs of related birds' nests. In 14 cases, relatives nested in different satellites. In the 23 cases where relatives nested in the same colony they appeared to avoid each other as 22 pairs nested "far" from each other. Only two of 56 pairs of relatives ever nested "near" each other; we conclude that relatives avoid each other's nests.

Why should Pinyon Jays avoid nesting near their relatives? This question is especially perplexing when we consider the co-operative mobbing of predators by jays nesting in the same colony. Nesting near a relative would seem to allow those sharing genes to cooperate in the expulsion of predators from each other's nesting areas. There must be considerable resistance for such a potentially beneficial trait not to evolve. The searching techniques of aerial nest predators may provide this resistance. Visually acute predators such as crows and ravens may use two sophisticated weapons to better their odds of finding jay nests. First, they are likely to hunt in an area-restricted fashion. That is, once a prey item has been found, the area yielding the prize is thoroughly searched for more prizes. Second, as more and more nests are found, these predators may key in on specific attributes of nests or nest locations and form "search images" (Tinbergen 1960) for these attributes that allows even better detection of the prey. Aerial predators are not the only ones likely to use search images to aid in the detection of nests. Our colleagues that had spent much time locating nests could correctly discriminate photographs of nest trees from non-nest trees (Gabaldon 1978)! They had formed specific search images for nest trees. If nest predators use area-restricted searching and employ search images to find nests, then even if a jay's nest is not initially spotted by a raven it may be discovered after the raven spots its neighbor's nest and focuses its hunting around that nest for other similar-looking meals.

We have several lines of evidence suggesting that neighboring Pinyon Jay nests suffer similar fates. First, consider the chances of a nest being preyed upon given the fate of its neighboring nest. From 1981 to 1987 nearly three out of every four jays whose neighbors' nests were preyed upon also had their nests preyed upon. In contrast, jays had only a 50:50 chance of having their nest preyed upon if their neighbors' nests fledged young (Table 19). These results indicate that the probability of a nest being preyed upon increases if its neighbor is preyed upon. This result would be expected if successful predators search nearby areas for more nests (i.e. if predators forage in an area-restricted pattern and have a search image).

**Table 19.** *Number of nests preyed upon or fledging young in relation to the fate of their nearest neighboring nest. Neighbors' fates influence a nest's fate; 72% of nests whose neighbors were preyed upon were also preyed upon, but only 51% of nests whose neighbors were fledged young were preyed upon. Fate of nest and fate of neighboring nests were positively associated (chi-squared test of association, $\chi^2 = 8.32$, DF = 1, P < 0.01).*

| Fate of neighboring nest | Fate of nest | |
|---|---|---|
| | Preyed upon | Fledged young |
| Preyed upon | 88 | 35 |
| Fledged young | 35 | 34 |

In order to determine whether all nest predators are equally capable of finding adjacent nests and to investigate the influence of colony structure on the fate of neighboring nests we set up a series of experiments. We collected abandoned jay nests, filled them with two domestic quail eggs, and wired them into trees as "experimental colonies" for predators in the Town Flock's home range to discover. After we placed our colonies we recorded the number of days until a nest and its neighbor were preyed upon. There were strong positive correlations between the days until a nest was preyed upon and the days until its neighbor was preyed upon suggesting that nearby nests were taken by the same predator at nearly the same time, as would be expected if predators were using area-restricted search techniques (Table 20).

The strength of this pattern, however, depended upon the colony's configuration and the type of predator. Ravens and crows which remove eggs from nests and soar above colonies looking for nests were most efficient at finding neighboring nests in large colonies, in colonies with a majority of their nests exposed, and in colonies with their nests either closely clumped together or spaced far apart (Table 20). These predators may always search around discovered nests for more food, but our results suggest they do not find neighboring nests when colonies are small or nests are well hidden. In these cases searching around discovered nests is unlikely to be rewarding as more nests cannot be located (concealed colonies) or they are very rare (small colonies). Steller's Jays prey on only a few Pinyon Jay nests, but distinguish themselves by pecking holes in the eggs they prey upon. These predators seem to search for nests as they flit between trees; they do not soar above trees looking for nests. Accordingly they are efficient at finding neighboring nests only in colonies with nests packed closely together. In such colonies, flits between trees may often reveal another nest to prey upon.

**Table 20.** *Ability of egg-eating predators (crows and ravens) and egg-pecking predators (Steller's Jays) to locate neighboring Pinyon Jay nests. Correlation coefficients relate the timing of predation on one nest to the timing of predation at its neighbor's nest. Strong negative correlations (those close to −1) indicate that neighboring nests were found many days apart. Strong positive correlations (those close to +1) indicate that neighboring nests were found at nearly the same time. Positive correlations are starred according to increasing significance (\* = P <0.05, \*\* = P <0.01, \*\*\* = P <0.001). Correlations are reported separately for experimental clumped (internest distance = 30 m), spaced (internest distance = 100 m), large (7, 10 nests), small (1, 3, 5 nests), exposed (all nests on outer edges of tree's foliage), and concealed (all nests in inner portions of tree's foliage) colonies. Correlations are partial correlation coefficients between days to predation of a nest and days to predation of its nearest neighbor controlling for colony configurations (spacing, size, or exposure) not investigated.*

| Colony configuration | Eggs removed | | Eggs pecked | |
|---|---|---|---|---|
| | Number of nests | Correlation | Number of nests | Correlation |
| Clumped | 85 | +0.32*** | 30 | +0.49** |
| Spaced | 31 | +0.35* | 19 | −0.54 |
| Large | 97 | +0.33*** | 39 | +0.19 |
| Small | 21 | +0.28 | 10 | −0.30 |
| Exposed | 91 | +0.39*** | 24 | +0.11 |
| Concealed | 27 | +0.18 | 25 | +0.13 |

Let's put the evidence together. Nests of the Town Flock are heavily preyed upon by crows and ravens which are very efficient at finding neighboring nests. Therefore, a jay's chances of having its nest preyed upon rise sharply if its neighbors' nests are preyed upon. As we have argued, relatives should cooperate to propagate the genes they share. In the present case, nesting far from a relative may be the best way to attain this goal, albeit indirectly by not aiding predators in their search for nests. The last thing a jay should do is help a raven find its relatives' nests.

### Advantages and disadvantages of colonial nesting

It is doubtful that any single factor is responsible for the evolution of a nesting dispersion pattern (e.g. solitary *versus* colonial nesting). If this were the case, nest predation by ravens and crows should be sufficient to produce widely spaced or solitary nesting in Pinyon Jays. Instead, many factors may each give some advantage to colonial nesters which taken together create a benefit great enough to outweigh the costs of nesting in close proximity to many potential competitors. Here we discuss many potential costs and benefits of colonial nesting in order to understand better why jays do not try to foil predators by nesting solitarily.

*Costs of coloniality*   The benefits of colonial nesting must be substantial because the costs are potentially severe. Coloniality makes competition for nest materials, food, and mates likely if these resources are in short supply. Nest materials may only rarely be in short supply (after heavy snows) and food (at least for the Town Flock) is unlikely to be in short supply. However, we have shown that a shortage of mates is typical for many flocks of jays. Males commonly outnumber females within a flock and compete for mates. Colonial nesting brings many competing males in close proximity when their mates are in reproductive condition. Colonial males may therefore have to guard their mates so that they do not sneak any "extra-pair copulations". Males and females certainly stick close together during nest-building.

Coloniality is confusing. Each hour when males return to feed their mates, pandemonium breaks out. The usually quiet stretch of forest explodes into a noisy mêlée of begging, flapping, and flying Pinyon Jays. Confusion may work in favor of the nesting birds if it makes it difficult for a predator to isolate a particular nest because of all the activity going on at the same time. However, confusion may be costly because it can result in misdirected parental care such as the feeding of another bird's mate or offspring. Colony members may co-operate at some levels but they are (except relatives) genetic competitors and aiding unrelated birds decreases their fitness. Misdirected parental care must therefore be kept to a minimum. Confusion may also lengthen the time eggs and nestlings are exposed to inclement weather or predators. Females that respond to every colony member will spend unnecessary time off their nests.

Male Pinyon Jays negate the confusion cost of coloniality by announcing their arrival in the colony. Upon returning to the colony, they perch high in trees close to their nests and give Near calls. Females recognize and respond only to their mate's Nears. John Braly discovered this in 1931 and we investigated this intriguing observation further in 1986 with a simple playback experiment (Marzluff 1988b). We recorded males calling before they visited their nest and played these calls back to brooding and incubating females. We waited until a male had been away from his nest for half an hour and then broadcast his calls (alternately with his neighbor's calls) to his female. We "talked" to 11 females in this way and all of them responded more strongly to their mates' calls than to the calls of their neighbors. Six females left their nest during playback of their mate's calls and four of these approached to within a few meters of the speaker we were using. A female recognizes her mate's voice and exposes the nest contents only when he signals his presence. We have never seen a male feed the wrong female, but this does not appear to be mediated by vocal communication, as females do not vocalize their location. Instead, males may simply memorize the location of their nests and care for the female and young at that site. Males do recognize their mates and their offspring, but this is not necessary to allocate parental care correctly in the colony. As we will see in Chapter 9, however, it is needed and used in the creche.

Coloniality may also increase the ease with which diseases and ectoparasites are spread. Coloniality forces individuals into contact with potential carriers so that a pathogen infecting one jay can quickly spread to all the jays in a colony. We have no evidence of this occurring in Pinyon Jays, but evidence from other colonial birds suggests it can be a significant cost. Barn Swallows, for example, carry higher parasite loads and lose more nestlings to blood-sucking fly larvae as colony size increases (Shields *et al.* 1988). Pinyon Jays have similar parasites, but we do not know if they are more common in large colonies than in small ones. In any case, jays are much closer to each other during other parts of the year so the spread of many diseases is possible without colonial nesting.

Increased risk of predation is perhaps the cost of coloniality most relevant to Pinyon Jays. Predation is the major source of nest failure in our study population and, as we have argued above, nesting in large colonies may enhance predation by allowing predators to forage in an area-restricted fashion and use a search image of nest sites. Our results with experimental colonies suggests that coloniality has great potential for increasing the level of predation. In 1983 and 1984 we set out abandoned nests with quail eggs in differing colony configurations and monitored the rate of predation on nests. The visibility of nests, the number of nests in a colony, and the distance between nests influenced the rate of predation at artificial colonies. Colonies in which we placed all nests high in trees or far out on the tips of branches were preyed upon faster than colonies in which we concealed nests deep in the tree's foliage (compare open and hatched bars in Fig. 65). Large colonies (seven and ten nests), especially exposed ones, were preyed upon faster than small colonies (one, three, and five nests). Small, spaced (100 m between nests) colonies had slow rates of predation, regardless of whether their nests were exposed or concealed. Overall, large colonies with their nests clumped (30 m between nests) suffered the fastest predation and small, spaced colonies suffered the slowest predation (compare dotted bars in Fig. 65).

Two factors determine the fast rate of predation at large, clumped colonies relative to slower predation at small, spaced colonies. As we argued above, small colonies have slower rates of predation because the primary predators appear to have greater difficulty locating neighboring nests in such colonies. In addition, small colonies are harder to locate initially than large colonies.

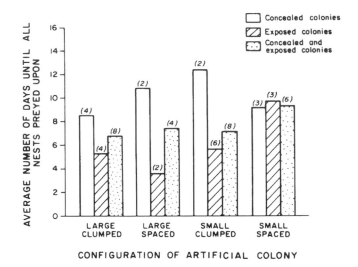

Figure 65   *Rate of predation on artificial colonies differing in configuration. Open bars are concealed colonies. Hatched bars are exposed colonies. Dotted bars are both concealed and exposed colonies lumped together. Numbers above bars are the number of colonies of each configuration. Three-way analysis of variance indicated a strong effect of exposure ($P = 0.02$) and a relatively strong interaction between colony size, spacing, and exposure ($P = 0.10$).*

Egg-eating predators take four times longer to discover small colonies than large colonies (Fig. 66A).

We must, however, point out that the potential cost of greater predation in large colonies is dependent upon the type of predator searching for nests. In our study area, ravens and crows (egg-eating predators) are the primary nest predators. These birds actively search wide areas for nests and appear to detect large colonies more rapidly, and find subsequent nests therein more easily, than in small colonies. Other predators, such as Steller's Jays (egg-pecking predators), snakes, and squirrels, have more limited foraging ranges and grouping nests in colonies may not affect their ability to find and destroy nests. In our study area, however, egg-pecking predators also select against colonial nesting. We argued above that clumping nests enhances these predators' abilities to find subsequent nests. They are also able to locate large colonies more rapidly than small colonies (Fig. 66B).

All things being equal, our evidence so far suggests that nesting together as a colony increases the risk of predation but, of course, all things are not equal! The critical factor we removed in our experiments was the activity of parental jays. Our experiments indicate that the presence of many nests, *per se*, near your nest is detrimental but perhaps the presence of many nest-owners is not so detrimental. Maybe it is even beneficial. As we will now discuss, colony members could help you find food, discourage predators, or baby-sit your nestlings.

*Foraging benefits of coloniality*   Osprey and Cliff Swallows that nest in colonies

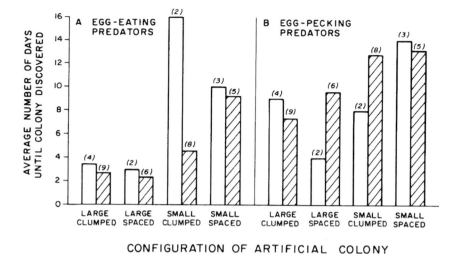

Figure 66 *Rate at which egg-eating (A) and egg-pecking (B) predators discover concealed (open bars) and exposed (hatched bars) artificial colonies. Sample sizes are given above bars. Three-way analysis of variance indicated highly significant size (P = 0.001), exposure (P = 0.005), and an interaction between size, spacing, and exposure (P = 0.03) effects for egg-eating predators. Size (P = 0.03) was the only significant effect for egg-pecking predators.*

benefit by being able to keep track of changing locations of abundant foods (Brown 1986, Greene 1987). Colony members watch each other return from hunting and ascertain whether or not an individual succeeded in catching a meal. Watchers then follow successful hunters back to rich feeding grounds. Colonies may thus serve as "information centers" where parents learn the best areas for feeding each day (Ward and Zahavi 1973). Pinyon Jay colonies are unable to function as information centers, however, because all parents forage together as a group and thus *together* discover rich hunting grounds.

Coloniality may enhance foraging on patchily distributed food by minimizing travel between the nest and the food. This would be beneficial because the time and energy saved could be used to procure more food, tend to more young, increase the care of young, or enable parents to rest. Ligon (1978) pointed out that insects, which are an important food for nestlings, are found in patches that vary in their productivity from day to day. Nesting as a colony in the center of such patches assures jays that they will always be relatively close to food and they will never have to travel far in search of a rich patch of insects. However, solitary nesters, dispersed evenly throughout the foraging area, may have to travel extreme distances to find a productive feeding spot one day, but be lucky enough to have a rich patch next to their nest another day. On average, dispersed nesters will have to travel farther than colonial nesters in search of mobile bonanzas thereby favoring the evolution of coloniality (Horn 1968).

The importance of cached pinyon pine seeds during the nesting season may also favor colonial nesting. Cached seeds are localized, but they are not mobile in the sense that insects are mobile. Nesting as a colony, adjacent to cached seeds assures each jay of a minimal commute from its nest to its caches. Moreover, as colony members forage together, coloniality allows each jay to make sure its neighbors are not raiding its food caches. We have no evidence that jays defend their caches. Lack of defense after several months of harvesting seems possible only if flock members recover caches as a group. Coloniality may be most important for its synchronizing effects that allow each individual to monitor the behavior of the other breeders in the flock. Monitoring cache recovery is one example of the potential importance of synchronized breeding.

*Is predation higher in colonies?*   Synchronized breeding in a colony also promotes co-operation which may negate the potential for high predation suggested by our experiments. As breeders return to tend their nests they first perch high in trees throughout the colony and survey the area for predators. If one jay detects a predator it emits a raucous Multiple Rack and many colony members quickly converge to dive at and scold the predator (Chapter 4). This group mobbing is effective in deterring many would-be predators. The ability to spot predators and muster an army sufficient to drive them away should increase with increasing colony size. If co-operative mobbing is effective, we should see a reduction in the probability of predation as colony size increases.

Indeed, an individual's nest has a slightly lower probability of being preyed upon if it nests in a large ($\geq$ 7 nests) rather than a small ($<$ 7 nests) colony. We observed the fates of nests in 16 large and 24 small colonies from 1981 to 1987. To index the risk of predation we calculated the proportion of nests in each colony that were preyed upon. In large colonies an average of 59% of nests were preyed upon (N = 16, SD = 29.1), but in small colonies 69.5% of nests were preyed upon (N = 24, SD = 36.1). This difference is not statistically significant, but this is a striking *reversal* of our results in experimental colonies where large colonies were *more* susceptible to predation. We do not claim that nesting in a large colony is beneficial because it reduces predation, but we can safely conclude that it is not costly. The many parents in a large colony are able to negate the higher risk of predation indicated by our experiments to be associated with increasing the number of nests in a colony. Parents may reduce predation by co-operative mobbing and by reducing colony visibility by concealing and spacing out their nests. On average, 41.3% of nests in a colony are well concealed (N = 39 colonies, SD = 30.6) and the average distance between neighboring nests is 110 m (N = 40 colonies, SD = 95 m, Appendix 1). These average values of real colonies are close to our "concealed" and "spaced" experimental colonies which suffered relatively low rates of predation.

### Coloniality as a selective factor

Our best guess is that coloniality evolved because it minimizes the costs of foraging on unpredictable (insects) and predictable (cached seeds) clumps of food. Coloniality, however, is potentially costly and as such is a powerful selective factor. For example, widely spaced nests (especially relatives' nests), synchronized behaviors, and individually unique vocalizations are attributes of Pinyon Jay colonies that we feel have evolved to counter the potential detriments of coloniality. These counter-adaptations ensure that: 1) predators encounter synchronized group defenses, 2) all eggs and young are vulnerable to predators for the same, and therefore minimum, length of time, 3) potential cache raiders find themselves in the company of cache owners, and 4) nestlings or incubating females trying for a free meal encounter discriminating parents and mates. Counter-adaptations to colonial life mitigate many costs thereby allowing foraging benefits to dictate the evolution of colonial nesting.

NEST PLACEMENT WITHIN A TREE

We have emphasized the tight link between Pinyon Jays and pinyon pines throughout this book, but this link does not always carry through to a jay's decision of where to build its nest. Both flocks we studied in detail are unusual in inhabiting high elevation ponderosa pine forests. This pine forest is a virtual monoculture and thus the vast majority of nests were placed in ponderosa pine trees (Fig. 67). In the lower elevations of Arizona and New Mexico, most nests are placed in pinyon pine or juniper trees. We have found nests in virtually every part of a tree from the lowest branches where you can look down at the developing young to the uppermost reaches of mature trees over 30 m tall where even the slightest breeze feels like a hurricane as you measure and band young. Nests are placed on every compass bearing around trees and may touch the tree's trunk or be placed at the tips of its farthest reaching branches. This variability in nest placement is not meant to imply that nests are randomly

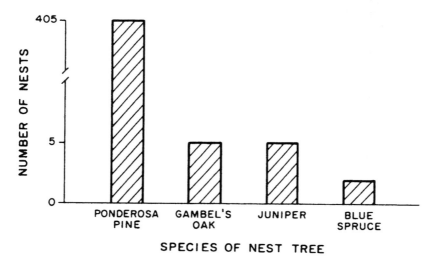

Figure 67 *Species of trees used as nesting sites by jays in the Town Flock in 1981–1987. Use of trees reflects approximate abundance of tree species in the study area.*

placed in trees. To the contrary, jays are very choosy about where they will nest and this finicky nature may enhance their fitness. Much of the initial work on nest placement by Pinyon Jays was done by Diana Gabaldon as part of her doctoral research on the Town Flock. Here we summarize her findings and test many of her ideas with more recent data we collected on nest placement by jays in the same flock.

Jays in the Town Flock appear to nest in particular types of ponderosa pine trees. They prefer trees with high foliage density and tall trees surrounded by shorter trees that are none the less higher than the nest. Diana Gabaldon speculated that the tall nest tree serves as a lookout post from which the surrounding area is scanned for predators prior to, and after, visiting the nest. The shorter trees may act as a screen to intercept gusty winds before they reach the nest. Jays avoid trees with abundant pine cones perhaps because these cones directly attract some nest predators (e.g. Abert's Squirrels) and may attract other predators as they hunt other birds feeding on cones (Gabaldon 1978).

Town Flock jays also prefer particular sites within their chosen tree. Choice of a bearing for nest placement illustrates a common problem faced by nesting birds: different environmental factors often favor different nest sites. The majority of jays orient their nests toward the south throughout the nesting season (Fig. 68). Southerly nesting, especially early in the season, would benefit jays because solar insolation warms females in such sites. Cannon (1973) calculated that females nesting on the south sides of trees absorb 40% more solar energy than females nesting on the north sides of trees. Such passive solar heating reduces a female's need to burn her own energy stores to warm the nest contents. South-facing nests have some drawbacks, however, especially when the wind blows. Prevailing winds during breeding are from the southwest. These winds cool south-facing sites and may destroy nests placed on the edges

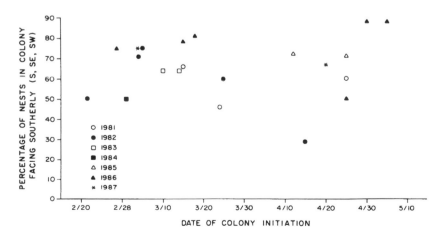

Figure 68    *Pinyon Jay nests in northern Arizona typically face south, regardless of when they are initiated. This figure includes data on 229 nests in 21 colonies built in 1981–1987.*

of the foliage. Pinyon Jays appear to solve this conflict by nesting on the south side of trees that are sheltered from wind by neighboring trees.

Choice of a beneficial nest height provides another example of conflicting environmental influences on nest placement (Marzluff 1988b). Nests near the tops of nest trees are frequently preyed upon by aerial hunting nest predators such as American Crows and Common Ravens. However, these sites are warmer so that snow falling on them during spring storms melts rapidly. Lower nests are better concealed from predators, but concealment shades females and nestlings. The cooler microclimate of a shaded site may increase an incubating female's energy expenditure by as much as 10% (Cannon 1973) and/or may lead to abandonment after early spring snow storms and freezing temperatures (Table 21).

**Table 21.**   *The relative height (nest height divided by tree foliage height) of nests has a strong influence on the fate of the nest. High nests are preyed upon, nests in the middle of the tree are abandoned after snow storms, and nests slightly higher than the center of the tree's foliage fledge young. Relative occurrence of each fate is the percentage of all nest failures attributed to a fate. Data are for the Town Flock in 1981–1986, from Marzluff (1988a).*

| Fate of nest | Number of nests | Relative occurrence of each fate | Average relative height of nest |
|---|---|---|---|
| Preyed upon | 138 | 48.9% | 70.3% |
| Abandoned after snow | 32 | 11.3% | 49.7% |
| Fledged young | 77 | 27.3% | 61.3% |
| Total | 247 | 87.5% | |

This simple conflict in where to nest should be easy for jays to solve. Just nest in a compromised spot about 60% of the way up in the tree to minimize predation without incurring excessive shading and increased energy expenditure. Jays choosing such sites should out-reproduce those nesting in more visible sites or in colder sites thereby allowing natural selection to produce compromised nest placement as a fixed or constant aspect of a jay's behavior. The evolution of a relatively fixed response such as this depends, however, upon the behavior having relatively constant costs and benefits. This was apparently the reason most jays choose southerly sites that are also protected from the wind. The costs and benefits of nest exposure are not constant within the Town Flock's home range and therefore cannot favor consistent compromised nest placement. Well concealed nests have constant costs and exposed nests have constant benefits because snowfall and cold temperatures have presumably been important sources of nest failure and energy drain ever since jays nested in high elevation ponderosa pine forests. The costs to exposed nests and benefits to concealed nests imposed by the jays' black, predacious cousins are not constant. The numbers of ravens have increased dramatically in Flagstaff over the last ten years (Fig. 69). Town Flock jays are therefore in a rapidly changing environment in which nearly half of all nesting attempts from 1981 to 1987, especially those in high, warm sites were destroyed by nest

*Female Pinyon Jay incubating in a shaded location after a spring snowstorm. The snow has already melted from more exposed portions of the tree, but remains around the rim of the nest concealing all but the female's head. (J. Marzluff).*

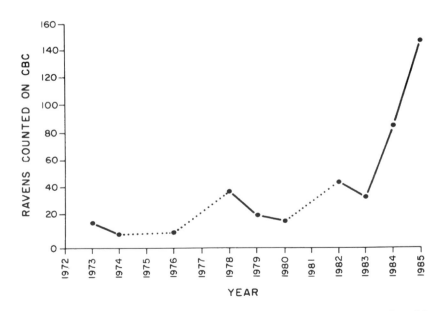

Figure 69    *Recent rise in the abundance of ravens in the Town Flock's home range. Dotted lines indicate years when censuses were not conducted. Ravens were censused each December during the Audubon Society's traditional "Christmas Bird Counts" (CBCs).*

predators (Table 21, Fig. 70). In the early years of our study (1973–1977) only 20% of nests were preyed upon (Fig. 70). Natural selection probably has not had long enough to select for intermediate nest height in this rapidly changing environment.

Long-lived species often first adapt to novel environments by learning to modify their behavior. If environmental conditions persist relatively

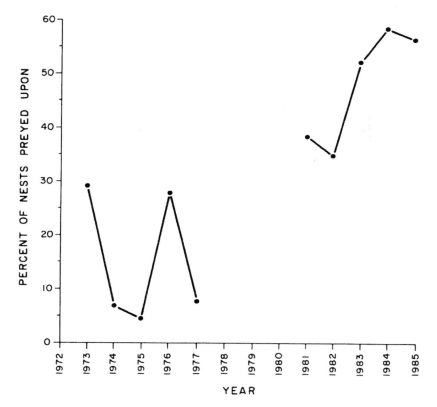

Figure 70   *Annual variation in the percentage of Town Flock nests preyed upon. On average pre-dation is over twice as frequent in the 1980s as in the 1970s.*

unchanged for generations, then specific morphological and/or behavioral adaptations to them may later evolve. Initially, however, a general adaptation such as the ability to learn associations between the environment and the hazards it brings may allow a species to survive in variable and/or changing environments. Pinyon Jays appear to learn to associate their nest location with the fate befalling it and later nest in locations less susceptible to those fates. For example, when exposed nest sites (those in the upper 30% of the tree) are preyed upon, eight of every ten jays re-nest in more concealed locations which, on average, are 27% lower in the nest tree. This drop appears to be a result of predation because when exposed sites were not preyed upon, jays subsequently nested only 10% lower in the nest tree. Similarly, jays only exhibited significant shifts away from concealed sites after nests in such sites failed from snowfall. Following snowfall, subsequent nests were 25% higher in the nest tree. In summary, Pinyon Jays appear to learn to associate their nest location with its chances of fledging young, failing due to snowfall, and being preyed upon. Jays may put this information to good use by: 1) nesting in more concealed (lower) locations after their exposed nests are preyed upon, 2) nesting in warmer (higher) locations after shaded nests fail from snowfall, and 3) continuing to nest at heights from which young have successfully fledged.

Black-billed Magpies also may use prior experience to modify their nest placement and lower the chance of failure (reviewed by Dhindsa *et al.* 1989). This corvid faces conflicting biotic and abiotic pressures on where to nest; cold weather favors exposed nesting and human disturbance favors concealed, high nests. Thus, weather and "predation" (by humans) affect magpies in nearly opposite ways as they affect Pinyon Jays. Magpies compromise their nest placement to minimize losses to both forces and may quickly increase the height at which they nest after humans begin to disturb low nests.

Learning where to nest is a process that occurs gradually throughout a jay's life. Males appear to select the nest site and as they age they experience predation at exposed sites. In response to having their exposed nests preyed upon,

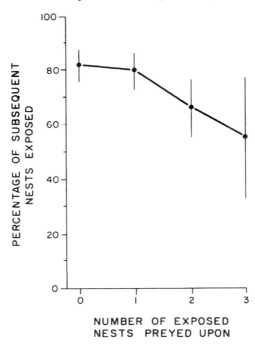

Figure 71    *Decline in the use of exposed nest sites as a function of experience with predation. Mean percentage of attempts per individual that were exposed is plotted. Bars show one standard error above and below the mean. From Marzluff (1988a).*

males gradually shy away from such sites; after having their first experience with predation at exposed sites males still place 80% of future nests in exposed locations, but this drops to 55% after their third experience with predation at an exposed site (Fig. 71). This gradual avoidance of exposed sites leads to an inverse relationship between average nest height and male age; older jays nest lower than younger jays (Fig. 72). Old jays may still get cold while tending their concealed (low and shaded) nests, but being experienced foragers they

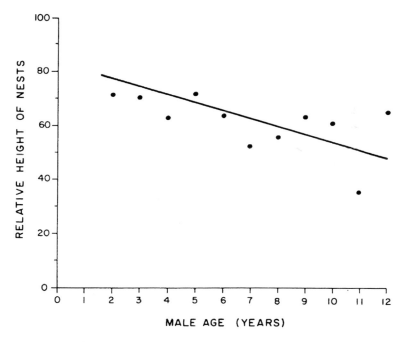

Figure 72    *Gradual decline in relative nest height as males age. Predation at high nests through an individual's life may result in selection of lower nest sites by old males. The average nest height of males of a given age is plotted. This inverse relationship is statistically significant (r = −0.60, N = 11 ages, P = 0.03). The plotted line is that best fitting the points determined by least-squares estimation. The number of males used to calculate each average was 26, 29, 38, 23, 18, 19, 13, 12, 5, 3, 6 for two-, three- four-, five-, six-, seven-, eight-, nine-, ten-, 11-, 12-year-old-males.*

may be able to handle the additional energetic demands of concealment. Perhaps in an effort to keep warm and yet keep out of the ravens' view, old jays nest low throughout the season but nest relatively far from the trunk early in the season. Nesting farther from the trunk may allow for some solar warming without greatly increasing the nest's visibility. As the season progresses and the threat of snow subsides, old jays conceal their nests further by selecting sites closer to the tree trunk (Fig. 73).

Imprinting as well as trial-and-error learning may influence nest placement. Choice of the nest tree may reflect the type of nest tree a jay was

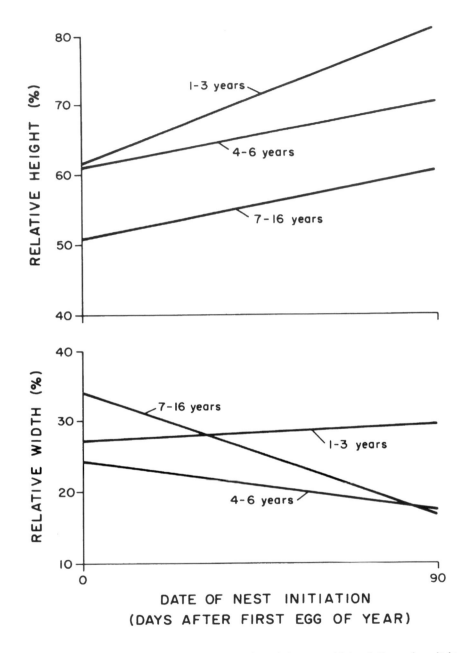

Figure 73 *Age-related changes in nest placement through the season. Males of all ages show slight increases in nest height as the season progresses (top graph). Youngest males (one to three years old) also place their nests slightly further from the nest trees' trunks later in the season (bottom graph). Relative width of the nest site equals distance from trunk to nest divided by distance from trunk to tip of foliage at nest height). Older jays, especially seven- to 16-year-olds, begin the season by nesting relatively far from the tree's trunk and gradually nest closer to the trunk later in the season. From Marz-luff (1988a).*

raised in. Parents and their young select trees similar in distance to nearest tree, nearest tree height, growth condition of the tree, and proximity of the tree to roads (Gabaldon 1978). Choice of a site within a tree is affected less by imprinting of parental preferences and may be affected primarily by the relative abundance of nest predators and an ability to learn to associate nest placement with the fate befalling the nest.

Although learning may play a major role in selecting a nest site, the construction of the nest is apparently more innate. Hand-reared birds construct complete nests shortly after they are released into the wild (Nancy Stotz, pers. comm.).

BRACING FOR THE CHILDREN

Now that we have investigated the reasons jays nest at particular times in particular places, and the consequences these decisons may have on productivity let's consider how Pinyon Jays rear their young. If parents nested at a good time in a good place they will confront a brood of hungry nestlings totally dependent upon them for survival. In the next chapter we investigate the behavior of parental jays and examine how factors other than nest placement and timing influence success.

CHAPTER 9

# Parental jays

*"From three to five eggs are laid to a set, those numbering four being the most often found. Incubation lasts about sixteen days. The Pinyon Jays are close sitters and, like Clarke's Nutcracker, are devoted parents. The young are able to leave the nest in about three weeks, and may easily be distinguished by their somewhat duller plumbeous blue color. They at once form in flocks and rove about from place to place in search of food."*
Bendire (1895)

In the game of evolution, the one who dies leaving the most genes wins. Reproduction is the key to attaining this goal as it allows organisms to replicate their genes by packaging them in the form of offspring. In the preceding chapters we have discussed how Pinyon Jays prepare for reproduction. Here we discuss reproduction beginning with egg-laying and continuing until the production of independent yearlings. One of the most consistent findings to emerge from long-term studies of animals is that not all breeders are equally prolific. This is evident in the Town Flock as many jays fail to leave surviving descendants but others leave an abundant genetic legacy (see also Chapter 10). Our goal in this chapter is to understand what makes some jays so successful as parents. First, we describe the timing of events following nest construction. We then describe the productivity and behavior of the "average" jay couple and seek correlations between environmental and parental attributes and reproductive success. We conclude with a description of an atypical, but very interesting variant of parental behavior: communal breeding.

The basic breeding biology of Pinyon Jays has been known for some time. In the opening quote for this chapter from Major Bendire, we learn that 95 years ago naturalists were aware of the Pinyon Jay's clutch size, durations of incubation and nestling life, and creching habits. The first recorded finding of nests and eggs of Pinyon Jays was by Mr C. E. Aiken on 13 May, 1874 near Colorado Springs, Colorado (Bendire 1895). Many nests were described during the early 1900s with authors consistently noting the insulating quality of nests, but describing a great variety of nest locations and degree of nest concealment (Bent 1946). Our long-term observations add little to these early depictions of nesting behavior. We describe only some of the variation on the basic themes defined by the early, astute observers. The power of our approach to studying Pinyon Jays comes from our ability to observe uniquely marked individuals. Here we can greatly expand on the works of early observers by describing the behavior of particular Pinyon Jay parents as they rear successive broods of young throughout their lifetimes.

## THE REPRODUCTIVE CYCLE

### Eggs and nestlings

Once the nest is constructed, pairs generally lay their first egg within three days. Another egg is laid each day, usually until a clutch of four is completed. Egg laying occurs in mid-morning after the nesting birds feed as a loose flock for about 2.5 hours. Pairs fly to the nest tree and the female enters the nest. A reliable sign that the female is laying an egg is the presence of the male sitting silently on a high, exposed perch nearby. Early in the season we often see females sitting on nests with incomplete clutches, even during mid-afternoon. These females were not incubating, but appeared to be shielding their eggs from freezing temperatures.

Pinyon Jay eggs in our region have a pale blue background color which is spotted chocolate brown to reddish brown as accurately described by Bendire (1895). The size of eggs within a clutch varies considerably, however their color pattern (background hue, color of blotches, position of blotches, density of blotches) was qualitatively similar within a clutch. This pattern may allow females to recognize their own eggs, as we found out the hard way. In order to measure the temperature of eggs during incubation we removed an egg from each of three nests and replaced them with eggs from other nests which we fitted with small thermometers. We learned nothing about egg temperature at these nests because pairs deserted in two cases and rejected the foreign egg in the third. In the latter case the female continued to incubate even though our thermometer egg dangled on the end of the thermometer cable outside the nest! When thermometers were placed in eggs belonging to a given clutch (N = 4), females readily accepted them. These observations lead us to believe that females recognize their own eggs and can distinguish them from eggs of other females in the flock.

Bent (1946) claims that Pinyon Jay clutches occasionally contain six eggs. However, in all our years of checking nests we have never found more than five eggs in a clutch. We once met an oologist (a person who studies eggs) who proudly informed us that he possessed a six-egg clutch of Pinyon Jay eggs. Given the rarity of such a clutch we requested a viewing. It came as no big surprise to find five of the eggs contained very similar markings, whereas one egg was colored quite differently. The sixth egg was probably added to the nest by a previous collector rather than by the mother jay. Bent's large clutches may have received similar supplements!

Clutches of six eggs could also result if Pinyon Jays occasionally parasitized each other's nests. That is, jays could occasionally dump extra eggs into a neighbor's nest thereby letting an unsuspecting flock mate share in the work of rearing a large brood. Nest parasitism is common among some ducks and in a few birds, such as European Cuckoos and Brown-headed Cowbirds, it is the only mode of reproduction. It is easy to see how such a strategy could evolve as the benefits of letting another bird do the work of rearing your nestlings must be immense. However, the ability to recognize your own eggs would be an adaptation to thwart such parasitism. Clearly, Pinyon Jays in our study flock are able to recognize foreign eggs and remove them from their nests. We have peered into hundreds of nests in northern Arizona and never suspected parasitism. The difference in color patterns between eggs of different nests or unusually large clutches of eggs would have been conspicuous to us. However, this may not be a universal Pinyon Jay trait. Researchers suggest that nest parasitism may occur in Pinyon Jays that live in Idaho because they observed a few very large clutches and noted eggs being moved between nests (Trost and Webb 1986). Is it possible that Pinyon Jays in Idaho lack the ability to recognize their eggs? Until researchers there try to measure egg temperature we may never know!

Incubation lasts 17 days and normally begins with the laying of the third egg, even in clutches of four or five eggs. This means that in larger clutches there are usually three nestlings of equal age and size and one or two smaller, younger ones. Such asynchronous hatching is common in birds of prey and may facilitate brood reduction. Brood reduction usually occurs when food supplies dwindle and older siblings monopolize food deliveries or forcefully eject younger siblings from the nest. Asymmetrical competitive ability enables parents to feed only the winners instead of spreading little food among many equally competitive nestlings. Raising a few strong nestlings may be better than raising many weak nestlings with poor chances of future survival. We rarely see brood reduction in the Town Flock, perhaps because it has supplemental food at feeders. Brood reduction may be more common in unsupplemented flocks foraging on patchily distributed foods that may be inaccessible for days following unpredictable spring snow storms. Alternatively, hatching asynchrony may be an indirect result of natural selection favoring quick concealment of eggs to reduce their conspicuousness to predators (Bollinger *et al.* 1990).

During incubation females rarely leave the nest. They are fed by their mates at a rate of about once every 73 minutes. However, this is highly variable and may be determined in part by the ease with which males can procure food. During cold weather with snow covering the ground four incubating females were fed as infrequently as only 3.8 times per day. When conditions are exceptionally harsh females desert the nest, sometimes eating their young before departure (Balda and Bateman 1976). Such cannibalism occurs in other birds as well as in chimpanzees (Goodall 1986) and may be a way for females to recover some of the energy they already have invested in reproduction.

Unmolested females are fed while on the nest or standing on the nest rim. On rare occasions, usually during the warmer afternoon hours females fly to their approaching males and are fed 300–400 m from the nest. Females then relax, stretch, preen and defecate before returning to the nest. Females also leave the nest prior to feeding if potential predators are near. Following these feedings, both the male and female scold the source of danger. During food transfer the female, if on the nest, gives soft Chirrs or a melodious Sub-song accompanied by a quivering of the wings. If she is off the nest, she gives loud raucous Chirrs with violent wing beating – similar to the behavior of fledglings when begging for food from their parents.

Males of incubating females often forage together as an all-male foraging flock. Foraging forays often take this sub-unit of the flock up to 2 km from the nests. After intense foraging both on the ground and in foliage the males assemble high in one or two trees after a series of rhythmic contact calls are given. Males then fly silently as a group back to the nesting area where they announce themselves and feed their mates (Chapter 8). Only mates were seen feeding females during incubation. Feeding of females lasts no more than 45 seconds, often much less. As males return to feed their mates the silent nesting area is transformed into a noisy and busy restaurant. When the males depart, silence returns.

The integration and co-ordination of the males during this time is striking and may have two important functions. Group foraging may be beneficial in finding food and increasing vigilance for predators, thus benefiting the males directly and the females indirectly. The synchronous return to the colony means all females are fed at the same time. Predators thus have little time to use vocalizations to find nests and females. Mate recognition enables females to prepare for an efficient and rapid transfer of food.

At hatching the young are pink-skinned, bare, and unco-ordinated. By the age of four days they can raise their heads and gape, exposing a bright orange-red mouth-lining. During these early days any harsh, rhythmic noise or vibration of the nest elicits a gaping response. Eyes begin to open seven days after the young hatch. Until this time nestlings produce soft "squeaks" when being fed. After the seventh day these become much louder and are accompanied by wing-flapping. Females brood attentively during the first ten

*Developmental stages of young Pinyon Jays. An egg gives way to a naked, pink nestling. The nestling gets darker and darker and more physically coordinated as it ages and its feathers develop. The first nestling is approximately two days old, the second nestling is eight days old, the third nestling is fifteen days old and the last image shows a recently fledged jay.*

days, only rarely leaving the vicinity of the nest. Thus males, which perform as during incubation, are now responsible for feeding the female and offspring.

Pinyon Jay nestlings are fed a variety of arthropods as well as pine seeds that parent birds recover from caches made the previous fall (Table 22). One of the first comments on the diet of nestling jays comes from Mrs Wheelock in 1904. She noted that pinyon pine seeds were fed to nestlings and that grasshoppers were fed only after having had their wings and legs removed (Bent 1946). We have often noted the same. Nestlings receive similar foods in northern Arizona and in New Mexico. However, there appears to be a greater use of pinyon pine seeds in New Mexico. Parents hull seeds and feed them whole or in pieces to the nestlings. Reliance on pinyon seeds in New Mexico is likely to be typical of Pinyon Jays as this study flock resides in dense pinyon-juniper woodland (Ligon 1978). The wide range of food items delivered to nestlings is a tribute to the foraging efficiency of adults during low temperatures and snow cover. Parents approach nests with their esophagi distended and bob and shake their heads to pass food items into the bill. Begging mouths of nestlings are then filled with food covered with copious amounts of a clear mucilaginous secretion. (To find out what nestlings get to eat, we put pipe cleaners loosely around the nestlings throats so they couldn't swallow. After the parents fed them we rushed over and removed the food. This trick was only a short-lived inconvenience for baby jays, however, as we fed them a nutritious meal that we had prepared.)

**Table 22.** *Food of nestling Pinyon Jays in northern Arizona and New Mexico. The percentage of total items belonging to each class is entered in the table. Data for northern Arizona are from Bateman and Balda (1973). Data for New Mexico are from Ligon (1978).*

| Food Item | Percentage of Items in Diet | |
|---|---|---|
| | northern Arizona | New Mexico |
| Plants | | |
| Pinyon pine seeds | 8.8 | 32.0 |
| Ponderosa pine seeds | 2.2 | 0.0 |
| Invertebrates | | |
| Spiders | 15.8 | 4.0 |
| Insects | | |
| Grasshoppers | 37.1 | 33.0 |
| True bugs | 2.0 | 3.0 |
| Cicadas | 0.4 | 0.0 |
| Lacewings | 0.6 | 2.0 |
| Beetles | 11.9 | 7.0 |
| Butterflies | 15.2 | 21.0 |
| Flies | 4.3 | 0.0 |
| Bees and wasps | 1.0 | 0.0 |
| Vertebrates | | |
| Western fence lizard | 0.6 | 1.0 |

Individual nestlings begin to thermoregulate (maintain a constant body temperature) at five days of age by shivering and they can maintain high body temperature despite low ambient temperatures when they are 12 days old

(Clark and Balda 1981). Brood size appears to be very important in determining when a group of nestlings can effectively thermoregulate. Broods of three can regulate temperature at eight days of age but broods of two do not regulate temperatures until nestlings are ten days old. Thus, females with small broods may have to invest more time brooding than females with large broods. This may explain why females with only two young often abandon their nests during harsh weather, but their neighbors with larger broods do not (Clark and Gabaldon 1979).

When nestlings reach 14–15 days of age a number of significant behavioral changes occur. First, nestlings beg loudly for food which is brought by both parents. These calls become quite complex and may actually differ between males and females. Second, the number of approach calls given by adults increases dramatically. Third, young have a full complement of feathers on their heads, necks, backs and wings. Their belly feathers are just starting to emerge and their tail feathers are short and still encased in protective sheaths. Belly feathers can appear last because the nest provides ample insulation, while tail feathers remain short to reduce wear in a compact nest. Fourth, both parents feed away from the nest, leaving the nestlings unattended. Nestlings can survive outside the nest if they happen to leave the nest after they are 15 or 16 days old (Marzluff 1985). Normally, however, nestlings do not fledge until they are 21 or 22 days old. Just before fledging, young sit on the nest rim and nearby branches where they flap their wings, preen themselves and each other, sleep, and view the world around them.

During the latter stages of nestling life an interesting behavior occurs during the warm mid-day periods. Groups of adults occasionally fly slowly and quietly through the colony, stopping and peering into active nests. Sometimes they even preen or feed the young they visit. During these visits the young are often crouched in the bottom of the nest but occasionally they erupt into loud begging. These "nursery visits" could serve to assess density and age of nestlings in the colony as well as the reproductive competence of the parents. Maybe these visits give jays the information they need to avoid mating in the future with a poor parent.

Young Pinyon Jays are naturally very inquisitive and as they become independent from their parents they begin to explore the world around them. They examine in great detail everything they encounter. For example, we have seen young birds pause and stare into holes, crevices, and cracks of all shapes and sizes. They seem attracted to movement and carefully watch branches and leaves waving in the breeze. Perhaps a young jay's attention is drawn to some of the holes and crevices by the movement of insects within them.

A wide variety of loose objects on the ground are investigated and manipulated by young jays. They pull needles, leaves, twigs and other objects off trees and bushes and handle them with their bills. They carry sticks, stones, pine cones, and numerous other items around in their bills and present these to other young birds. Two or more jays may get into tug-of-war competitions over especially prized items. Young Common Ravens also engage in such contests. Nancy Stotz watched one young Pinyon Jay drag a meter-long piece of string along the ground for about two minutes and then apparently lose interest, only for another bird to come along and pick up where the first jay left off!

These acts all suggest play behavior, which may function as practice needed to perfect behavior used later to survive and reproduce. String is often used in nest lining and many of the other items prized as "toys" are important parts of the birds' diets. For instance, large beetles seem to hold a particular fascination for young jays as they will pick them up and carry them, drop them, pick them up again, roll them over and over, peck at them, and sometimes even eat them. This is much like a cat playing with a prized mouse. Such play behavior may give the young the experience necessary to distinguish between edible and inedible food items, in addition to practice with food-handling techniques.

Nancy Stotz watched young Pinyon Jays in the field and laboratory for two years and discovered that birds just three weeks out of the nest cache objects such as pebbles and old seed hulls. Caged jays are fond of hiding any object in the nooks and crannies of their cages. Occasionally, they repeatedly cache and recover a single object in one site or in a series of sites. Because these behaviors appear without the obvious influence of an adult jay, it appears that at least the act of sticking objects into the substrate and recovering them is inborn, or instinctive, in these birds. John and Colleen Marzluff reared nestling Common Ravens and observed instinctive caching by these corvids as well. Ravens are not dependent upon caches for over-winter survival to the degree that Pinyon Jays are, so the tendency to hide and later recover potential food items must be a primitive characteristic in corvids, not a derived character that has evolved in a seed-caching specialist.

### Creching and parent-young recognition

After leaving the nest, at about 21 days of age, young Pinyon Jays gather together (by hopping and weak flight) into a creche that often extends over an area of about 1 hectare and contains between 20 and 60 birds. Creche formation is done strictly by the young with no enticement or encouragement from the parents. We determined this by playing a tape of loud begging calls to young three days out of the nest. Like rats after the Pied Piper, they followed us in an attempt to locate and contact the source of the begging calls. We believe the loud, harsh begging calls, indicative of a feeding visit, stimulate fledglings to search for and join the begging birds.

Even after fledging, nest-mates normally perch near one another to sleep and preen while waiting for their parents to appear with food. During the creching stage many (but not all) parents and helpers feed as a flock and return to feed their young together. Other adults often sit silently at the periphery of the creche, apparently acting as sentinels. Adults with food approach their young giving soft Nears. Fledglings respond by becoming alert, peering about, stretching wings and legs and giving soft Kaws. As parents draw closer young begin begging loudly exposing their bright red mouth-lining and flapping their wings vigorously. At times crechlings are so active they knock each other or their parents off the perch.

Young in the creche are fed about once every 45 minutes and feedings take about 3.5 minutes to complete. Young not fed for three hours beg from other species of birds and from butterflies flying overhead as well as from one another. Begging young are occasionally (13% of 313 feedings) fed by adults other than their parents (Balda and Balda 1978). These non-filial feedings occur after filial feedings and are only performed if the young gave the complete begging behavior as described above. The vast majority of these feedings are by adult males. We speculate that non-filial feedings are not accidents but intentional acts, perhaps serving to curtail begging vocalizations in an attempt to keep the creche as inconspicuous as possible.

Our knowledge of life in the creche is limited because of the incessant noise crechlings make. One can maintain sanity for only a few hours when surrounded by this cacophony! A creche of 30 to 40 birds can be heard from a distance of 1 km and as surely as it gives us a headache, it must attract predators. In 1974 we observed a Northern Goshawk return to a creche on four consecutive days removing a single young jay each day. (This was the hawk we described in Chapter 4 that killed a mobbing adult jay.) As we discuss later in the chapter, creche life is perilous and most fledglings do not survive their first few months out of the nest.

Crechlings at feeding time appear disoriented as they noisily chase adults for a nourishing mouthful. Watching banded birds revealed a method to this madness, however: parents and their young quickly group together, suggesting that parents and young recognize each other. To test for this we captured two crechlings of known parentage and held them overnight without feeding them. The next morning we placed each hungry young bird in a heavy paper sack and returned them to the edge of the creche. Within minutes the young started calling and were located by their calling parents which flew to the ground and approached the bag. Many other adult birds were seen in the area but none called out or interacted with the bagged young. Apparently parents are able to recognize their fledglings by listening to their begging calls.

When does parent-young recognition first appear in nestling Pinyon Jays, what form does it take and what are the consequences for this highly social species? These questions were partly answered by Pat McArthur (1979, 1982) with a series of controlled observations, manipulations, and experiments.

Pat determined that mutual parent-young recognition gradually develops during the latter stages of nestling life. Parents and young do not recognize each other for the first two weeks, but during the third week of nestling life parents learn to recognize their offspring's calls and young begin to learn their parents' calls. In the early 1970s we moved nestlings 1 km from the nest and tethered them at a feeding station known to be frequented by their parents. Father recognized his offspring on his first visit to the feeder and commenced feeding them. These young were 15 days old when the transfer occurred. When we did the same with seven-day-old nestlings no recognition occurred. Mutual recognition appears to develop just before the young leave the nest to mingle with others in the creche.

Mutual recognition of parent and young is obviously advantageous in the creche as it allows parents to care preferentially for their own genetic endowment. Recognition during nestling life is also important as parents, who have a wider view of their environment, can report prevailing conditions to their young and co-ordinate their activities. At a vocal signal from approaching parents, nestlings become alert and prepared to accept food allowing feedings to be efficient and swift. When another signal is given by parents, young become silent and motionless to reduce their conspicuousness to potential predators.

## BEHAVIOR OF PARENTAL JAYS

Let's take a closer look at how Pinyon Jay parents care for their nestlings. Our main objective here is to investigate how parents work together and how or whether they adjust their behavior to prevailing conditions. Specific conditions we will examine are the timing of breeding, size and age of the brood, and age and experience of the parents. Parents are expected to invest enough time and energy in rearing nestlings so that fledging and future survival is successful. In 1972, Robert Trivers formalized the idea that parents should invest neither too little nor too much in their offspring. Too little investment may jeopardize offspring survival. Too much investment may tax the parents to the extent that investment in other offspring is compromised (Linden and Møller 1989).

We are also interested in determining whether parental behavior influences the chances of a nest being preyed upon. As we argued in the last chapter, nest concealment is the primary means by which jays reduce predation. Behavior at the nest may provide an alternative, or additional means of reducing predation.

We spent over 800 hours in a small, too cold, or too hot blind (hide) observing 36 nests in the Town Flock from 1981 to 1984. We will first discuss nests tended only by a pair and later discuss four nests tended by parents and helpers. As soon as a pair visited their nest we timed and recorded their movements as follows. When a parent entered the nest tree, we recorded the time it spent in the tree before reaching the nest. When parents were at the nest we timed the length of their feeding and cleaning bouts. Feeding could be distinguished from cleaning behavior because of the violent head bobbing described above which accompanies feeding. We include probing of the nest for parasites and fecal sac removal in cleaning behavior. Parents either come to the nest individually, which we term a single visit, or together, which we term a double visit. We calculated the percentage of the observation period that at least one parent attended the nest and recorded how long parental activities made the nest appear conspicuous to us. Behaviors at the nest which we included as conspicuous were the parents' approach and departure from the nest, their feeding, and their cleaning. We did not include the time when a female was quietly sitting on the nest as conspicuous.

### Average parental behavior

The greatest challenge in recording nesting behavior of Pinyon Jays is staying awake between visits to the nest! The reason is that nest visits are very infrequent, on average occurring 1.3 times per hour or once every 46 minutes. The average pair (Table 23, Fig. 74) remains in the nest tree just over 30

**Table 23.**  *Average behavior of Pinyon Jays tending nests from 1981 to 1984.*

| Behavior | Number of pairs | Average | Variance |
|---|---|---|---|
| Number of feeding trips per hour | 32 | 1.3 | 0.12 |
| Percentage of visits by one parent | 30 | 47.8 | 482.70 |
| Total time feeding per hour (seconds) | 29 | 54.8 | 286.00 |
| Percentage of time at least one parent was tending the nest | 32 | 15.6 | 359.10 |
| Percentage of time behavior at the nest was conspicuous | 32 | 6.1 | 10.20 |

seconds before reaching the nest and feeds their young for nearly one minute. The nest is probed and fecal sacs collected for nearly two minutes before the parents depart. Two minutes of cleaning doesn't sound like much but it does the trick. Pinyon Jay nests are very neat, never soiled with feces like those of hawks and eagles. One reason parents are able to clean the nest quickly is that they often collect feces right from their nestlings' anuses. Parents touch the young near the anus and in response nestlings deposit a mucous-encased fecal sac into the prodding parental bill. Nature isn't always pretty, but it is efficient! About half of the visits to the nest are by one parent and half by both parents. The end result of nest visits is that the average nest is attended by one parent about 15% of the time and is conspicuous only about 6% of the time.

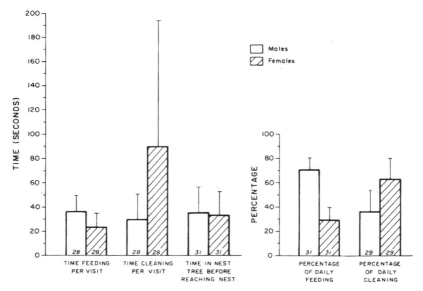

Figure 74  *Behavioral roles of males and females. Height of open bars give averages for males. Height of shaded bars give averages for females. Light vertical lines indicate one standard deviation. T-tests indicated that males and females differed significantly (P < 0.01) with respect to all behaviors except time spent in the nest tree before reaching the nest (P = 0.70). Number of pairs used in each comparison is given at the bottom of each histogram. An average value was calculated for each individual throughout the nesting cycle so that each parent was used only one time in this analysis.*

Males and females show significant division of labor while caring for nest-lings (Fig. 74). As in the classic human stereotyped sex-roles, male jays are the providers and females are the domestic engineers! Male jays feed more and clean less than female jays. In terms of time, on average over 70% of the feed-ing is done by the male and over 60% of the cleaning is done by the female.

*A ten-year-old male Pinyon Jay pauses on the way to feed his nestlings; his throat is bulging from the food he carries. (J. Marzluff).*

### Parental flexibility

Average behaviors make nice summaries of parental care but variation in care may help us understand what motivates jays to tend their broods. We will begin our exploration of factors influencing parental care by examining three major factors: date of nest initiation, number of nestlings, and age of nestlings. We expect these factors profoundly to influence parental care because they influence the requirements of young jays. Energetic requirements should be highest early in the season when snow storms and freezing temperatures are common, and should increase with brood size and nestling age as more and bigger nestlings demand more food.

Parents attend nests early in the season differently than they attend nests late in the season (Fig. 75). As the season progresses, both parents increase the time per visit spent cleaning, but decrease their time per visit spent feeding. The reduction in feeding per visit is compensated for by an increase in the number of visits per hour as the season progresses so that the average time spent feeding young per hour remains constant through the season. Parents appear to dole out food in more frequent, but smaller amounts as the season

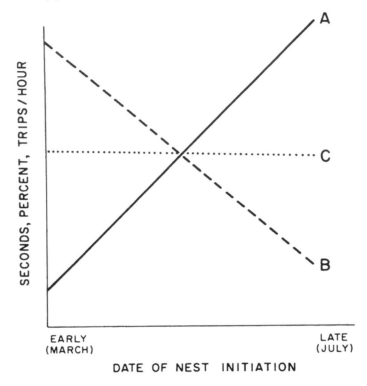

Figure 75  *Changes in parental behavior associated with the time of nesting. Three trends are evident: (A) Males and females clean for more seconds per visit, make more feeding trips per hour, and are in attendance at the nest and conspicuous at the nest for greater percentages of time at nests initiated late in the season. (B) Males and females feed for fewer seconds per visit at nests initiated late in the season. (C) Parents feed a similar amount of time per hour regardless of when nests are initiated. Relationships indicated in (A) and (B) are statistically significant correlations between date of nest initiation and average behavior during the time nestlings were 10–20 days old for N = 32 pairs.*

progresses. This may represent a very fine-tuned adjustment to energetic costs of brooding young. The major cost of brooding is re-warming the nest and young after they have cooled while the female is away from the nest for feeding and cleaning (Vleck 1981). Fewer, longer visits to the nest early in the season when cold temperatures make re-warming costly will reduce the number of times young need to be re-warmed thereby lowering parental energy expenditure. Increased cleaning later in the season reflects an increase in probing of the nest, especially by females. Females clamber all about the nest probing it inside and out apparently removing blood-sucking fly larvae (family Calliphoridae). Adult flies lay eggs in the nostrils of baby jays. As the fly larvae mature they drop into the nest lining and pop up from time to time to suck a blood meal from the naked belly of a nestling jay. As temperatures warm up, fly larvae increase their activity which may explain why females probe more at

late season nests than at early nests. Increased cleaning activity later in the season results in late season nests being attended and conspicuous a greater percentage of the time than early nests.

Parents appear to be very sensitive to the greater energy demands of a large brood (Fig. 76). Changes in parental behavior with respect to brood size revolve around one major shift: parents spend more time feeding larger broods than small ones. This is hardly surprising, but it has interesting ramifications affecting other parental activities. Most importantly, females with big broods spend less time cleaning and probing for parasites. Interestingly, males increase their contribution to cleaning big broods, perhaps to fill the void left by females now busy feeding. Big broods are also attended by parents for a

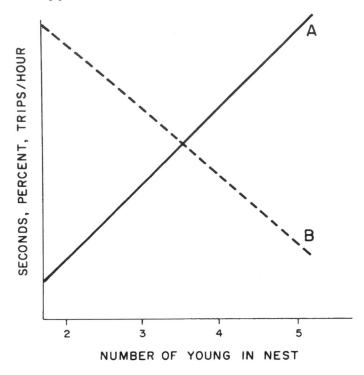

Figure 76    *Changes in average parental behavior when nestlings were 10–20 days old that were associated with differences in brood size. Two statistically significant trends for N = 32 pairs are evident: (A) As brood size increases males and females spend more seconds per visit feeding, males spend more seconds per visit cleaning, parents make more feeding trips per hour and together feed their brood longer each hour, and the percentage of time the nest is conspicuous increases. (B) As brood size increases the percentage of time at least one parent is at the nest declines and females reduce the time they spend cleaning.*

lower percentage of time each day than small broods. This may be another ramification of greater demands for food as parents must spend more time away from the nest gathering food for large broods than for small broods. Alternatively, it may be that parents are free to spend more time away from large broods because big broods can thermoregulate at an earlier age than small broods.

As nestlings grow they change in two important ways: they are able to co-ordinate their movements and they are able to control their body temperatures. Parents respond to these changes in a variety of ways (Fig. 77). First, older nestlings are fed more frequently, but for shorter durations than younger nestlings. This is partly mechanical because older, more physically co-ordinated nestlings take food faster from their parents. Total time spent feeding nestlings per hour remains constant as nestlings grow. However, we suggest older nestlings actually receive more food during feedings because of their

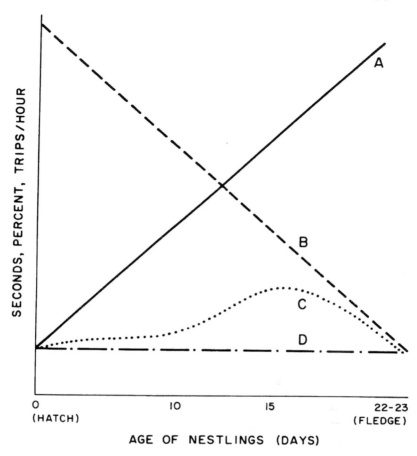

Figure 77 *Changes in parental behavior associated with growth of nestlings. Four trends are evi-dent: (A) As nestlings grow parents make more feeding trips per hour to the nest. (B) As nestlings grow males feed them for fewer seconds per visit, fewer visits to the nest are by single parents, and the percentage of time at least one parent is at the nest declines. (C) The time per visit that males and females clean and the percentage of time parents are conspicuous at the nest peaks at day 15. (D) Total time per hour parents feed their young is similar regardless of the nestlings' ages. Trends in (A) and (B) are statistically significant correlations for N = 32 pairs.*

improving motor control. Second, females are able to reduce the time spent brooding older nestlings. Reduced brooding liberates the female to forage for her young. As a result, nest attentiveness declines. Parents also visit older nest-lings together more frequently than singly. This reduction in the percentage of single visits implies that parents co-ordinate their activities when both are away from the nest. This co-ordination may be one reason Pinyon Jay nests are conspicuous for such a short percentage of the day.

We mentioned earlier that nestlings undergo several major changes when they are about two weeks old. Parental cleaning and, as a result, nest conspicuousness also peaks at this time (Fig. 77). Females may be more free to probe the nest for long periods, especially from underneath it, at this time because they no longer need to brood. Increased cleaning may also reflect greater parasite activity two weeks after the young hatch because by this time the fly larvae have left the nestlings' nostrils and taken up residence in the nest lining.

To summarize, Pinyon Jays are very flexible parents. They adjust their rates of feeding and cleaning to the needs of their young and to prevailing environmental conditions. This should help them meet the changing demands of their growing nestlings in a variable climate. Parents may also adjust their behavior to complement their mate's behavior. Increased cleaning by males with big broods may be one such adjustment as their mates are not able to clean as much as females tending small broods.

### How much should parents invest in the care of their offspring?

Parental flexibility suggests that parents invest just as much care as is necessary to raise their offspring successfully. This investment is considerable and the fact that it is reduced when possible suggests that it may be costly. As we will discuss in Chapter 12, parental investment may put jays in deadly situations. However, without substantial care by both parents it is unlikely that Pinyon Jays could rear any offspring (Chapter 7). Pinyon Jay parents are in a cruel dilemma; they must invest to reproduce, but this investment could kill them. The important relationship that determines how much care parents should invest is the correlation between investment and breeder mortality. In Pinyon Jays this is not likely to be a monotonic relationship; increasing parental care is unlikely to directly translate into an increased risk to breeders. Unless females lengthen their incubation and brooding times or unless males greatly increase their rate of food delivery, the types of breeding season mortality we discuss in Chapter 12 are unlikely to be increased by greater parental care. Therefore parents can invest heavily in their broods without greatly reducing their abilities to breed in the future. According to Trivers (1972), this would favor the evolution of extensive parental care.

A population's sex ratio may further influence the evolution of parental care. Randy Breitwisch (1989) has convincingly argued that male-biased sex ratios, such as we documented for Pinyon Jays (Chapter 5) and which are so common in monogamous birds, may enable females to demand substantial parental care from their mates. Rare females can get away with these demands because they hold an ace up their sleeves in the form of an ability to desert their current mate and quickly obtain a new and more co-operative one. However, female Pinyon Jays, as well as other corvids (Dunn and Hannon 1989), may not be able to desert. Remember from Chapter 7 that desertion spells certain doom for a brood of nestlings and therefore would not be an evolutionarily stable strategy (Maynard Smith 1977). It therefore seems unlikely that the demands of females influence the extent of parental care in Pinyon Jays.

*Influence of parental behavior on predators*

Biologists rarely are lucky enough to be in the right place at the right time to witness a predation event. We consider our time cramped up in the hide adequately repaid because we witnessed five instances of predation; four by crows or ravens and one by a Bullsnake. Crows and ravens simply flew to the nest, reached into the cup and extracted one nestling at a time. When parents caught these predators they drove them away from the nest and resumed caring for their reduced brood. Parental deterrence was ultimately ineffective, however, as the marauders returned, despite parental protest, to eat the rest of the brood. The Bullsnake preyed upon a relatively low nest (3 m) by climbing directly up the tree trunk and onto the top of the young. As more of the snake's 1.8 m long body entered the nest the young were constricted, suffocated, and swallowed one by one. The parents returned to their snake-filled nest after three of four nestlings had been swallowed. Evidently it is difficult to scold with a full throat because both parents emptied their mouths before they surrounded the snake and mobbed it. For all their efforts the snake did not even move, except to swallow their remaining nestling.

Our observations indicate that parents can do little to dissuade predators after their nest has been located, but can the parents' behavior at the nest influence its chances of being discovered in the first place? Apparently so. There were three significant differences in parental behavior at nests preyed upon *versus* nests not preyed upon (Fig. 78). At preyed upon nests, males fed longer, females cleaned longer, and as a result there was a parent at these nests nearly three times longer than at nests not preyed upon. Parents at preyed-upon nests also approached the nest more directly, remaining in the nest tree for slightly shorter periods of time than parents at nests not preyed upon. Spending more time in the nest tree before reaching the nest may enable parents to spot potential predators before revealing the location of the nest. Our quantification of conspicuousness was not greater at preyed-upon nests. Either what we felt was conspicuous was rarely detected by predators or was not equivalent to what predators viewed as conspicuous.

*Do experienced pairs behave differently than inexperienced ones?*

In the previous chapter we proposed that older, experienced jays selected concealed nest sites because they learned to associate exposure with a greater risk of predation. If old jays can learn that trick we would expect them to learn to behave carefully around their nest to further reduce the chances of predation. Based on our observations at preyed upon nests we expect experienced jays to reduce their time at the nest and scan longer for predators when in the nest tree before approaching the nest.

Experienced pairs do exactly that (Fig. 79). Experienced pairs (those with four or more collective years of breeding experience) are at their nests only half as often as less experienced pairs. This reduction is accomplished by a reduction in male feeding and female cleaning, yet experienced pairs do not reduce the total time they feed their young. How can males reduce feeding without

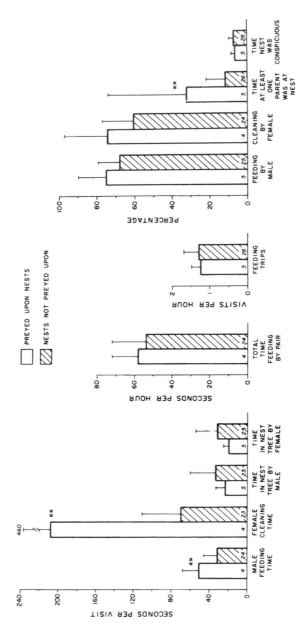

Figure 78　Behavioral differences in nests preyed upon (open bars) and nests not preyed upon (shaded bars). Stars above histograms indicate preyed upon nests differ statistically in behavior from nests not preyed upon (two-tailed probability, ** = P < 0.05). Height of bars indicate average value of behavior. Narrow vertical lines indicate one standard deviation above the average. Number of nests in each category are given at bottom of each bar.

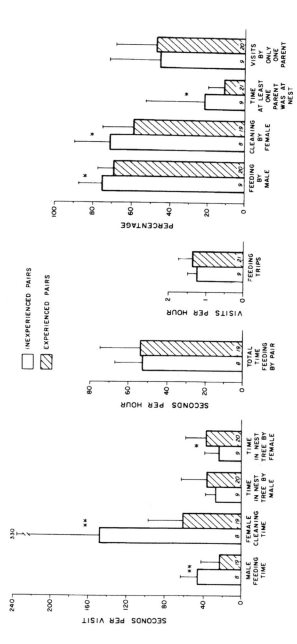

Figure 79  *Behavioral differences between experienced and inexperienced pairs. Inexperienced pairs (open bars) have fewer than four years of breeding experience between them. Experienced pairs (shaded bars) have four or more collective years of breeding experience. Stars above histograms indicate that inexperienced pairs differ significantly from experienced ones (\* = P < 0.10, \*\* = P < 0.05, one-tailed tests of predictions in text). Sample sizes are given at base of bars. Thin vertical lines show one standard deviation above the average.*

their young suffering? First, experienced pairs visit their nests slightly more often than inexperienced pairs. More importantly, experienced pairs divide their duties more equally than inexperienced pairs, an attribute that has been tied to greater reproductive success in Herring Gulls (Burger 1987). Experienced males still do the majority of the feeding and experienced females still do the majority of the cleaning, but these majorities are less extreme than found in inexperienced pairs. Experienced males and females also approach their nests more slowly than inexperienced parents. The greater time this allows for locating predators may compensate for an experienced pair's slightly more frequent nest visits.

REPRODUCTIVE SUCCESS

*Productivity through the nesting cycle*

We have summarized annual productivity at various stages of the nesting cycle in Appendix 2. During a nesting attempt, an average jay is likely to incubate four eggs, rear two nestlings, fledge one nestling, and produce 0.4 yearlings. Looking at average productivity another way, 55% of eggs survive the incubation period and hatch, 56% of nestlings fledge, 32% of fledglings survive their first few months in the creche, and 51% of crechlings survive their first winter. Muliplying these probablities indicates that only 5% of eggs ever become yearlings! Ligon's study (1978) suggests that Pinyon Jays in New Mexico are similar in productivity to the jays we studied.

Survival is clearly not constant throughout the nesting cycle. Fledglings appear to be especially vulnerable when they first leave the nest to join a creche. Young are poorly co-ordinated fliers at this time and are not always under adult supervision. As such they may be easy targets for predators (such as the Northern Goshawk we reported earlier in the chapter). Florida Scrub Jays provide an interesting comparison. Fourteen per cent of their eggs produce yearlings and survival is roughly equal across all stages of the nesting cycle (Woolfenden and Fitzpatrick 1984). Fledgling Florida Scrub Jays do not wander in creches immediately after leaving their nests, remaining instead in the relative safety of their natal territory. Pinyon Jays, at least relative to Scrub Jays, do not seem to enjoy relaxed predation pressure because of their creching habits. Perhaps the primary importance of creching is to facilitate socialization rather than thwart predators.

*Annual variability and environmental correlates*

A striking feature of Appendix 2 is the large amount of year to year variation in all measures of productivity. Here we try to understand reasons for this variation by examining correlations between productivity and aspects of climate and pinyon pine seed production. Those factors significantly correlated with breeding productivity in the Town Flock are listed in Table 24.

Average annual clutch size varied from barely three eggs in 1976 to well over

**Table 24.** *Significant correlations between environmental attributes and breeding performance of Town Flock jays. Pearson correlation coefficients are given in parentheses. Stars indicate level of statistical significance (\* = P < 0.05, \*\* = P < 0.01, \*\*\* = P < 0.001).*

| Measure of Breeding Performance | Number of Years Studied (N) | Significant Environmental Correlates |
|---|---|---|
| Clutch size | 11 | Spring precipitation ($-.47$)\*, Spring snowfall ($-.46$)\*\* |
| Number of nestlings | 7 | Annual precipitation ($-.57$)\* Fall pinyon crop ($+.67$)\*\* |
| Number of fledglings | 7 | Summer precipitation ($-.74$)\*\*\*, Summer temperature ($+.53$)\*, Fall pinyon crop ($+.60$)\*\* |
| Number of crechlings | 11 | Annual precipitation ($-.59$)\*\*\*, Fall pinyon crop ($-.47$)\*\* |
| Number of yearlings | 11 | Annual precipitation ($-.73$)\*\*\*, Fall pinyon crop ($-.39$)\* |
| Percentage of eggs hatching | 7 | Winter temperature ($+.61$)\*\*, Fall pinyon crop ($+.61$)\*\* |
| Percentage of nestlings fledging | 7 | Summer precipitation ($-.60$)\*\*, Winter snowfall ($+.56$)\*, Fall pinyon crop ($+.50$)\*[1] |
| Percentage of fledglings color-banded | 8 | Spring snowfall ($-.56$)\*\* |
| Percentage of crechlings overwintering | 12 | Summer temperature ($-.44$)\*, Winter temperature ($+.48$)\*[2] |

[1] *Partial correlation holding summer precipitation and winter snowfall constant*
[2] *Temperature in winter following birth*

four eggs in 1974. Spring precipitation, in particular snowfall, appears to be an important correlate of clutch size: clutches were typically smallest following snowy springs. This is probably a causal relationship as we have observed females curtailing laying and initiating incubation at the onset of heavy storms. Females remain on their nests throughout the storm and if storms hit as eggs are being laid many females may settle on incomplete clutches.

The importance of spring snowfall may be peculiar to the Town Flock because it inhabits an especially snowy area. Lower elevation flocks, such as those in New Mexico, may rarely face snow during egg-laying. Clutch size in these flocks is likely to be influenced by the abundance of pine seeds. Large clutches are more common after bumper pine crops in such areas (Ligon 1971). Town flock jays also lay more five-egg clutches and fewer two- and three-egg clutches after large pinyon crops than after poor crops. However, the influence of pine crop is often weakened by the intervening effects of snow storms (Marzluff and Balda 1988a). In most locations abundant food stores can be expected to be converted into an extra "bonus baby".

Precipitation and the size of the fall pine crop are important correlates of hatching and fledging success. Dry years and those following large pine crops and warm, snowy winters were associated with high productivity and high survival of eggs and nestlings (Table 24). Warm, dry summers also were associated with high fledging success.

It seems intuitive that abundant seed stores would lead to greater productivity during nesting because pine seeds can be converted into eggs, but why should precipitation lower success? The influence of precipitation may be partly cumulative, that is, the number of nestlings that a jay has in its nest is lower partly because of the detrimental influence of precipitation on the number of eggs laid. In addition, summer rains in the southwestern mountains come in the form of violent thunderstorms which may preclude foraging for several hours each afternoon and require parents to shelter their nestlings. Summer rains should benefit parents by enhancing insect productivity, but this benefit would be delayed until young are no longer in the nest. Grasshoppers, a summer favorite of the jays, perish when their bodies get too wet, thus one food source is negatively affected by summer rains (P. Price, pers. comm.).

Correlations between events in the fall and winter preceding breeding and spring breeding performance suggest that energy reserves of parents may influence success. Parents may be in great shape after feasting on pine seeds all fall and winter especially if winters are mild. This may explain why productivity is high following such conditions. Healthy females may be able to incubate their eggs and tend their young more closely thereby reducing the likelihood of eggs freezing or being eaten by predators such as squirrels or Steller's Jays. Males in good shape may further their mates' efforts by provisioning them at high rates.

The negative influence of precipitation on productivity continues through to the production of crechlings and yearlings. Again, spring snowfall is the likely culprit as it was associated with low survival of fledglings in their first few months out of the nest. This may be a cumulative effect of reduced clutch size, but the impact of snowfall on survival of fledglings suggests a more direct impact. We suggest that nestlings reared during snowy springs may be weaker because of heightened energetic costs and smaller amounts of food eaten. Weak young may be especially susceptible to dangers during their first few weeks out of the nest. Being in the safe confines of the nest may mask the relative strength or weakness of the young. Overwinter survival is greatest when crechlings encounter cool summers and warm winters.

The big mystery is why crechling and yearling productivity was low after large pine crops. This is the reverse of nestling productivity and seems counterintuitive. Pine crop was not correlated with survival of fledglings or crechlings but only with the number raised per pair each year. However, as we will discuss in Chapter 12, when we look at survival for the larger sample of all juveniles (crechlings) born in the Town Flock (not just those from nests that we located as in Appendix 2) a positive effect of pine seeds is again evident. The lack of a conclusive result in the present analysis is likely the result of determining crechling survival from an incomplete sample of birds. This incomplete sample could be biased if our actions around nests influenced their fates. However, Diana Gabaldon (1978) found no significant differences in productivity from nests she observed *versus* nests she did not observe. It is still possible that some productive parents are good at concealing their nests from us. We rarely located the nest of one pair that consistently brought their young to our feeders. Such pairs would not enter our incomplete sample of known nests, but would be used in our larger sample of survival within the flock. If more successful pairs elude our searches in bumper pine crop years than in poor crop years we would fail to discover a positive relationship between crop size and productivity. Clearly we need more observations of fledglings during their first few months of life to determine if this is the case. At the moment we must put our faith in our most complete sample which suggests that pine crops benefit productivity throughout the nesting cycle. Ideally, future students will experimentally add pine seeds to some flocks and use unsupplemented flocks as controls to determine if pine seed abundance causes higher productivity.

One other dramatic change in annual productivity of crechlings and year-lings deserves mention. Productivity during the 1970s was twice as high as productivity during the 1980s (Appendix 2). Jays reared an average of 1.21 (SD = 0.40) crechlings and 0.55 (SD = 0.15) yearlings from 1973 to 1979. In contrast, jays averaged only 0.55 (SD = 0.42) crechlings and 0.24 (SD = 0.09) yearlings from 1981 to 1985. Three factors could be responsible for this plung-ing productivity. First, observers differed in the 1970s and 1980s. Diana Gabaldon and Pat McArthur made the majority of field observations in the 1970s and John Marzluff made the observations in the 1980s. Additionally, Gene Foster and Jane Balda kept daily records of the jays in the 1970s and early 1980s. Productivity may be lower in the 1980s because slightly more nests, especially unsuccessful ones, were found each year (Fig. 80A) and because daily records of survival were less precise. Second, Flagstaff has doubled in size over the course of our study and urban sprawl may be profoun-dly reducing jay productivity. Many nesting areas now contain apartments, schools, and roads but more importantly growth of the city has resulted in a garbage boom that feeds hundreds of crows and ravens. These corvids turn their attention from dumps to jay nests each spring which has resulted in a dramatic increase in the percentage of nests annually preyed upon (Fig. 80B). It is possible that our activities compounded this problem by alerting some of these abundant predators to the location of a tasty meal. We were careful to climb nest trees only when no predators were in sight. However, on a few occa-sions ravens flew by when we were in nest trees. They nearly always stopped in mid-air or circled to gain a better look at us. Invariably, nests receiving such attention were preyed upon a short time later. Third, spring snow storms could have been more destructive in the 1980s than in the 1970s, but this is not borne out in the data (Fig. 80C).

It is difficult to sort out the many strong correlations between environ-mental factors and productivity. Partial correlation analyses enable us to untangle some correlations. This analysis allows us to hold weather factors statistically constant while we investigate the influence of pine crop (and *vice versa*). Partial correlations strengthened our conclusion that large pine crops increase nestling productivity. When we held weather factors constant we found that pine crop became significantly correlated with survival of nestlings. In other cases partial correlations did not aid our understanding of the relative importance of environmental factors because many factors are highly corre-lated with each other. We will require many more years to sort out the relative importance of environmental factors to breeding success.

In general, even the strongest correlations between the abiotic environment and productivity were unimpressive. Squaring a correlation coefficient

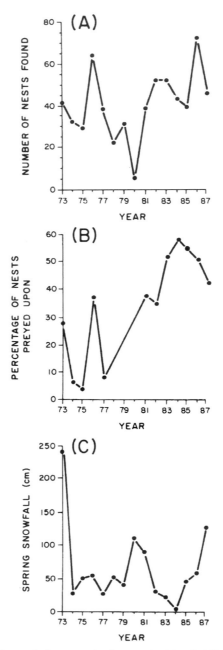

Figure 80    *Annual changes during our study of nest productivity. (A) Number of nests found from 1981 to 1987 is slightly larger than the number found in earlier years. (B) The percentage of nests preyed upon is substantially higher in the 1980s than 1970s. (C) The amount of snow falling during the onset of nesting (15 March–15 February) is quite variable from year to year.*

indicates what percentage of the variation in productivity is attributable to variation in an environmental correlate. Our best relationship (the negative impact of summer rain on fledgling production) explains only 55% of the variation in productivity. Most relationships in Table 24 explain far less than half of the variation in breeding productivity. A likely reason is that parents influence productivity regardless of weather and seed crop. We now turn to the importance of parental attributes on reproduction.

### Parental influences on productivity

We have already discussed the influence of several parental attributes on productivity. We showed that pair bond duration had little impact on productivity (Chapter 7). Additionally, dominant parents may have high productivity (Chapter 6), and decisions of parents such as when and where to nest may profoundly influence productivity (Chapter 8). Earlier in this chapter we suggested how parental behavior may influence productivity and showed that experienced pairs behaved differently from inexperienced ones. Here we tie these ideas together by focusing on the influence of age and experience on productivity.

Many studies of birds have shown close ties between age, experience, and productivity (e.g. Coulson and White 1958, Ryder 1981). In general, researchers report that older, more experienced parents raise more offspring than young, inexperienced parents. The reasons for greater success of old timers are myriad; they may select better nest sites, nest earlier in the season, be more proficient hunters, or enjoy high dominance. We expect old, experienced Pinyon Jays to out-reproduce young, inexperienced ones because experienced parents nest in sites less accessible to predators (Chapter 8).

We present a summary of age-specific productivity through the nesting cycle in Table 25. Young and old jays alike lay about the same number of eggs per nesting attempt. However, ignoring the small sample of very successful yearling males, the youngest jays produce fewer nestlings and fledgings than older jays (t-tests comparing productivity by one- and two-year-olds to the oldest class of jays are significant at $P < 0.05$). Age has less of an effect on productivity of crechlings and yearlings (t-tests are not significant). These offspring are more independent than fledglings and many factors out of their parents' control influence their survival. What about the successful yearling males? Their success may be an artifact of a small sample size, but it is not entirely unexpected if, as we suggest in Chapter 11, yearling male breeders are the "top guns" of their class.

Increased productivity of fledglings is especially consistent and closely related to parental age if we total productivity for a pair over the season (Table 25). Part of the reason for the greater success exhibited by older jays is their ability to re-nest successfully after nest failure (Table 26). If old jays do not succeed in fledging young, they try again and again and eventually succeed over half of the time. Young jays also try again after failure. However, they are rarely successful (less than 20% of one- and two-year-olds successfully re-nest; Table 28). As we discussed in the previous chapter, this ability of old jays to succeed after failure is likely because they learn from their mistakes and pick better nest locations after they fail.

More striking to us than the low productivity of young jays is the high variability in productivity among jays of a given age. In many cases, standard deviations in Table 25 are as great or greater than averages indicating extreme inconsistency in the productivity of equal-aged birds. An obvious reason for

**Table 25.**    *Age-specific fecundity for pairs breeding in the Town Flock from 1972 to 1987. Table entries are means, sample sizes, and standard deviations in that order.*

|  | Clutch size | Number of Nestlings per Nest | Number of Fledglings per Nest | Number of Fledglings per Year | Number of Crechlings per Year | Number of Yearlings per Year |
|---|---|---|---|---|---|---|
| Male Age |  |  |  |  |  |  |
| 1 | 3.75,  4, 0.96 | 1,80,  5, 2.49 | 2.44,   9, 1.94 | 2.44,  9, 1.94 | 0.69, 13, 1.11 | 0.36, 11, 0.92 |
| 2 | 3.50, 18, 0.71 | 0.88, 34, 1.55 | 0.87,  61, 1.44 | 1.23, 43, 1.57 | 0.98, 55, 1.25 | 0.57, 46, 0.83 |
| 3 | 3.72, 39, 0.76 | 2.12, 42, 1.97 | 0.86,  62, 1.63 | 1.15, 46, 1.80 | 1.02, 46, 1.37 | 0.38, 45, 0.68 |
| 4 | 3.82, 39, 0.68 | 2.71, 42, 1.77 | 1.36,  61, 1.82 | 1.84. 45, 1.89 | 0.91, 45, 1.41 | 0.47, 38, 0.89 |
| 5 | 4.00, 17, 0.79 | 2.03, 29, 2.01 | 1.29,  41, 1.75 | 1.56, 34, 1.81 | 1.10, 30, 1.09 | 0.44, 32, 0.72 |
| 6 | 3.50, 18, 0.99 | 1.75, 28, 1.97 | 1.37,  30, 1.88 | 1.78, 23, 1.98 | 0.61, 28, 0.99 | 0.30, 23, 0.64 |
| 7+ | 3.58, 71, 1.01 | 2.06, 85, 1.85 | 1.26, 107, 1.79 | 1.96, 69, 2.19 | 1.13, 70, 1.44 | 0.52, 58, 0.82 |
| Female Age |  |  |  |  |  |  |
| 1 | 3.86, 14, 0.66 | 1.50, 26, 1.88 | 1.07,  41, 1.65 | 1.29, 34, 1.73 | 0.76, 38, 1.15 | 0.23, 30, 0.57 |
| 2 | 3.69, 29, 0.71 | 1.35, 37, 1.77 | 0.70,  57, 1.40 | 1.08, 37, 1.75 | 1.00, 48, 1.35 | 0.43, 42, 0.70 |
| 3 | 3.93, 40, 0.53 | 2.50, 44, 1.96 | 1.27,  63, 1.78 | 1.91, 42, 1.89 | 1.07, 45, 1.25 | 0.57, 42, 0.89 |
| 4 | 4.03, 30, 0.72 | 2.82, 33, 1.76 | 1.67,  43, 2.00 | 2.18, 33, 2.02 | 1.03, 37, 1.17 | 0.56, 34, 0.89 |
| 5 | 3.73, 15, 0.80 | 2.29, 28, 2.05 | 1.53,  34, 1.81 | 1.86, 28, 1.84 | 0.85, 26, 1.26 | 0.50, 22, 0.74 |
| 6 | 3.41, 17, 1.18 | 1.95, 22, 1.94 | 1.31,  29, 1.85 | 2.00, 19, 1.97 | 0.79, 19, 1.03 | 0.43, 14, 0.65 |
| 7+ | 3.79, 19, 0.71 | 2.11, 27, 1.89 | 1.88,  43, 2.10 | 2.46, 33, 2.30 | 1.19, 43, 1.52 | 0.47, 38, 0.76 |

**Table 26.**    *Probability of Pinyon Jays successfully re-nesting within a season. Number and percentage of Town Flock jays that failed during nesting and later re-nested and successfully fledged young are given as a function of their age.*

|  | Total Number Re-nesting | Number Successfully Re-nesting | Percentage Successfully Re-nesting |
|---|---|---|---|
| Male Age |  |  |  |
| 1 | 1 | 1 | 100.0 |
| 2 | 24 | 4 | 16.7 |
| 3 | 17 | 7 | 41.2 |
| 4 | 16 | 4 | 25.5 |
| 5 | 7 | 2 | 28.6 |
| 6 | 8 | 1 | 12.5 |
| 7+ | 47 | 27 | 57.4 |
| Female Age |  |  |  |
| 1 | 8 | 2 | 25.0 |
| 2 | 24 | 2 | 8.3 |
| 3 | 20 | 12 | 60.0 |
| 4 | 11 | 1 | 9.1 |
| 5 | 10 | 3 | 30.0 |
| 6 | 12 | 7 | 58.3 |
| 7+ | 18 | 11 | 61.1 |

this variation is that we lumped productivity for several years into each age class. As we showed above, productivity is not equal across years; we address the annual influence of age and experience on productivity below. A bird's family legacy may also influence productivity and we discuss this factor in the last section of this chapter.

In order to investigate annual variation in the influence of age and experience on productivity we correlated male age, male experience, female age, and female experience with productivity through the nesting cycle each year. We summarize these correlations in Figure 81.

Age and experience appear most influential on fledging success (Fig. 81). Only a few strong ($P < 0.05$) or moderate ($P < 0.10$) correlations exist at other stages. It makes sense for parental attributes to be most influential on fledging success because this is the end of complete nestling dependence on parents. Any rights or wrongs during egg-laying, incubation, and tending young combine to influence the number of young that parents fledge. After fledging, parental attributes have less of an influence on success presumably because, relative to sessile nestlings, mobile young are more in control of their own destiny and less responsive to every parental command.

Female age and experience were especially highly correlated with fledging success. This sex bias, however, was not consistent through the cycle. There were 24 strong or moderate correlations; 12 included male attributes and 12 included female attributes.

The number of seasons a jay has nested may be a better correlate of productivity than its age. Fifteen of the 24 strong or moderate correlations involved experience; nine involved age. Experience may be the biological reason that age influences productivity as experience and age are highly correlated.

The influence of age and experience on productivity was not equal every year. We began an in-depth study of this influence in 1981 (Table 27). Three distinct patterns are evident. In 1981, young, inexperienced birds consistently

**Table 27.** *Yearly impact of parental age and experience on production of fledglings. Negative correlations suggest that young, inexperienced parents produced more fledglings than old, experienced parents. Positive correlations suggest that old, experienced parents produced more fledglings than young, inexperienced ones. Conclusion is our interpretation of the relative advantage of an age class based on the consistency and strength of correlation tests. There are four correlation tests each year: male age, female age, male experience, and female experience correlated with number of fledglings. The number of negative and positive correlations is listed in the table. The number of significant correlations ($P < 0.05$) are given in parentheses.*

| Year | Number of correlations | | | | Conclusion |
|------|----------|---------------|----------|---------------|------------|
|      | Negative | (Significant) | Positive | (Significant) |            |
| 1981 | 4 | (3) | 0 | (0) | young advantage |
| 1982 | 0 | (0) | 4 | (4) | old advantage |
| 1983 | 3 | (0) | 1 | (0) | no advantage |
| 1984 | 1 | (0) | 3 | (2) | old advantage |
| 1985 | 0 | (0) | 4 | (1)* | old advantage |
| 1986 | 1 | (0) | 3 | (0) | no advantage |
| 1987 | 3 | (0) | 1 | (0) | no advantage |

* *The significant correlation in 1985 had a P-value of 0.06.*

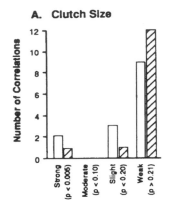

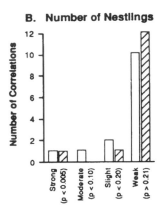

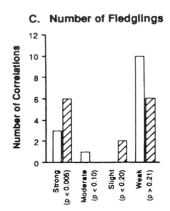

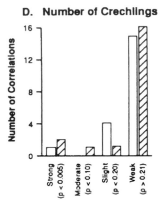

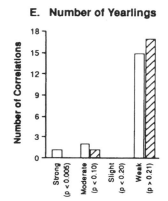

and significantly fledged more young than older, experienced birds. Three years (1982, 1984, 1985) provide striking contrasts as older, experienced birds were most productive. Finally, age and experience had slight and mixed effects in 1983, 1986, and 1987. In these years neither young nor old jays had higher fledging success. Within each year the pattern through the nesting cycle is as we discussed above. When there were effects they accumulated and had their strongest effects at fledging, then dropped. In years without effects the lack of influence was equal throughout the cycle.

The annual variation in the influence of age and experience on reproductive success suggests that there are disadvantages, as well as advantages, to being an old Pinyon Jay. These are probably related to nest placement and the timing of nesting as these attributes are influenced by age and are important to productivity. Remember that older jays nest slightly earlier and in more concealed locations than younger jays. Accordingly, we would expect old jays to suffer lower rates of predation, but higher losses from spring snow storms, than young jays. This holds for females as those with nests preyed upon had 1.7 years of experience, but those with nests failing from snow had 2.8 years of experience (F = 2.9, P = 0.09). Using this logic, we can make three predictions: 1) young jays should be at an advantage in snowy years with little predation because they nest later and avoid some of the snow, 2) old jays should be at an advantage in years with little spring snow but heavy predation, and 3) parental age should not be correlated with success when snow is heavy and predation extreme.

Our predictions are generally upheld. We graph the combined influence of snowfall and predation on the importance of age and experience to success in Figure 82. Young jays out-reproduced old jays in 1981, a year characterized by low predation and moderate snowfall. Older jays out-reproduced young jays only when snowfall was minimal regardless of predation. Neither age out-reproduced the other when snowfall and predation were high.

Other species that inhabit harsh and unpredictable environments are likely to exhibit variable influences of age and/or experience on productivity. To date only one other species (Lesser Snow Goose, Rockwell *et al.* 1985) has been examined in this respect. It also showed annual fluctuations in the relative productivity of old geese. Female geese five years old and older always laid more eggs than two-year-old geese. However, the relative magnitude of this advantage varied three-fold from 1974 to 1982. The proximate factors controlling the relative advantage of old geese remain unknown. An important

---

Figure 81   *Associations between productivity through the nesting cycle and male age, female age, male experience, and female experience. This figure is a summary of the results of correlation tests between productivity and age/experience each year. Four correlations (one for each measure of age and experience) are produced each year for each measure of productivity. The height of the bars indicates how many of these correlations were strong (P < 0.05), moderate (P < 0.10), slight (P < 0.20), or very weak (P > 0.21). If parental age and experience have marked effects on productivity we expect many correlations to be strong, moderate, or slight and few to be weak. Conversely, an abundance of weak correlations indicates parental age and experience has little influence on productivity. The influence of male age and experience (open bars) and female age and experience (hatched bars) is shown separately.*

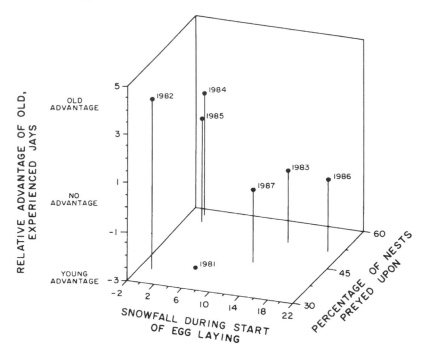

Figure 82    *Influence of snowfall at the initiation of egg laying and the intensity of predation on the relative advantage of known-age breeders. Our measure of relative advantage is the number of significant correlations beween parental age/experience and the number of fledglings each year (data in Table 27). Many significant positive correlations indicate old, experienced pairs fledged more offspring than young, inexperienced pairs (1987, 1984, 1985). Many significant negative correlations indicate young, inexperienced pairs fledged the most offspring (1981). Few or no significant correlations indicate fledging success was similar for old, experienced and young, inexperienced pairs (1983, 1986, 1987).*

conclusion of this study and ours is that models of population productivity using average age-specific fecundity are likely to be oversimplistic for species living in variable environments.

## COMMUNAL BREEDING

In our opening chapter we introduced the intriguing observation that some individuals help others breed rather than, or in additon to, breeding themselves. Breeding systems characterized by such auxiliaries (or helpers at the nest) are termed communal breeding systems (Brown 1987). Typically, young of the previous few breeding seasons are the helpers and their role is to help their parents feed, clean, and guard the current nestlings. The important observation which has broadened our understanding of this apparent altruism is that helpers often benefit by increasing their indirect fitness. That is, by helping, helpers allow their parents to raise more young than would otherwise be possible. These "bonus babies", being related to the helper, are the helper's indirect genetic payment.

Our observation that helpers in Pinyon Jays are the small males, losers in the battle for mates (see Chapter 11) suggests that helpers may not provide much help. As we will show, they do not. We knew this after a few years of study but felt compelled to find some benefit to helping, fearing without one our colleagues would never completely accept our story and assume we missed something. After all, if helpers don't help then they do not receive inclusive fitness benefits and would seem to be better off not helping. We will expand our previous arguments (Marzluff and Balda 1990) that in Pinyon Jays some parents allow their weaker sons to "help" as a form of extended parental care that buffers weaklings from the rigors of life on their own. This may allow the helpers to later attain breeding status equal to their larger, more dominant competitors.

*Helping is a family affair*

Any discussion of helping in Pinyon Jays is complicated by the observation that some families appear more prone to have helpers than other families. We have studied 41 extended families of jays and have observed helping within only nine (22%) of them. Moreover, helping is relatively frequent among male descendants in these families (65% of 34 sons helped). Helping occurs periodically through the generations of these families (see Appendix 5 1971 L-M). The rest of the families in the Town Flock have never had helpers although 25 sons lived long enough to potentially help. We will refer to the members of families having helpers as H-lineage birds and refer to other jays as NH-lineage birds. It is important to realize that many H-lineage birds have not actually been helpers nor received help; many are relatives of helpers and are classified as H-lineage birds by birthright (Table 28).

Helping may have a genetic component; however, a nesting transfer experiment suggests that helping is cultural. In 1978 Pat McArthur switched a male nestling from an NH-lineage nest into an H-lineage nest (see Appendix 5 1978

**Table 28.** *Methods by which Town Flock jays are classified as H-lineage birds. Individuals are only included in one category. Numbers and percentages are for data collected from 1972 to 1982. From Marzluff and Balda (1990).*

| Method | Number of Individuals | Percentage |
|---|---|---|
| Had a helper | 23 | 24.7 |
| Birthright | 36 | 38.7 |
| Helped parents at the nest | 17 | 18.3 |
| Was helped as a nestling | 10 | 10.8 |
| Was helped as a nestling and later helped parents at the nest | 5 | 5.4 |
| Was helped as a nestling and later was aided by a helper | 1 | 1.1 |
| Helped parents at the nest and later was aided by a helper | 1 | 1.1 |

E and 1978 Z). As luck would have it this male survived and helped his foster parents the next year!

H-lineage and NH-lineage families differ in three important aspects that may explain why helping occurs. In a nutshell, H-families are extraordinarily fit relative to NH-families and H-families produce a preponderance of sons that are less likely to breed as yearlings than NH-lineage sons. Let's investigate these differences in more detail.

H- and NH-lineage pairs produce different sex ratios of yearlings. H-lineage pairs produced over five times as many male yearlings as female yearlings (33 males: 6 females) whereas NH-lineage pairs produced only twice as many males as females (25 males: 13 females). (These differences are not quite statistically different, P = 0.07.) H- and NH-lineage males also pursued different options as yearlings. Most H-lineage sons helped; few bred and few floated (Table 29). In contrast, most NH-lineage sons floated and many bred as yearlings (Table 29).

**Table 29.**    *Family lineage differences in behaviors of Town Flock male yearlings. From Marzluff and Balda (1990).*

| Lineage | Number of Sons | Helped Parents | | Bred as a Yearling | | Non-breeding, Nonhelping, Floaters | |
|---|---|---|---|---|---|---|---|
| | | N | % | N | % | N | % |
| H-lineage | 34 | 22 | 65 | 4 | 12 | 8 | 24 |
| NH-lineage | 25 | 0 | 0 | 8 | 32 | 17 | 68 |

A wide variety of fitness components indicate that members of H-lineage families are more fit than members of NH-lineage families (Table 30). H-lineage pairs produce more yearlings per year than NH-lineage pairs. Further, more H-lineage birds successfully attain breeding status in their natal flock or as emigrants in nearby flocks than NH-lineage birds. H-lineage individuals have an average life span nearly a year longer than NH-lineage individuals. The major difference in survivorship and production of yearlings between lineages occurs during the first year of life. H- and NH-lineage pairs behave

**Table 30.**    *Differences in fitness components for members of H-lineage and NH-lineage families. Lineages differ significantly in all these comparisons (Marzluff and Balda 1990). Sample sizes are given in parentheses below averages.*

| Lineage | Annual Fecundity | Individual Longevity | Production of Emigrants | Production of Breeders in Natal Flock | Percentage of Families going Extinct |
|---|---|---|---|---|---|
| H | 0.64 yearlings/ pair/year (64 pairs) | 3.02 years (128 indivs) | 0.89/pair (9 pairs) | 3.7/pair (9 pairs) | 0 (9 families) |
| NH | 0.37 yearlings/ pair/year (84 pairs) | 2.13 years (506 indivs) | 0.09/pair (32 pairs) | 1.2/pair (32 pairs) | 56 (32 families) |

similarly at their nests (the only significant differences are tendencies for H-lineage parents to visit the nest together more than NH-lineage parents and for H-lineage males to delay their approach to the nest more than NH-lineage males; Table 31). Similarities in behavior may explain why H- and NH-lineage pairs had equal success up through the production of crechlings

**Table 31.** *Comparison of breeding behavior by H-lineage and NH-lineage pairs. F-statistics are from one-way analyses of variance testing for differences in behavior attributable to lineage. P-values are significance levels for F-statistics; values less than 0.05 are considered statistically significant.*

| Behavior | H-lineage Pairs | | NH-lineage Pairs | | | |
| | Mean | Number of Pairs | Mean | Number of Pairs | F | P |
|---|---|---|---|---|---|---|
| Time males feed (seconds/visit) | 41.7 | 10 | 35.9 | 15 | 1.20 | 0.09 |
| Time females clean (seconds/visit) | 132.4 | 9 | 71.6 | 15 | 1.70 | 0.21 |
| Number of times parents feed nestlings (trips/hour) | 1.2 | 12 | 1.4 | 15 | 0.80 | 0.38 |
| Time parents feed (seconds/hour) | 57.5 | 10 | 54.2 | 15 | 0.20 | 0.66 |
| Percentage of observation period nest was conspicuous | 6.7 | 12 | 5.4 | 15 | 1.00 | 0.33 |
| Percentage of observation period at least one parent attended nest | 21.8 | 12 | 13.3 | 15 | 1.17 | 0.29 |
| Percentage of feeding trips made by only one parent | 31.2 | 10 | 56.3 | 15 | 9.86 | 0.01 |
| Time male is in nest tree before visiting nest (seconds) | 46.5 | 11 | 27.2 | 15 | 4.56 | 0.04 |
| Time female is in nest tree before visiting nest (seconds) | 39.0 | 11 | 27.9 | 15 | 1.91 | 0.18 |
| Percentage of feeding done by male | 74.8 | 11 | 69.9 | 15 | 1.40 | 0.25 |
| Percentage of cleaning done by female | 66.7 | 10 | 64.1 | 15 | 0.14 | 0.71 |

(Fig. 83). Survival of crechlings, however, is another matter; 47% of H-lineage crechlings survived their initial fall and winter, but only 37% of NH-lineage crechlings did so (Fig. 84). Annual survivorship of young H-lineage birds was also less dependent upon vagaries of the climate than was NH-lineage survivorship (Table 32). Lastly, none of the H-lineage families we monitored from 1972 to 1986 became extinct, but over half of the 32 NH-lineage families were extinct by 1986. The bottom line of the fitness differences has been that the composition of the Town Flock has gradually shifted to include more H-lineage birds. The proportion of breeders belonging to H-lineage families rose sharply from 1973 to 1979 and has declined slightly in recent years (Fig. 85). Overall, we have seen a significant increase in H-lineage breeders in our study flock.

Now we can see the origins of helping in a better light. Helpers come from families that are exceptionally productive, especially in terms of the number of sons they produce. Parents in these families also may be especially attentive to

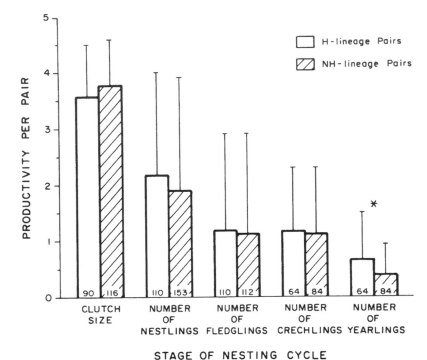

Figure 83    *Productivity of H-lineage (open bars) and NH-lineage (hatched bars) pairs through the nesting cycle. Top of bars indicate mean values. Lines above bars show one standard deviation. Sample sizes are given inside base of bar. Data on clutch size through fledgling success are from 1981 to 1987. Means of these variables were derived by averaging all nests found during this time. Crechling and yearling data are from 1972 to 1987. Means of these variables were derived by averaging per pair average productivity. Each pair only contributes one value to these means, but may contribute a value each year to productivity at earlier stages. The star (\*) indicates that H-lineage productivity is significantly (P < 0.05) greater than NH-lineage productivity (t-test).*

**Table 32.**    *Lineage-specific correlations between annual survivorship of young jays and annual weather factors. Winter temperatures are averages from December through February. Spring temperatures are averages from March to May. Summer temperatures are averages from June to August. Significant correlations (P < 0.05) are evidence suggesting that survivorship is influenced by weather. Non-significant correlations (P > 0.05) suggest that survivorship does not depend on the weather factor considered. Pearson correlation coefficients (r), sample sizes (N), and significance (P) are listed for each lineage. From Marzluff and Balda (1990).*

| Weather Factor | Age | H-lineage | | | NH-lineage | | |
|---|---|---|---|---|---|---|---|
| | | r | N | P | r | N | P |
| Winter temperature | Juvenile | −0.18 | 10 | 0.31 | +0.61 | 10 | 0.03 |
| Spring temperature | Yearling | −0.26 | 10 | 0.24 | +0.45 | 10 | 0.09 |
| Summer temperature | Yearling | −0.09 | 12 | 0.39 | −0.67 | 12 | 0.01 |

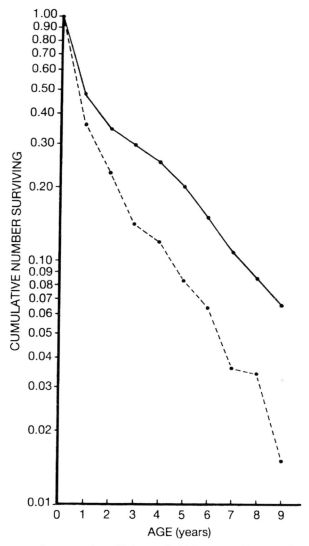

Figure 84  *Age-specific survivorship of H-lineage (solid line) and NH-lineage (dotted line) birds. Survivorship of males, females and birds of unknown sex born in 1972–1982 was followed in 1972–1986. H-lineage birds survived at significantly greater rates than did NH-lineage birds. From Marzluff and Balda (1990).*

their offspring as they gain independence because their young have exceptional survivorship during the first tenuous year of their life, regardless of prevailing weather. Now the interesting question is: are H-lineage families successful because they have helpers or do they have helpers because they are successful? To answer this we will investigate the role of helpers and the impact of their help to breeders, to the nestlings receiving their aid, and to themselves.

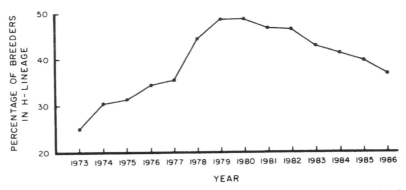

Figure 85    *Annual variation in percentage of breeders in the Town Flock that belonged to the H-lineage. The proportion of H-lineage breeders has increased significantly over 14 years (r = 0.57, DF = 12, P < 0.05). Number of breeders censuses per year from 1973 to 1986 were: 28, 36, 32, 46, 42, 34, 31, 33, 45, 41, 42, 39, 38, 38. From Marzluff and Balda (1990).*

## The behavior of helpers

As the nesting cycle progresses to the hatching stage a small, but variable number of nests in the Town Flock each year are attended by parents and one or, rarely, two helpers (Fig. 86). Helpers do not aid in the construction of the nest, in feeding of the female, nor in care of the eggs. In fact they rarely

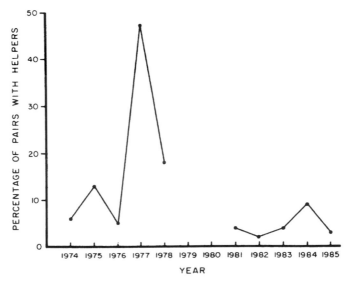

Figure 86    *Annual variation in the percentage of Pinyon Jay pairs that had helpers in the Town Flock. No data are available for 1979, 1980, 1986, 1987 because nests found were not adequately observed. Sample sizes of nests censused from 1974 to 1985 are: 31, 23, 40, 15, 22, 26, 52, 50, 43, 39. From Marzluff and Balda (1990).*

approach the nest until the eggs hatch, after which they participate fully in the feeding, cleaning, and guarding of the nestlings. Helpers either approach the nest alone or in the company of one or both parents (Fig. 87). Helpers behave as expected for males: their activity at the nest is more paternal than maternal as they contribute more to feeding than to cleaning. We recorded activities at

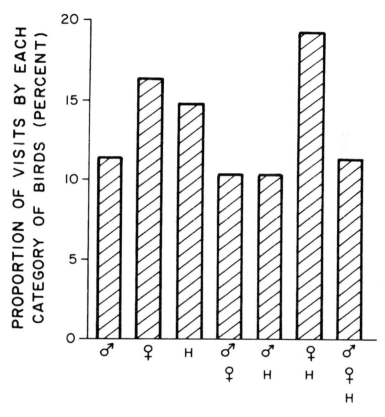

IDENTITY OF BIRDS VISITING NESTS

Figure 87    *Composition of birds attending four nests with helpers from 1981 to 1983. Visits to nests were by fathers (♂), mothers (♀), or helpers (H) alone; by fathers and mothers (♂,♀), fathers and helpers (♂, H), or mothers and helpers (♀, H); or by fathers, mothers, and helpers together (♂, ♀, H). From Marzluff and Balda (1990).*

four nests with helpers and found that helpers accounted for 30% of the time nestlings were fed and 14% of the time the nest and nestlings were cleaned (Table 33). This is a substantial amount of care which tends to increase the overall feeding rate at nests with helpers relative to nests without helpers (Table 33). With our small sample it is hard to say whether parents reduce their workload because of this help. At two nests (1 and 4, Table 33), parents

**Table 33.**  *Nesting behavior of pairs with helpers and pairs without helpers. All observations were conducted on nests with four young from 1982 to 1984. From Marzluff and Balda (1990).*

| Number of Hours Observed | Number of Feedings per Hour | Time (s) Feeding per Hour | | | | Time (s) Cleaning per Hour | | |
|---|---|---|---|---|---|---|---|---|
| | | Total | Male | Female | Helper | Male | Female | Helper |
| Nests with helpers | | | | | | | | |
| 1    32.5 | 2.6 | 57.2 | 26.5 | 10.7 | 20.0 | 16.6 | 148.8 | 21.3 |
| 2    10.8 | 1.8 | 72.8 | 44.6 | 15.2 | 13.1 | 38.0 | 68.8 | 5.5 |
| 3    6.0 | 2.5 | 83.8 | 44.0 | 19.2 | 20.7 | 15.7 | 15.7 | 3.2 |
| 4    7.0 | 1.6 | 49.0 | 19.7 | 7.7 | 20.9 | 29.3 | 39.6 | 30.0 |
| Average at nests without helpers | | | | | | | | |
| 321.2 | 1.4* | 53.3 | 39.2 | 14.0 | – | 35.4 | 76.4 | – |
| | 0.3** | 12.1 | 13.3 | 6.6 | – | 23.4 | 66.1 | – |

* *Mean of 17 nests*
** *Standard deviation of 17 nests*

with helpers worked well below the average, but at the other two they worked above the average. Suffice it to say that helpers make a tangible contribution to the nesting attempt.

### Do helpers help?

Given that helpers are full participants in the nesting effort we might expect that the number or quality of young produced with help would be greater than without help. This does not appear to be the case. We can most directly test the influence of helpers on productivity by comparing the number of young a pair produces with help to the number of young the same pair produces without help. As breeding experience may influence productivity this comparison would only be strictly valid if pairs differed only in the presence of helpers and not in their experience. We were able to find 16 pairs that had statistically equal average experience when helped *versus* when not helped. These pairs did not produce more crechlings nor more yearlings in years when they were helped compared to years when they were not helped (Table 34). Within a year helpers also do not increase a pair's productivity (Table 35). What about the quality of the young produced with the aid of helpers? If any differences in quality existed, they were short-lived as nestlings reared with help did not survive their first year of life any better than those reared only by parents, nor did helping correlate with nestlings' overall life spans (Table 34).

Perhaps helpers help their parents by reducing their workload and enhancing their survival. As we showed in Table 33 this may not be a general benefit, but it may help some parents (i.e. ones like those at nests 1 and 4). Parents in poor condition might be in special need of help so we predicted that helped parents (especially the smaller, subordinate females) should be lighter in weight than unhelped parents. This was not the case. The median weight of ten females with helpers was 98.9 g and median weight of 15 NH-lineage females was 99.4 g. Additionally, we measured three females in years they had

**Table 34.** *Impact of helpers on annual productivity and longevity of breeders, helpers, and helped offspring. From Marzluff and Balda (1990).*

| | Average Crechlings per Breeding Attempt (N) | | Average Yearlings per Breeding Attempt (N) | | Average Life span (years) of Yearlings (N) | | Average Body Weight (g) of Mate (N) | |
|---|---|---|---|---|---|---|---|---|
| Pairs with helpers | 1.2 | 16 | 0.72 | 16 | | | | |
| Same pairs without helpers | 1.3 | 16 | 0.76 | 16 | | | | |
| Males that were helpers | 0.78 | 16 | 0.20 | 16 | 5.2 | 21 | 99.0 | 10 |
| H-lineage males that lived 1 year in their natal flock and did not help | 0.82 | 9 | 0.37 | 9 | 6.1 | 8 | 98.3 | 11 |
| Male nestlings that were tended by helpers | | | | | 5.1 | 15 | | |
| H-lineage male nestlings that were not tended by helpers | | | | | 4.8 | 19 | | |

**Table 35.** *Annual productivity by H-lineage pairs having helpers versus H-lineage pairs not having helpers. Within a year, having a helper does not result in higher productivity (paired t-test comparing average productivity within years: crechlings $t = 0.16$, $N = 6$, NS; yearlings $t = 0.41$, $N = 6$, NS).*

| | Average production of crechlings | | | | | | Average production of yearlings | | | | | |
|---|---|---|---|---|---|---|---|---|---|---|---|---|
| | Pairs with helpers | | | Pairs without helpers | | | Pairs with helpers | | | Pairs without helpers | | |
| Year | N | Average | SD | N | Average | SD | N | Average | SD | N | Average | SD |
| 1973 | 1 | 4.0 | – | 7 | 1.0 | 1.5 | 1 | 2.0 | – | 7 | 0.57 | 0.98 |
| 1974 | 2 | 1.0 | 1.4 | 8 | 2.6 | 1.1 | 2 | 1.0 | 1.4 | 8 | 1.1 | 0.83 |
| 1975 | 3 | 1.0 | 1.0 | 4 | 0.75 | 0.50 | 3 | 0.67 | 1.2 | 4 | 0.75 | 0.50 |
| 1976 | 2 | 1.0 | 1.4 | 12 | 1.0 | 1.13 | 2 | 1.0 | 1.4 | 12 | 0.42 | 0.67 |
| 1977 | 7 | 1.6 | 1.1 | 5 | 1.8 | 1.8 | 7 | 0.86 | 1.1 | 5 | 1.0 | 1.0 |
| 1978 | 1 | 0.0 | – | 4 | 1.5 | 1.3 | 1 | 0.0 | – | 4 | 0.75 | 0.50 |
| 1981 | 1 | 2.0 | – | 20 | 1.2 | 1.4 | 1 | 2.0 | – | 20 | 0.35 | 0.67 |
| 1982 | 1 | 1.0 | – | 14 | 1.1 | 1.3 | 1 | 0.0 | – | 14 | 0.43 | 0.85 |
| 1983 | 2 | 1.0 | 1.4 | 18 | 0.06 | 0.24 | 2 | 1.0 | 1.4 | 18 | 0.06 | 0.24 |
| 1984 | 4 | 0.0 | 0.0 | 17 | 0.18 | 0.53 | 4 | 0.0 | 0.0 | 16 | 0.06 | 0.25 |
| 1985 | 1 | 0.0 | – | 10 | 0.30 | 0.67 | 1 | 0.0 | – | 10 | 0.20 | 0.42 |

help and in other years they were not helped. Two weighed more (by an average of 3 g) when they were helped. Regardless of their condition females do not appear to gain a survival benefit from having help. We recorded the death or disappearance of nine females that had helpers. Eight died in years they were not helped; this is suggestive of help but it is not statistically significant because these females were not helped 75% of years. We are forced to conclude that helpers help neither their parents' productivity nor their survival.

Maybe helpers help themselves. Helping might benefit helpers in several ways: 1) by providing breeding experience that would raise their future reproductive success, 2) by removing them from potentially stressful competition and conspicuousness in the non-breeding flock of yearlings, and 3) by enabling them to gain status in the eyes of potential mates. Our data do not support these hypotheses, however (Table 34). Breeding males that helped and males that did not help produced equal numbers of crechlings and yearlings per breeding attempt. Helping did not further the life span of yearling males. Lastly, helping did not allow males to obtain high quality mates. Remember that heavy female Pinyon Jays are preferred by males and males mated with heavy females are very successful (Chapter 7). Males that helped their parents did not obtain larger females than males who did not help.


*Why has communal breeding evolved in Pinyon Jays?*

We are left with the conclusion that helpers do not really help, either directly or indirectly, any of the participants in communal breeding. This allows us to reject three common hypotheses for the evolution of communal breeding. First, we can reject hypotheses proposing that helpers gain an indirect fitness benefit: Pinyon Jay helpers do not enhance the number or quality of young their parents raise. Second, we can reject hypotheses proposing resource benefits to helpers: Pinyon Jay helpers do not improve their acquisition of the most important limited resource – mates. Third, it is unlikely that helping evolved through reciprocity. In some species, such as Green Woodhoopoes, nestlings reared by helpers reciprocate the good deed by later aiding helpers successfully emigrate into neighboring flocks (Ligon and Ligon 1978). In our system, reciprocity is unlikely as helpers do not emigrate and helped nestlings rarely become helpers (Table 28).

Four observations suggest that communal breeding in Pinyon Jays is a form of extended parental care (termed Parental Facilitation by Brown and Brown, 1984) whereby parents enhance the survival of some sons by shielding them from stress and exposure in the non-breeding yearling flock. First, as we suggest in Chapter 11, helpers are likely to be subordinates that cannot obtain mates. These sons could benefit most by extra parental care. Second, communal breeding is facultative in that its frequency is linked to mate availability. This is consistent with the parental facilitation hypothesis because weak sons need extra care only when they face severe competition for mates. Third, parents that have helpers are exceptionally productive. This may allow them the freedom to provide extra care for their weak sons even if it means fewer young will be raised that year. Fourth, even though helpers are small they eventually attain longevity and fecundity equal to that obtained by their larger yearling competitors. Parental facilitation thus appears to be beneficial for parents and their young. This flexibility in allocation of parental care may be one reason H-lineage families are so consistently successful at producing yearlings despite annual fluctuations in abiotic (weather) and biotic (mate availability) factors. To answer our initial question, helpers are not likely to be responsible for the success of H-lineage jays. Instead, only successful jays may be able to provide the extra parental investment that helpers require.

Lineage differences in parental behavior suggest that there may be two parental strategies among Pinyon Jays. One strategy exemplified by NH-lineage jays is the production of a few males and a few females. These offspring must all compete for breeding opportunities and losers remain on their own, either dispersing or floating. The second, and currently more successful strategy is exhibited by H-lineage birds. They produce many males and hedge their bets by allowing very large sons to try for a quick reproductive pay-off as yearlings in the mating lottery and simultaneously allowing smaller sons to practice their breeding skills while garnering protection necessary for survival and successful reproduction in the future.

A number of intriguing questions remain. As we mentioned earlier, during the 1970s and early 1980s there was a definite increase in the proportion of H-lineage birds in the Town Flock (Fig. 85). This means that NH-lineage birds must have been relatively successful in the past. What caused this balance to shift during our years of study? We have no answer to this complex question. Maybe the effect of feeders or the increase in crows and ravens is somehow involved? We will never know.

CHAPTER 10

# Lifetime reproductive success

In previous chapters we have examined many factors that influence the productivity of Pinyon Jays within a given breeding season. Analyzing reproductive success within a breeding season or comparing success between seasons is the most straightforward way of determining the influence of most factors on breeding success because many factors change during the course of a jay's life (e.g. pair-bond duration, mate compatibility, parental behavior, and dominance) while other factors' influence varies from year to year (e.g. snowfall, pine crop, and parental age and/or experience).

Long-term studies such as ours also produce another measure of reproductive success. By following the activities of marked individuals from year to year we begin to glimpse the total reproductive output of individuals during their lifetimes. We call this *lifetime reproductive success* (LRS). During our study of Pinyon Jays we have measured LRS, in the form of the number of surviving yearlings, for 97 jays (48 males and 49 females).

Lifetime reproductive success is an especially seductive measure of productivity because it enables us to eavesdrop on the process of natural selection. Quantification of variability in LRS allows us to see which individuals (and their associated phenotypes) are currently leading and which are trailing in the evolutionary marathon. Additionally, the degree of variation in LRS among individuals is an index of the opportunity natural selection has for modifying a population (Arnold and Wade 1984a,b). The relative distance between racers in a marathon is analogous to the relative reproductive success of individuals. When this difference is great, then the leaders' traits can quickly increase their representation in the population, relative to the representation of the losers' traits. In other words, natural selection can quickly cause the population to be comprised primarily of leading traits. If heritable traits that correlate with this variability can be identified we have good reason to conclude that the combination of traits associated with high LRS will become more common in future generations. LRS is not equivalent to "Darwinian Fitness" and should not be viewed as an "infallible oracle for answering important evolutionary questions" (Grafen 1988). It is, however, an important tool that we will use to time the evolutionary race among jays in the Town Flock.

In this chapter we will quantify LRS for our sample of male and female Pinyon Jays. We will focus on variation in LRS and compare our results with published accounts of LRS (Clutton-Brock 1988). Our major aim is to examine the influence of lifespan, annual fecundity, and crechling survivorship on the number of yearlings produced in a jay's lifetime. Determination of the relative importance of these components allows us to suggest where natural selection has the greatest opportunity to act. Lastly, we investigate the influence of age of first reproduction, body size, and lineage of descent on LRS. These are the only characteristics of individuals we have measured that remain fixed during a jay's lifetime thereby enabling us to correlate them with LRS.

## VARIATION IN LIFETIME REPRODUCTIVE SUCCESS

The average male and female Pinyon Jay have similar lifetime reproductive success (Table 36). A breeding male lives an average of 5½ years, breeds during four of these, and fledges nearly one young each year of which half survive to provide him with 2.7 yearlings during his lifetime. Likewise, an average breeding female lives five years, breeds during four of these, fledges one young a year, and produces 2.9 yearlings during her lifetime. The only significant difference between males and females is a tendency for females to spend a greater proportion of their life span breeding. This occurs because, as we have seen previously (Fig. 48), females initiate breeding at an earlier average age than males.

All Pinyon Jays are not created equal. Our sample of 97 breeding jays reveals considerable variation in all components of LRS (Figs. 88, 89). Perhaps the most striking feature of this variation is the realization that many jays completely fail to reproduce. Over a third (34 of 97) of the breeding jays in our

**Table 36.**    *Components of lifetime reproductive success for 97 Pinyon Jays. All of these jays were observed breeding at least once. Unsuccessful birds as well as those successful at producing crechlings are included in this table.*

| Component of LRS | Males N = 48 | | Females N = 49 | |
|---|---|---|---|---|
| | mean | variance | mean | variance |
| Life span of breeders | 5.65 | 13.64 | 5.20 | 10.58 |
| Number of years breeding | 4.19 | 9.56 | 4.27 | 7.95 |
| Proportion of life span breeding* | 0.71 | 0.03 | 0.82 | 0.04 |
| Average number of crechlings/year | 0.92 | 0.64 | 1.04 | 0.85 |
| Total number of crechlings/ lifetime | 4.23 | 25.97 | 4.82 | 20.82 |
| Average annual survival of crechlings | 0.51 | 0.09 | 0.44 | 0.06 |
| Number of yearlings/lifetime | 2.74 | 5.49 | 2.86 | 3.70 |

* *Females spend a significantly greater proportion of their life spans reproductive than males (Kruskal Wallis = 8.7, P = 0.003).*

population failed to produce any yearlings. Males and females have similar distributions for each component. However, males tend to exhibit more extremes in productivity than females. The record productivity was set by a male (Dark Green-Red) who lived 16 years and produced 28 crechlings and ten yearlings during his glorious 14-year career. Another male (Mr Green) was also quite successful, raising eight yearlings in just five years. This male, unlike the first, had only one mate and she (Red/White-White) was the most prolific female during our study.

Early theories and empirical studies suggested that males should have greater variation in reproductive success than females and that species forming harems or defending territories should have greater variation in LRS than monogamous, non-territorial species. These claims rested on the notion that female reproduction was limited by physiological constraints but male reproduction was not. Females, after all, cannot continue to conceive while pregnant and their productivity is constrained by clutch size. Males, on the other hand, do not face such physiological barriers and are limited only by their ability to inseminate females (an easy task according to theory). Stud males able to assemble large harems or defend choice territories could be very successful (in attracting and inseminating females) in comparison with duds thus generating large variance in male (especially polygynous male) LRS.

Recent studies have not upheld all these claims (Clutton-Brock 1988), but the early expectations for monogamous, non-territorial animals are confirmed in Pinyon Jays. As expected for monogamous species, male and female Pinyon Jays have similar variation in LRS (Tables 36, 37). The variation in LRS shown in Figures 88 and 89 may seem large, but it is in fact quite small in comparison to most other monogamous birds (Table 37). Pinyon Jays have variation in LRS similar to House Martins, European Sparrowhawks, Kittiwake Gulls, and Northern Fulmars. Kittiwakes, Fulmars, House Martins, and Pinyon Jays share an important characteristic which may lessen their variation in LRS: they are colonial nesters with a minimum of territorial

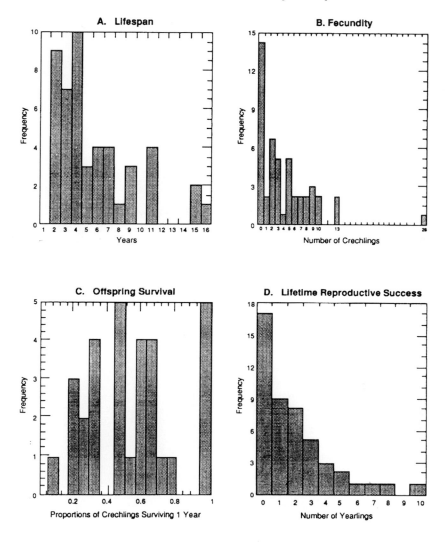

Figure 88   *Individual variation in fitness components for 48 male Pinyon Jays in the Town Flock.*

defense (Table 37). Lack of territoriality precludes stud males from claiming super-territories which could enhance their fitness. Variable territory quality therefore cannot be a factor inflating variation in LRS for these species. All of the more variable monogamous species in Table 37 are territorial. Coloniality and lack of territoriality may be important, but they cannot be the only determinants of low variation, as European Sparrowhawks both have low variation in LRS and defend classic territories.

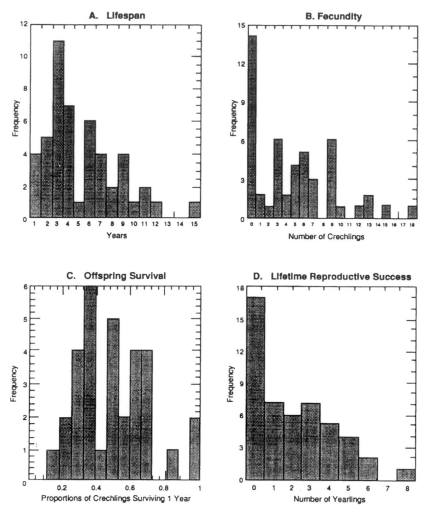

Figure 89    *Individual variation in fitness components for 49 female Pinyon Jays in the Town Flock.*

## Components of lifetime reproductive success

Lifetime reproductive success is composed of several multiplicative components. A typical model of LRS includes life span or the number of years breeding (L), annual fecundity, usually meaning the number of eggs laid (F), and the survival of eggs or young to maturity (S). Thus LRS = L × F × S. Representing LRS as a product of components enables us to ask how important each component is to overall LRS. In this section we will use two procedures to investigate this question. First, we will investigate correlations

**Table 37.**   *Variation in lifetime reproductive success for monogamous birds with different nes-ting habits. All results are from Clutton-Brock (1988). Standardized variances are defined in Table 39 and allow for comparisons between species with different mean LRS. Sex is indicated in parentheses (m = male, f = female).*

| Species | Original Data | | Standardized | Nest |
| | mean | variance | variance | Dispersion |
| --- | --- | --- | --- | --- |
| Great Tit (m, f) | 1.09 | 2.36 | 3.34 | Territorial |
| Song Sparrow (f) | 1.27 | 2.66 | 2.51 | Territorial |
| House Martin (m) | 8.69 | 54.97 | 0.83 | Colonial |
| House Martin (f) | 7.02 | 23.06 | 0.51 | Colonial |
| European Sparrowhawk (f) | 6.01 | 19.39 | 0.38 | Territorial |
| Bewick's Swan (m) | 5.15 | 33.32 | 1.26 | Territorial |
| Bewick's Swan (f) | 5.36 | 35.64 | 1.24 | Territorial |
| Kittiwake Gull (m) | – | – | 0.83 | Colonial |
| Kittiwake Gull (f) | – | – | 0.69 | Colonial |
| Northern Fulmar (f) | 3.23 | 6.32 | 0.60 | Colonial |
| Florida Scrub Jay (m, f) | 1.42 | 2.71 | 1.25 | Territorial |
| Pinyon Jay (m) | 2.74 | 5.49 | 0.51 | Colonial |
| Pinyon Jay (f) | 2.86 | 3.70 | 0.34 | Colonial |

between the components of LRS. Second, we will partition the variation in LRS into portions accounted for by each component. We have information on several components of Pinyon Jay LRS and here investigate life span, the number of years breeding, annual productivity of crechlings, survival of crechlings for one year, and lifetime production of yearlings. In our multiplicative model of LRS we use breeding lifespan as L, annual production of crechlings as F, and crechling survival as S. We are unable to use egg production, the usual measure of fecundity, because we lack complete clutch size data for birds during their lifetimes. Additionally, the number of eggs laid by a pair is a poor measure of fecundity because pairs re-lay eggs after nest failure, which may happen three or four times in a season. Production of crechlings is the most appropriate measure of fecundity for Pinyon Jays because it quantifies total productivity during the period when offspring are dependent upon parental care (crechlings are still dependent upon their parents for food and predator detection). We should expect our measure to be more variable than usual measures because the myriad of factors influencing egg and nestling survival have already contributed to the variation in the number of crechlings produced. The product L × F × S equals the lifetime production of yearlings, our measure of LRS.

Life span and the number of years breeding are the primary correlates of the lifetime production of yearlings by male and female Pinyon Jays (Table 38). Annual production of crechlings is correlated with LRS to a lesser extent. Survival of crechlings is the least important correlate of LRS. Partitioning variance in LRS also suggests that variation in breeding life span is the prime reason that LRS is variable (Table 39). This procedure does, however, suggest that variable survival of crechling jays is more important to variation in LRS than is annual fecundity. In females, variable crechling survival is equal in

**Table 38.**   *Pearson product moment correlations between components of Pinyon Jay lifetime reproductive success. Successful and unsuccessful breeders are included in this analysis. Correlations for males are given above the diagonal, females below the diagonal. Sample sizes are given below each correlation.*

| Component of LRS | L | BL | F | S | LRS |
|---|---|---|---|---|---|
| Lifespan (L) | | 0.94 | 0.19 | −0.15 | 0.66*** |
| | | 48 | 48 | 34 | 48 |
| Number of years breeding (BL) | 0.95*** | | 0.15 | −0.14 | 0.67*** |
| | 49 | | 48 | 34 | 48 |
| Average number of crechlings/year (F) | 0.32*** | 0.15 | | 0.02 | 0.58*** |
| | 49 | 49 | | 34 | 48 |
| Annual survival of crechlings (S) | −0.18 | −0.07 | −0.08 | | 0.27** |
| | 35 | 35 | 35 | | 34 |
| Total number of yearlings/lifetime (LRS) | 0.63*** | 0.62*** | 0.54*** | 0.52*** | |
| | 49 | 49 | 49 | 35 | |

** $P < 0.01$
*** $P < 0.001$

importance to breeder life span. Analysis of variation is perhaps a bit more illuminating than correlation analyses in this context because it enables us to look at the influence of each component in isolation from the other components. Thus we get a better feel for the individual importance of each component using analysis of variance (see Brown 1988 for more details). Regardless of which method we employ, the importance of breeder life span is evident. Simply put, the lifetime reproductive success of a Pinyon Jay is a direct consequence of how long it lives.

Failure to produce any crechlings may also influence variation in LRS.

**Table 39.**   *Original and standardized measures of lifetime reproductive success for Pinyon Jay breeders that produced at least one crechling during their lifetime. Standardized variances for single components equal the variances in the original data divided by the square of the original data mean. Standardized variances of products are calculated after each component in the product is divided by the product's average. Percent of LFS refers to the percentage of variation in lifetime reproductive success accounted for by variation in each component or product of components. These percentages are computed using the upward progression method of Brown (1988) and are discussed in the text. Negative percentages for products indicate strong correlations between their constituent components.*

| | Original Data | | | | Standardized Variance | | | |
|---|---|---|---|---|---|---|---|---|
| | Males N = 34 | | Females N = 35 | | | Percent | | Percent |
| Component of Lifetime Reproductive Success | mean | variance | mean | variance | male | of LFS | female | of LFS |
| Number of years breeding (L) | 4.97 | 10.51 | 5.11 | 7.99 | .426 | 83.86 | .306 | 89.97 |
| Average number of crechlings/year (F) | 1.30 | 0.40 | 1.46 | 0.57 | .237 | 47.05 | .267 | 79.65 |
| Annual survival of crechlings (S) | 0.51 | 0.09 | 0.44 | 0.06 | .346 | 66.14 | .310 | 90.86 |
| Product of L and F | 5.97 | 26.27 | 6.74 | 16.02 | .629 | − 7.09 | .288 | −84.66 |
| Product of L and S | 2.39 | 4.35 | 2.22 | 2.69 | .682 | −15.75 | .524 | −26.25 |
| Product of F and S | 0.66 | 0.28 | 0.63 | 0.21 | .652 | 15.16 | .499 | −23.30 |
| Total number of yearlings/lifetime (LFS) | 2.74 | 5.49 | 2.86 | 3.70 | .508 | −89.37 | .339 | −26.25 |

However, Brown's (1988) analysis of variance method is restricted only to successful breeders. Brown does provide a formula that enables us to calculate an overall variance, including non-breeders, and then determine what proportion of this variance is accounted for by successful *versus* unsuccessful jays. Doing this for our sample of 34 successful males, 35 successful females, and 14 unsuccessful jays of each sex yields an overall variance of 5.4 for males and 4.3 for females. Roughly two-thirds of this variation is attributable to variation among the successful breeders (71% of the male variation and 61% of the female variation). Putting this result together with those in the previous paragraph we can conclude that the primary source of variation in LRS is variation among successful breeding birds and this results from their variable life span and variable survival of their crechlings.

We can gain some insight into how natural selection may influence the breeding behavior of Pinyon Jays by examining the relative importance of the components of LRS. Currently, natural selection should be favoring long-lived jays that produce crechlings with a high chance of survival because such jays have the highest LRS. Maximizing annual productivity of crechlings is less likely to profoundly increase LRS and in fact could reduce it if the extra energy demands of a large brood reduce crechling or breeder survival. This suggests that the typical Pinyon Jay strategy of producing a few offspring each year may have evolved because it enables parents to extend their care of these offspring thus increasing their chances of survival without overtaxing parents and shortening their life-span.

## CORRELATES OF LIFETIME REPRODUCTIVE SUCCESS

The documentation of lifetime reproductive success and its variation is interesting because it provides a window through which we can peer in at evolution in action. Natural selection can lead to evolutionary change only if genetically or culturally heritable attributes of individuals are correlated with variation in LRS. We have shown that variation in LRS exists and now we attempt to correlate properties of individuals with that variation. A strong correlation between a property and LRS suggests that natural selection would be capable of shifting the frequency of that property through evolutionary time. Whether such evolution occurs or not is dependent upon the heritability of the property and the strength of the selective pressure. We lack rigorous data on heritability of many characters in Pinyon Jays. However, body mass is likely to be heritable as fathers and sons are of similar mass (Marzluff and Balda 1988b).

Body mass is correlated with several components of LRS. However, in neither males nor females is overall LRS significantly correlated with body size (Table 40). Therefore it is unlikely that natural selection is currently changing the average size of a Pinyon Jay (or the distribution of sizes). An overall advantage to large or small body size probably does not exist because, as we argued in Chapter 7, reproductive performance depends more on compatibility of mates' sizes than on absolute size. Large or small jays can succeed provided they have an appropriate-sized mate.

**Table 40.**    *Correlations between parental body mass and components of lifetime reproductive success. Correlation coefficients are listed above sample sizes throughout table. Significance of correlations is indicated by asterisks (\*P < 0.05, \*\*P < 0.01).*

| | Pearson Correlations With: | |
|---|---|---|
| Component of lifetime reproductive success | Female Mass | Male Mass |
| Average crechlings/year | −0.36** | 0.23* |
| | 33 | 38 |
| Annual survival of crechlings | 0.27* | −0.34** |
| | 30 | 29 |
| Number of years breeding | −0.04 | −0.04 |
| | 33 | 38 |
| Life span | −0.20 | 0.03 |
| | 33 | 38 |
| Proportion of life span breeding | 0.26* | −0.11 |
| | 33 | 38 |
| Total number of crechlings/lifetime | −0.32** | 0.15 |
| | 33 | 38 |
| Total number of yearlings/lifetime | 0.04 | 0.06 |
| | 33 | 38 |

Heavy males and heavy females appear to be maximizing different components of LRS. Male mass is positively correlated with annual fecundity but negatively correlated with crechling survival. Thus it appears that heavy males produce many, poor surviving crechlings. In contrast, female mass is negatively correlated with annual fecundity but positively correlated with crechling survival. This suggests that heavy females produce fewer, but better surviving crechlings. These contrasting strategies bring up two interesting speculations. First, remember that we suggested partners of similar size (big females mated to small males) were the most productive because they were more compatible than partners disparate in size (Chapter 7). Our results here suggest one reason why such partners are compatible: they work toward similar goals by producing a few offspring that have high survival. Partners more disparate in size (e.g. large males and large females) may not co-ordinate their reproductive efforts as effectively because they each attempt to maximize different components of LRS. Incidentally, although Table 40 shows no advantage to the strategy of producing a few, quality offspring, our earlier analysis suggests that maximizing offspring survival is more likely to increase LRS than is maximizing fecundity. Therefore larger females and smaller males appear to be allocating reproductive effort in similar ways and in a way that is most likely to increase LRS. This may explain why we find them to be so successful.

Our second speculation revolves around the correlation suggesting that heavy females breed for a greater proportion of their life spans than light females (Table 40). This correlation is expected given the laboratory and field results which suggested that heavy females were preferred by males (Chapter 7). Heavy females apparently breed earlier in life because they are quickly selected as mates and/or they out-compete lighter females for the opportunity to breed.

The age at which individuals first breed is not significantly correlated with LRS (Table 41). Birds that breed as yearlings breed on average one year longer than individuals which do not breed until they are three years old. These differences in breeding life span enable yearling breeders to attain the highest LRS, albeit only slightly higher than LRS of individuals delaying breeding. This occurs because of the low survivorship of crechlings produced by yearling breeders.

**Table 41.** *Influence of age of first reproduction, lineage of descent, and helping on components of lifetime reproductive success. Only birds attempting to breed are included in this analysis. Both sexes are combined. Significance of comparisons is tested with Kruskal-Wallis one-way ANOVAs and indicated by asterisks between means that are significantly different (\*P < 0.05, \*\*P < 0.01, \*\*\*P < 0.001).*

| Comparisons | N | Average number of crechlings/ yr | N | Average survival of crechlings | N | Average number of yearlings | N | Number of years breeding |
|---|---|---|---|---|---|---|---|---|
| First breed as yearling | 30 | 0.63 | 17 | 0.51 | 30 | 1.8 | 30 | 4.2 |
| *versus* | | | | | | | | |
| First breed at age 2 yrs | 45 | 1.00 | 32 | 0.50 | 45 | 1.4 | 45 | 3.3 |
| *versus* | | | | | | | | |
| First breed at age 3 yrs | 5 | 0.89 | | 0.53 | 5 | 1.0 | 5 | 3.2 |
| H-lineage breeders | 30 | 1.40 | 26 | 0.52 | 30 | 3.1 | 30 | 4.6 |
| *versus* | | ** | | | | *** | | |
| NH-lineage breeders | 67 | 0.82 | 43 | 0.45 | 67 | 1.5 | 67 | 4.1 |
| H-lineage sons that helped | 7 | 0.76 | 4 | 0.65 | 7 | 0.7 | 7 | 1.9 |
| *versus* | | | | | | ** | | * |
| H-lineage sons that did not help | 14 | 1.30 | 13 | 0.57 | 14 | 3.1 | 14 | 4.6 |

The striking differences in lineages occasionally having helpers (H-lineage) *versus* those never having helpers (NH-lineage) that we considered in the previous chapter are confirmed in our analysis of LRS (Table 41). Birds belonging to H-lineage families have significantly higher LRS than do birds belonging to the NH-lineage families. This result suggests that natural selection should currently be biasing our study flock toward H-lineage birds. This appears to be the case (Fig. 85). Over the course of our study the percentage of breeders in the flock that belong to the H-lineage families rose from a low of 25% in 1973 to a peak of 48% in 1979 and 1980. It dropped slightly to 35% in 1986. Again we must emphasize that helping, *per se*, is unlikely to be fueling the success of H-lineage birds; rather, helping is tolerated by successful parents. Accordingly, we reassert our claim that helpers are making the best of a bad situation because helpers end up breeding for relatively few years and have low LRS (Table 41).

In conclusion, it seems that much of the variation in LRS is unaccounted for by the individual properties we have measured. Only lineage of descent has profound influences on LRS, and as predicted, natural selection appears to be changing the lineage composition of our flock in favor of the extremely successful H-lineage families. Why this is happening at this time remains a curious mystery. Natural selection is unlikely to change the size of Pinyon Jays, possibly because intrapair compatibility of size is a more important determinant of reproductive success than is absolute size. Lastly, helping at the nest is likely to remain a last-ditch option as the consequential delay in breeding can dramatically lower LRS.

Our investigation of LRS has continually pointed to the importance of long life and the production of crechlings with a high probability of survival. We will address these important aspects of Pinyon Jay biology in our next chapters on the options pursued by young jays and survivorship.

CHAPTER 11

# Travels and tribulations of young jays

*"Growing up in animal societies is not an easy task. Apart from the obvious hardships imposed by predators and the physical elements, each juvenile must contend with a complex social environment in which other juveniles and their parents attempt to increase their inclusive fitness at its expense."*

(Rubenstein 1982)

If ornithologists could fly we would know a great deal more about birds. This statement is especially pertinent to birds such as Pinyon Jays that roam widely across large, undefended home ranges. Our single greatest frustration in studying Pinyon Jays is one we share with many field biologists: we cannot keep up with moving jays for long. As flightless mammals, ornithologists and cats focus on activities that occur at regular, easily monitored locations such as feeding stations and nests. This allows us to spend most of our time in the field in contact with our study organisms. The disadvantage of this approach is that it narrows our view of an animal's natural history. Most behavioral research on birds centers around breeding. Our work is no exception, but we will broaden our view of Pinyon Jays in this chapter by investigating the options pursued by young jays.

As the opening quote suggests, growing up is a trying time for social animals. Rubenstein's comments were made with respect to the socialization of young horses, but they aptly summarize the pressures imposed on growing Pinyon Jays. Young jays face several decision points during their first two years of life as they attempt to maximize their inclusive fitness. They must decide whether to remain in their natal flock or disperse from it. This decision rests on the relative costs and benefits of remaining at home and competing for a mate or traveling through unfamiliar terrain to join another flock and there obtain a mate. (Remember from Chapter 8 that these "decisions" are likely to be made by natural selection, not by conscious thought.) After a jay settles down it faces another point of decision: should it help its parents breed, secure a mate and breed itself, or just hang out with other unmated young jays in the non-breeding yearling flock? In this chapter we will argue that the availability of mates in a jay's birth (or natal) flock and its relative dominance within the flock are two important factors that influence dispersal and the options pursued by yearlings as they enter the breeding population.

## THE PATTERN OF PINYON JAY DISPERSAL

Regular trapping and banding of jays in the Town Flock from 1973 to 1982 provided us with two views of dispersal: immigration into the flock and emigration out of the flock. Most of our insight into dispersal comes from the former view. Each year we captured and marked all young produced by breeders in the Town Flock by early August. Soon thereafter we noticed many, relatively skinny, young birds associating with the Town Flock. Over the years we captured and banded 681 of these dispersing birds which we call *wanderers*. During this time span 383 young Town Flock birds also disappeared. An unknown proportion of these disappearances are attributable to death and the remainder to emigration.

A few more definitions will aid our discussion of dispersal. Wanderers that establish themselves as breeders in the Town Flock are referred to as *successful immigrants*. Likewise, we call Town Flock jays that leave the flock and establish themselves as breeders in other flocks *successful emigrants*. Town Flock jays that are absent for two or more months and then return to the Town Flock are called *returning emigrants*. We use the general term *disperser* to refer collectively to wanderers, successful immigrants, and successful emigrants.

The amount of dispersal varies from year to year by several orders of magnitude. In 1975, 1980, 1982, and especially 1978 wanderers were abundant in the Town Flock, but in the remaining years they were rare or totally absent (Fig. 90). Recall from Chapter 5 that we attributed much of this variation to climatic conditions; wanderers are especially prevalent in dry years. During

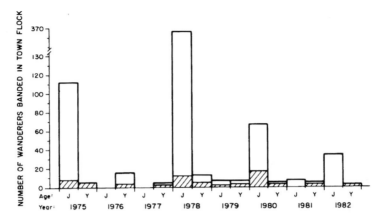

Figure 90 *Year-to-year variation in the number of wanderers banded in the Town Flock. Juvenile (J) and yearling (Y) wanderers are plotted separately. Fewer juvenile wanderers than yearling wanderers remained till December with the Town Flock (shaded portion of bars).*

these years wanderers appeared in "waves". On some days ten to 20 new birds would appear. Despite the occasional outbreaks of wanderers that enter the Town Flock, few ever stay more than several days (note shaded portion of Fig. 90). Juveniles and yearlings differ in this respect; only 7% of wandering juveniles compared to 51% of wandering yearlings remained with the Town Flock through the fall.

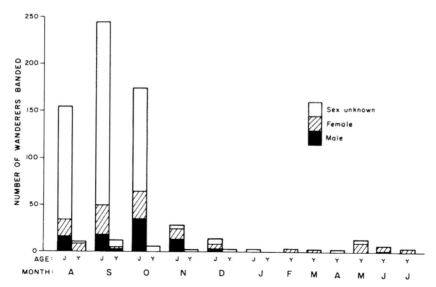

Figure 91 *Number of wanderers banded in the Town Flock per month from 1973 to 1982. Juveniles (J) and yearlings (Y) are plotted separately and their sex, when known, is indicated by shading. From Marzluff and Balda (1989).*

Dispersal is primarily a fall event whose participants are young males and females (Figs. 91, 92). Dispersal typically begins in August, peaks in September, and has ended by December. There is a secondary wave of dispersal during the spring and early summer. The two seasonal waves of wanderers

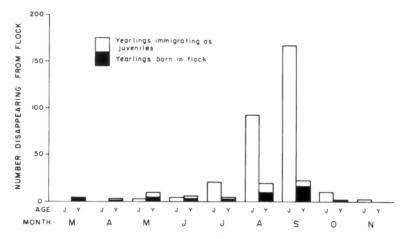

Figure 92    *Number of juveniles (J) and yearlings (Y) disappearing from the Town Flock per month from 1973 to 1982. Yearlings are separated into those originally born in the Town Flock and those joining the Town Flock as juvenile immigrants. From Marzluff and Balda (1989).*

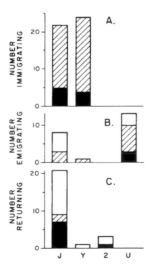

Figure 93    *Known fates of jays dispersing as juveniles (J), yearlings (Y), two-year-olds (2), or jays dispersing at an unknown age (U). Sex is indicated by shading as in Figure 91. (A) Number of jays successfully immigrating into the Town Flock. (B) Number of jays successfully emigrating from the Town Flock into the neighboring Country Club Flock. (C) Number of jays returning to their natal flock after emigrating from it. From Marzluff and Balda (1989).*

have different sex compositions. Fall wanderers include males and females in relatively equal abundance (51.6% female), whereas spring wanderers are primarily females (77.8% female). Dispersers are most commonly juveniles; juveniles comprise 90% of wanderers and 80% of the young birds (juveniles and yearlings) that disappeared from the Town Flock. Adult Pinyon Jays rarely disperse (Fig. 93). Occasionally, adult females emigrate into the Town Flock, apparently actively recruited as mates by recent widowers. These unknown females suddenly appeared with established Town Flock males shortly after their mates died.

Few wanderers associate for long with the Town Flock (Fig. 90), and even fewer establish themselves as breeders in the flock. Forty-six wanderers successfully immigrated and spent the remainder of their lives ($\bar{x}$ lifespan = 5.5 years) in the Town Flock. Most (80%) were females (Fig. 93A). Females established themselves in the flock at a higher rate (31.1% of 119) than did males (9.2% of 98). We documented 22 cases of Town Flock jays successfully emigrating. Successful emigrants also were primarily females (11 of 14; Fig. 93B).

Dispersing females quickly form lasting bonds with their adopted flocks. In one case, we captured a female yearling born in the Town Flock that had emigrated to a neighboring flock. We returned her to the Town Flock by releasing her at a feeding station when her father, mother, and brother were present. She was back in her adopted flock within 24 hours. Young females that disperse apparently prefer their newly acquired companions to their former flock mates and relatives!

Not all jays disappearing from the Town Flock emigrate or die; some return after prolonged periods of apparent wandering (Fig. 93C). Unlike successful dispersers, returning jays are usually males (eight of ten known-sex birds). We have documented 25 cases of returning jays and most (15) involved a juvenile wandering from the flock in the fall and returning five to ten months later during the breeding season. These returnees usually (13 of 15) spent the remainder of their lives in the Town Flock. The remaining ten cases involved birds leaving in the fall and returning the following fall or a year and a half later during the following breeding season. Four of these birds did not remain in the Town Flock. They stayed a few days then departed or died. The most unusual case involved a male that returned unpredictably three times over four years after his initial departure.

Most Pinyon Jays breed in the flock they were born in, and most that successfully emigrate breed in neighboring flocks (Fig. 94). We were able to determine the birthplace and breeding place for 134 Town Flock jays. Most (84%) bred in the Town Flock, and these were typically males (69.6% of 112). The other 22 bred in a neighboring flock and were typically females (Fig. 93B). Longer distance dispersal rarely occurs. The United States Banding Laboratory provided us with information on 170 subsequent sightings of jays originally banded throughout the western United States. Only six birds banded in one location were subsequently observed or found dead in a different location. Four of these were 160 km or less from the original location, but two were over 600 km away (Fig. 94). One of these long distance travelers was a juvenile we banded as a nestling in the Town Flock. During its

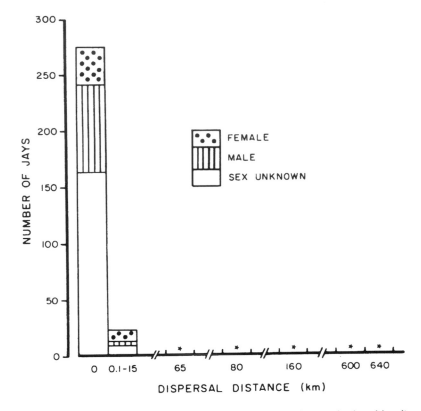

Figure 94    *Distribution of distances jays were known to disperse between birth and breeding. A distance of zero indicates breeding in the natal flock; a distance of 0.1–15 km indicates breeding in the flock neighboring the natal flock. Starred distances indicate that one jay of unknown sex was observed that distance from its place of birth (breeding uncertain). Data include jays banded in the Town Flock (N = 134) and all reported recoveries of jays banded throughout the western United States (N = 170). From Marzluff and Balda (1990).*

first autumn it traveled 600 km to eastern New Mexico where it was shot in the act of eating a farmer's grain. We cannot call this extensive wandering "successful dispersal", because the jay did not live long enough to breed. The other bird was banded in 1956 by a lifelong Pinyon Jay enthusiast, Dr N. R. Whitney, near Rapid City, South Dakota. It too was shot, five years later and 640 km away from Rapid City in the Bear Paw mountains of Montana. The age of this bird at banding was unknown. However, it presumably lived long enough to successfully emigrate into a distant flock.

We can summarize the pattern of dispersal in Pinyon Jays as one of female transfer between neighboring flocks. Males occasionally wander from their natal flock, but like prodigal sons they return and usually breed at home. Dispersal by young females is very common in birds (Greenwood 1980). However, consistent exchange of females between neighboring social groups and occasional male wandering and return is rare among social birds and mammals. In this aspect of their social organization Pinyon Jays closely resemble chimpanzees (Nishida 1979, Pusey 1979), gorillas (Harcourt *et al.* 1976, Baker 1978), and wild dogs (Frame and Frame 1976, Frame *et al.* 1980).

REASONS UNDERLYING PINYON JAY DISPERSAL

Let's try to understand the salient features of dispersal in Pinyon Jays by addressing three related questions. First, why is dispersal so common in young birds? Second, why are dispersers primarily females? Third, why do dispersers typically travel such short distances? In order to answer these questions we must consider the costs and benefits of dispersal and decide how they are influenced by an individual's age and sex.

Most current researchers, including ourselves, view dispersers as actively attempting to improve their fitness by increasing their access to one or more critical resources. Three common resources thought to influence an individual's decision to disperse or remain at home are food, space or territory, and mates. If an individual has only limited access to one or more of these resources at home its reproduction and/or survival may suffer and dispersal to a more bountiful location may be favored. Dispersal will be favored over remaining at home (philopatry) if dispersers average higher fitness in their new breeding locale than they would have attained at home. For dispersal to occur, dispersers must typically produce many more offspring than they would have produced at home because they must compensate for any costs of reduced survival expected by traveling through unfamiliar terrain.

In Pinyon Jays the resource most likely affecting dispersal is the availability of mates. Pinyon Jays are not territorial and they wander as a group in search of abundant food when crops on their home range fail (Chapter 1). Therefore food and space resources are unlikely to influence a Pinyon Jay's decision to leave its natal flock to the degree that they influence dispersal in species tied to small fixed territories such as Florida Scrub Jays (Woolfenden and Fitzpatrick 1984).

A Pinyon Jay's age and sex likely influence its ability to obtain a mate and therefore should influence its decision to disperse. Remember from Chapters 5 and 6 that Pinyon Jay flocks typically have a surplus of males and that individuals live within a dominance hierarchy where adult males dominate adult females. Further, adult females dominate juvenile males, who in turn dominate juvenile females. It seems reasonable to assume that a jay's ability to obtain a mate is determined by its dominance and/or desirability and the availability of opposite-sex birds. A flock's sex ratio and an individual's dominance status and desirability should therefore influence the decision to disperse. Our data support this theory, as dispersers are young (subordinate), immigration of females increases when flocks become exceptionally male-biased, and emigration of females increases when flocks become female-biased (Fig. 95). It appears that young birds, especially females, transfer from flocks with few extra males to flocks with many extra males.

Young males presumably disperse because as subordinates they are

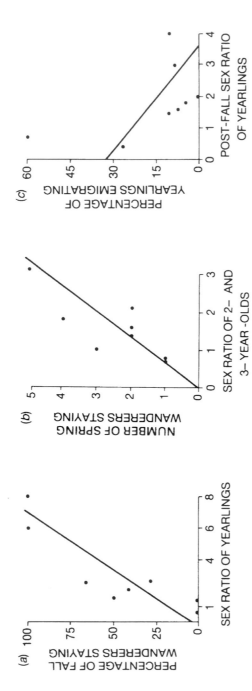

Figure 95   *Correlations between sex ratio (number of males per female) and various indices of dispersal. Dots represent sex ratio and dispersal per year from 1975 to 1982. Lines are least-squares regressions. (A) More wanderers banded in the fall remain with the Town Flock through December as sex ratios become more male biased. (B) More wanderers banded in the spring remain in male-biased flocks. (C) More yearlings disappear during the spring from female-biased flocks. From Marzluff and Balda (1989).*

unlikely to obtain a mate (or any other limited resources) before dominants, but why are most dispersers females? It would seem that males should disperse because jay flocks usually have more males than females. In this anomaly, however, may lie the answer; males may rarely benefit by dispersing because they are likely to encounter the same problem away from home, namely few females. This may explain why many dispersing males eventually return home. Females would be likely to benefit by dispersing because they could establish themselves in highly male-biased flocks. Emigration would be most beneficial for females when, as occasionally happens, females outnumber males. As expected, emigration is highest in such years (Fig. 95C). Emigration from slightly male-biased flocks to highly male-biased flocks may also be beneficial as females, especially subordinate ones, would broaden their pool of prospective mates. Remember that finding a compatible mate, not just any mate, is an important determinant of an individual's fitness. This may then explain why some females disperse from male-biased flocks. Mate acquisition cannot be the only reason jays disperse as a few males disperse each year. Perhaps this minor amount of dispersal represents a few very subordinate birds that find access to a variety of resources difficult in their natal flock. Their status and hence access to resources may be greater in a new flock.

An interesting aspect of dispersal in Pinyon Jays is its enormous year-to-year variation (Fig. 90). This would be expected if most dispersal was triggered by a factor, such as mate availability, that varies from year to year and from flock to flock. The same flock may act as a sink, importing many dispersers in some years, and as a source, exporting many dispersers in other years. Such a flexible disperal system ensures that the maximum number of jays breed in any year because one flock's surplus females can locate another flock's surplus males.

We should question any relief we may feel for providing data which are consistent with an adaptive hyothesis, such as dispersal allowing an individual to maximize its reproductive success, until we are able to rule out competing non-adaptive hypotheses. Stephen Gould and Richard Lewontin published an influential paper in 1979 which warned against uncritically accepting the adaptiveness of an organism's features and behaviors. It is easy to offer an adaptive reason for nearly all features, because such explanations seem so intuitive. However, many features may have no selective advantage or disadvantage, and others may have evolved indirectly in response to natural selection acting elsewhere. Take brain size in primates as an example. Gorillas and humans have larger brains than chimps, but should we conclude that large brains are an evolved adaptation in gorillas and humans, perhaps one that allows large primates to cope with more complex social environments? Before accepting this appealing adaptive hypothesis we should consider alternative hypotheses such as: brain size has increased indirectly as a result of selection for large body size in gorillas and humans. If brain size is larger than expected for a given body size, then we are more justified in proposing that selection has worked directly on brain size. (It apparently has in humans but not in gorillas.)

Let's consider the non-adaptive hypothesis that young female Pinyon Jays disperse because they get lost during the fall seed harvest. We tested this hypothesis by capturing jays and releasing them at distant locations. We displaced 12 adult jays and all returned, most within a day or two, to their home flock. We displaced 15 juvenile jays and their return was dependent upon the distance we moved them. Eight of ten (four of each sex) that were moved to their neighboring flock returned home. However, none of five (three males, two females) released farther from home returned. We once displaced eight juveniles 60 km from their natal flock and none returned. These results suggest that some long-distance dispersal may occur because young jays lose their way in the fall. However, this non-adaptive hypothesis does little to explain the prevalent pattern of female-biased dispersal between neighboring flocks, as males and females home with equal and high probabilities from neighboring flocks. We can now be more secure in our feeling that most dispersal by young females is an adaptation for finding high quality mates.

Let's next consider why most dispersal is across relatively short distances. This characteristic of Pinyon Jays is also found in most animals and plants and as such has attracted considerable speculation (Shields 1982). Two theories are probably relevant to Pinyon Jays. First, any travel costs associated with dispersal are reduced by short-distance dispersal (Bengtsson 1978). Second, breeding close to home ensures that mates will have developed in similar environments and therefore share genetic and cultural adaptations to these environments (Shields 1982). Not only would mates therefore be expected to have compatible backgrounds, but they would breed in environments similar to those they evolved in. This may be especially important in heterogeneous environments such as those inhabited by Pinyon Jays. If Town Flock jays dispersed to substantially lower elevations before breeding, their propensity towards delayed nesting or exposed nest placement in order to avoid damage from spring snows (Chapter 8) would be unnecessary and perhaps costly.

### OPTIONS OF YEARLINGS IN THEIR NATAL FLOCK

All young jays that do not disperse do not immediately breed in their natal flock. They pursue one of three options: they breed, they help their parents breed, or they float with other non-breeding, non-helping yearlings in a flock somewhat disjunct from breeding adults. In the Town Flock from 1974 to 1982, most of the yearlings floated (73%, SD = 16%), neither helping nor breeding. Only a few bred (14%, SD = 9%), or helped their parents (13%, SD = 11%). However, the number of yearlings pursuing each of these options varies greatly from year to year (Fig. 96).

Provided costs of breeding are not excessive, yearlings should strive to propagate their genes by reproducing as early in life as is possible. Thus, breeding should be the preferred option for yearlings. If an individual cannot reproduce because it is not dominant or sexy enough or because mates are unavailable, then it faces a less clear-cut decision: should it remain with other young jays in the non-breeding yearling flock or should it stay in the protective company of its parents and help them breed? Remaining in the non-breeding flock may be advantageous, especially for dominant yearlings, as alliances or consortships may be formed and dominance positions established. However, subordinates

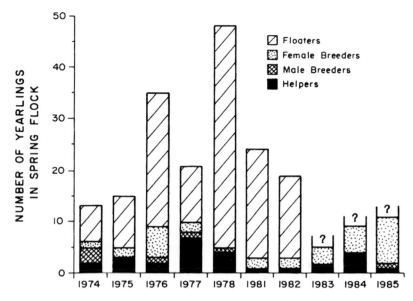

Figure 96  *Annual variation in the options pursued by yearling Pinyon Jays. Question marks in 1983, 1984, and 1985 indicate where we were unsure of the number of floaters.*

are unlikely to fare well in competition within the yearling flock and they may suffer from stress common to other subordinate animals. Subordinate Wood Pigeons, for example, show signs of stress such as lower feeding rates and enlarged adrenal glands because they are constantly alert for attacks from dominants (Murton *et al.* 1971).

If this scenario is correct, then mate availability and an individual's status should influence the options pursued by yearlings. We predict that as mate availability increases, the number of yearling breeders should also increase. As mate availability decreases, subordinates should not be able to successfully compete for mates so we expect the following relationships between dominance and the options pursued by yearlings: 1) dominant males and females should breed, 2) intermediate males and females should float and begin to establish pair bonds with each other, 3) subordinate females should disperse, and 4) subordinate males should help their parents. We lack dominance data for yearlings pursuing each of these options each year and therefore cannot test these predictions directly. We can test them indirectly if we assume dominance and size are positively correlated. Adult males with large bills dominate those with small bills (Chapter 6), so we may be on fairly stable ground. We chose to determine the relative size of birds when they were first captured as juveniles because at this time they interact frequently with each other as they begin to establish their position in the social hierarchy of the flock (Chapter 6). It seems reasonable to suggest that relative size of a juvenile might influence its ability to compete with other juveniles and therefore influence its status and future options within the flock. Our predictions can be modified in terms of size (measured as mass in grams) as follows: the largest yearlings each year will be most likely to breed, the smallest will most probably help or disperse, and those intermediate in size will be likely to float.

Availability of breeding opportunities appears influential in the decision to breed or help. As with the decision to disperse, two observations point to the importance of mate availability. First, a yearling's sex influences whether it will help or breed. Most (71%) yearling breeders are females, but only one helper was a female. Second, there is considerable year-to-year variation in the proportion of yearlings pursuing each option (Fig. 96) which is closely correlated with the flock's sex ratio. In general, when the flock has an excess of males, more males help and more females breed. This relationship makes sense because an abundance of males means better breeding opportunities for young females, but poorer opportunities for young males. However, when we look more carefully at this relationship, it appears that yearling males and yearling females are not equally influenced by surplus males. Yearling males are more apt to help when yearling males outnumber yearling females (Fig. 97A). Yearling females, however, are most apt to breed when two- and three-year-old males outnumber two- and three-year-old females (Fig. 97B).

Why should the sex ratio of yearlings influence males but the sex ratio of older birds influence females? Perhaps different processes influence yearling males and yearling females. Few male yearlings breed, thus most may be competing with other yearling males for mating opportunities. When male yearlings are abundant, competition for the rarer females should increase, and as a result more males may opt to avoid competition and instead help their parents. In contrast to males, breeding by yearling females is fairly common (see Chapter 7, Fig. 46). Yearling females commonly mate with two- or three-year-old males and as we have just seen, they appear to disperse actively to flocks with an abundance of these older males (Fig. 95B). Yearling females may be competing primarily with slightly older females for the ability to breed with two- and three-year-old males. As the abundance of the older females relative to the older males declines, then more yearling females are able to breed. As long as there are enough older females to breed with the older males, few yearling females will breed. No wonder yearling females disperse to flocks with surpluses of older males.

Our data are consistent with the hypothesis that a juvenile male's relative size, and presumably its status, influences its decision to breed, help, or float (Fig. 98). Given the many differences between families with helpers and those without helpers (Chapter 9) we tested our hypotheses within H-lineage and NH-lineage families separately (Fig. 98A *versus* 98B). Within H-lineage families, male breeders were larger than 86% of their potential competitors. Male floaters were larger than 79% of their competitors. Male helpers were smaller than breeders and floaters as they were larger than only 71% of their competitors. If we lump all the H-lineage males that did not help (N = 13 breeders, floaters, and those known not to help but not whether they bred or floated), then we find that "non-helpers" were larger than 87% of their competitors. This is statistically greater than the relative size of helpers reported above. Outsizing 70% of the competition may make helpers seem large, but we include all juveniles (males and females) in the pool of competitors. Males are

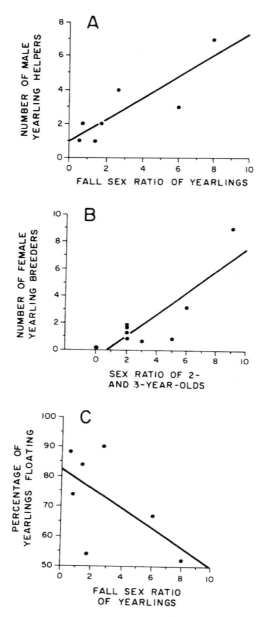

Figure 97   *Influence of sex ratio on options pursued by yearling Pinyon Jays. (A) The number of yearling helpers increased as the fall sex ratio of yearlings becomes more male-biased (y = 0.94 + 0.64x, r = 0.86, P < 0.05). (B) The number of female breeders increases as sex ratios of two- and three-year-olds (their usual mates) becomes more male biased (y = − 0.54 + 0.81x, r = 0.83, P < 0.05). (C) The percentage of yearlings floating declines as fall sex ratios of yearlings become more male biased (y = 82.6 − 3.3x, r = 0.60, P < 0.05). Lines are least-squares regressions.*

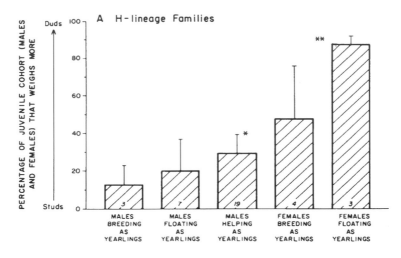

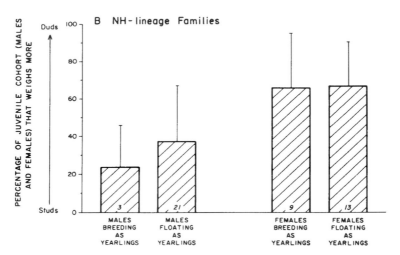

Figure 98    *Influences of a yearling's relative size on the options it pursues. The vertical axis indi-*
*cates the relative studliness of yearlings as measured by their size. Relatively large yearlings are*
*defined as studs and are low on the vertical axis because few other yearlings weigh more than they do.*
*H-lineage yearlings (A) and NH-lineage yearlings (B) pursue different options and are plotted*
*separately. Bars indicate the average relative size of yearlings in each category. Narrow lines indicate*
*one standard deviation above the mean. Sample sizes are at the base of bars.*

larger than females so even small males should be larger than about half of the
pool (likewise, even large females should be smaller than about half). These
results suggest that, relative to other juveniles within a year, males that breed
are bigger than those that float, which in turn are bigger than those that help.

H-lineage females show a similar trend. Females that bred as yearlings were
larger than 52% of their competitors whereas females that floated were only

larger than 12% of their competitors. Big females breed early in life! We expected this for two reasons: their greater competitive ability and their greater sex appeal (Chapter 7). We do not know the relative size of dispersers, but juveniles that disappeared (dispersed or died) were relatively small (only larger than 41% of juveniles). The fact that disappearing jays are larger (and presumably more dominant) than floating jays suggests that, for females, dispersal may be a better option than floating.

Non-helper lineage yearlings showed a similar, but less pronounced influence of size on their decision to breed or float. Male breeders were slightly more "studly" than male floaters. However, female breeders and floaters were of similar size (Fig. 98B). Although these data are slim, they suggest that competition may be stronger among yearling males than among females. This is consistent with our finding that males are more common than females.

Competition among yearlings should vary according to the bias in the flock's sex ratio. In Figure 98 we included yearling size in years of male bias and in years of female bias. This may have obscured some of the influence of competitive ability on yearling options. In Figure 99 we compare the relative size of yearlings pursuing various options when the flock is male-biased *versus* when it is female-biased. When the flock was female-biased, male floaters and male helpers were relatively smaller (less competitive) than when the flock was male-biased. This is exactly what we would expect if competition for mates drives the male yearlings' options. Competition should be relaxed when females are abundant, therefore allowing less studly males to breed and leaving only the "ultra-duds" to float and help. Conversely, competition should be

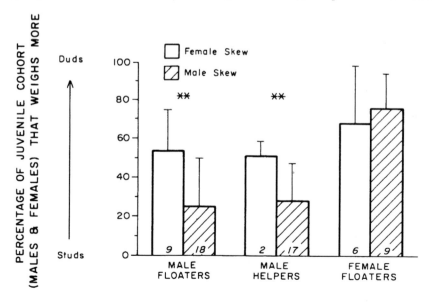

Figure 99  *Influence of flock sex ratio on the relative size of yearling floaters and helpers. Vertical axis and legend is explained in Figure 98. Stars indicate that yearling size differs according to the flock's sex ratio (** = P < 0.01).*

strongest when flocks are male-biased thus forcing many studs (as well as duds) to float and help. Unfortunately, we lack critical information on the influence of sex ratios on the size of male yearling breeders because so few breed in years of male-bias. Sex ratio appeared to have less influence on female yearling options again suggesting that yearling males are influenced by mate shortages more than yearling females.

In summary, our available evidence supports the idea that mate availability and status are important determinants of the option a yearling will pursue. We propose that yearlings strive to breed, but due to mate shortages, subordinates rarely achieve this goal. Many subordinate females disperse to flocks with better breeding opportunities and the rest join subordinate males in a non-breeding yearling flock. A few very subordinate males appear to remove themselves from the stress of competing in the non-breeding flock and seek refuge as helpers under their parents' protective care. Viewed this way the act of helping takes on a new function, perhaps it is a form of payment that yearlings offer in exchange for extended parental care.

CHAPTER 12

# A matter of life and death

We saw in Chapter 10 that life span is a major determinant of a jay's evolutionary fitness. The longer a jay lives the more young it contributes to future generations and the better its genes fare in the game of natural selection. The importance of lifespan to lifetime reproductive success demands that we understand why some jays live longer than others before we can fully understand the pressure natural selection imposes on Pinyon Jays. To that end, we begin this chapter by searching for social and environmental correlates of survivorship.

Survivorship is not only a key to individual fitness, it also plays a key role in determining the vigor of populations. Populations change in size when animals give birth, immigrate, emigrate, or die. We have already discussed the first three processes and here investigate the role of mortality as a force in

Pinyon Jay population dynamics. We first combine birth and death schedules into a life table as a summary of our population's health and as a predictor of strategies individuals might pursue. Lastly, we use key-factor analysis to identify the contribution of various mortality sources of fluctuations in population size and population regulation.

SURVIVORSHIP

*Methodology*

Our data on survivorship of Pinyon Jays come from censuses of marked birds in the Town Flock. Observers recorded the presence or absence of each jay every day that the flock fed at our feeders. These "roll call" counts were summarized each month into a list of birds present and birds absent. We produced an annual tabulation of survivorship by counting the number of birds alive during the last week of December and the first week of January each year. This provided a two-week census period to ensure that we knew a jay's status. In order to calculate survivorship we simply divided the number of birds surviving through a year by the number known to be alive during the previous year's census. Multiplication of this fraction by 100 gives the annual percentage of jays surviving. Survivorship is based on counts at the end of each year because flock membership is most stable at this time (fall dispersal is complete, spring dispersal and breeding have not begun) and because the flock regularly visits our feeders at this time of the year (visits are more erratic during the fall seed harvest and spring breeding season).

There is a cost to using a winter census. Most jays are born in March and April therefore juveniles have lived at most nine to ten months when we determine their first "year" survivorship. For simplicity we will ignore this and speak of birds as though they were born in late December. Accordingly, between birth and their first census we will call jays juveniles. Juveniles that live beyond their first census are called yearlings throughout the following year. Yearlings that survive the year and live beyond their second census are called two-year-olds throughout the following year. Surviving two-year-olds are called three-year-olds, and so on. All birds two or more years old are also collectively referred to as adults.

Dispersal disrupts calculations of survivorship. When a bird disappears from our flock we assume it has perished, yet we rarely find dead jays. Many of the young birds that disappear from the Town Flock presumably immigrate into other flocks and, because we did not regularly census surrounding flocks, elude our detection. Therefore our estimates of death are surely overestimates, at least for juveniles and yearlings, the primary dispersers. Dispersal is rare among adults so our estimates of adult survivorship should be more realistic.

We followed the survivorship of 708 jays color-banded as crechlings from 1973 to 1986. These jays represent birds from the 13 cohorts born from 1972 to 1984. We did not obtain complete censuses of crechlings in 1972 nor in 1983, so we have 11 years of data on juvenile survival. We have 13 years of yearling survival, but each older age class takes one more year to mature leaving us with 12 years of two-year-old survivorship, 11 years of three-year-old survivorship, and so on.

### The seasonal occurrence of mortality

Mortality is not equal throughout the year. It is most frequent during the spring and autumn months suggesting that breeding and havesting pine seeds are costly activities (Fig. 100). In contrast to our warm-blooded suspicions,

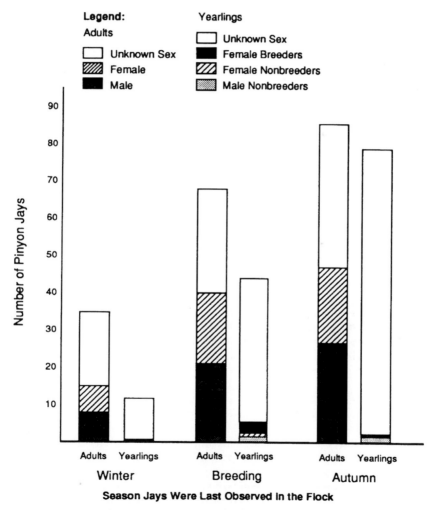

Figure 100   *Seasonality of mortality for adult and yearling Pinyon Jays in the Town Flock. Yearlings and adults differ in their seasonality of mortality ($\chi^2$ comparing losses in each season between adults and yearlings = 7.59, 2DF, P < 0.05). Winter is December–March. Breeding is April–July. Autumn is August–November.*

mortality is rarest during the cold and snowy winter months. The few recoveries of dead jays also point to the rigors of breeding. Incubating jays in Arizona and New Mexico have been killed by Great-horned Owls. In 1982, eight incubating Town Flock females were killed by a trigger-happy teenager. One jay was hit by a car while provisioning his mate and brood. Another was killed while mobbing a Northern Goshawk intent on eating crechling jays. Some of the autumn mortality may also reflect the accumulation of hardships during breeding. Jays may enter the fall in a weakened state after breeding and molting and then die as they range widely in search of ripe pinyon pine crops. Travel into unfamiliar terrain during the autumn may also directly increase mortality as foraging jays may be unaware of predators' hiding places. Hawks congregate in the pinyon-juniper woodland for migration during the autumn making jays likely to encounter many keen-eyed predators as they harvest, transport, and cache pine seeds.

The seasons do not pose the same problems to young and old jays (Fig. 100). Adults have greater winter and breeding season mortality than yearlings. Yearlings have greater autumn mortality than adults. Apparent differences in autumn mortality may, however, reflect differences in dispersal more than differences in mortality as autumn is the usual time for juvenile and yearling dispersal (Chapter 11). Breeding season differences, however, are probably due to real differences in mortality. Some yearlings disperse during the breeding season which would cause us to overestimate their mortality. The fact that adults still have greater mortality, despite yearling dispersal, suggests that yearlings fare better than adults over the breeding season. This is to be expected if breeding is costly because nearly every adult breeds, but most yearlings do not (Fig. 46).

Adult males and adult females have nearly identical seasonal occurrences of mortality (Fig. 100). Adult males have slightly greater autumn and slightly less winter mortality than females. Despite our suspicions that breeding is more costly for females than for males (because females do all the incubation and brooding) both sexes suffer equal mortality during the breeding season (39% of adult male mortality and 40% of adult female mortality). Males' costs of provisioning their mates and males' more blatant harassment of predators may bring their costs in line with females' costs of forming, laying, and incubating eggs.

*The influence of age on survivorship*

Very few eggs become free-flying Pinyon Jays. We present a standard way of graphing survival, known as a survivorship curve, in Fig. 101. This curve begins with 1000 eggs and uses our best estimates of survival in the Town Flock to predict how these eggs will fare as they age. From our calculations in Chapter 9 (Appendix 2) we expect a little more than half (55%) of the eggs will hatch, about the same proportion (56%) of nestlings will fledge, and about a third (32%) of the fledglings will survive to be color-banded as crechlings. Our survivorship curve therefore drops precipitously from 1000 eggs to 98 crechlings. In other words, only 10% of eggs become semi-independent crechlings.

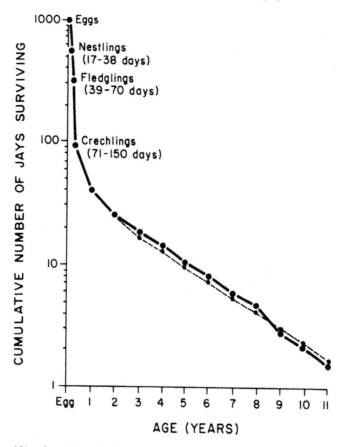

Figure 101   *Composite survivorship curve for the Town Flock. Calculations for survival of eggs, nestlings, and fledglings are based on data from 1981 to 1987. Calculations for survival of crechlings are based on data from 1974–1976 and 1981–1985. Calculations for survival of yearling and older birds are based on birds of known age born from 1972–1984 living over the period 1973–1986. The dashed line indicates expected survival if adult mortality was a constant 26% per year.*

Mortality continues at a high rate during a jay's first few years of life. Averaging each year's survival we calculate that 41.3% of juveniles survive their first nine to ten months and 62.1% of yearlings survive the next year. Adults fare much better as each year an average of 73.9% of jays two or more years old survive. Despite high adult survivorship, heavy mortality early in life assures that most jays will not survive into their golden years. Our sample of 1000 eggs is expected to yield only one 11-year-old jay! We have not observed a bird banded as a crechling survive beyond eleven years. However we know this is not the maximum age attained by wild Pinyon Jays. We followed the survival of 77 jays banded as adults in 1972. We know that these birds were at least two years old in 1972 and eight survived to be at least 11 years old in 1981. The longest-lived Pinyon Jay in the Town Flock was Dark Green-Red, a male who died in 1986 when he was at least 16. This is the same male that produced 23 crechlings and 10 yearlings as discussed in Chapter 10. The oldest female (Purple-Purple) died in 1985 when she was 14 (she was a yearling when we first banded her in 1972).

Adult Pinyon Jays have relatively constant survivorship regardless of their actual age. Using the average survivorship of 74% for jays aged two to ten years produces a survivorship curve that differs little from a curve based on the actual age-specific survival rates (Fig. 101). Among adults, six-year-olds and eight-year-olds have the lowest survivorship (69.8% and 60.9% respectively) while seven-year-olds have the highest survivorship (82.3%). The average survivorship of all other age classes is very constant, ranging only from 73.4% to 77.7%.

Constant adult survivorship suggests that senescence, or a general deterioration of life support systems among elderly organisms, does not occur in Pinyon Jays. With the possible exception of females banded as crechlings, we have been unable to document the sudden drop off in survivorship predicted if senescence was important (Fig. 102). For unknown reasons four of five seven-year-old females died, suggesting female senescence. However, this is more likely to be a chance event associated with our small sample of seven-year-olds than a general phenomenon. Support for this claim comes from our sample of females initially banded as adults in 1972. They exhibited constant survivorship over the following eight years (Fig. 102). Lack of senescence is also suggested by comparing the survival of known-aged adults to survival of birds banded initially as adults. Birds banded as adults of variable age surely include some very old individuals. If senescence occurs they are expected to have lower average survival than adults of known age (Woolfenden and Fitzpatrick 1984). Our sample of unknown-age adults actually survived slightly better than known-age adults (Fig. 102).

## Annual variation in survivorship

Year-to-year variation in survivorshp is pronounced (Appendix 3). Here we attempt to account for this variation by examining correlations between survivorship and environmental factors. This is the same approach we used in Chapter 9 to investigate annual variation in the productivity of breeders. Our findings in Chapter 9 are relevant here because survivorship of young jays was one of our measures of productivity. Moreoever, we might expect that years good for reproduction are also good for survivorship of breeders as is the case for Acorn Woodpeckers, another seed-storing, co-operative species (Koenig and Mumme 1987).

Breeding performance and breeder survival are not positively linked in Pinyon Jays (Fig. 103). If any trend is evident it is that years favorable for reproduction are poor for survival. This negative relationship occurs because cold, snowy springs wipe out many nests, thus lowering annual productivity, but such conditions do not kill adult jays. In fact, breeder survival is highest following moist springs, even if precipitation comes in the form of snow rather

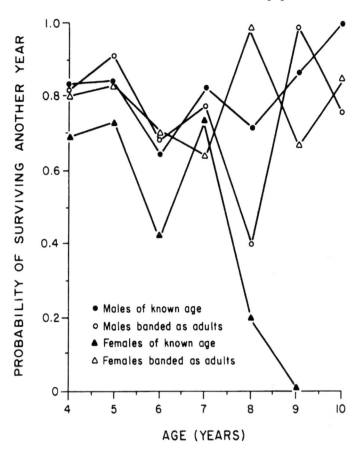

Figure 102  *Changes in the probability of survivorship as adults enter their golden years. If senescence occurs in Pinyon Jays we expect these probabilities to get lower and lower as jays age. (Sample sizes:* • *N = 65;* ○ *N = 26;* ▲ *N = 45;* △ *N = 30).*

than rain (Table 42). Adult survival also is not enhanced by large pinyon seed crops. Instead, believe it or not, survival is greatest when cone crops are of intermediate size (Fig. 104). This is a complicated relationship that we will return to later. What is important to our point here is that, unlike breeder survival, productivity of breeders is greatest following bumper pine crops. In Chapter 9 we found that eggs and nestlings fare best when reared after plentiful seed harvests. Our sample of nests suggests this benefit did not carry over to survival of independent young. However, our larger sample of cohorts strongly suggests that juvenile survival is augmented by abundant pine seeds (note the significant partial correlation between pinyon crop and juvenile survival in Table 42). One other factor also acts to nullify any positive association between breeder survival and productivity. Urban sprawl and the associated explosion of Common Raven and American Crow populations in Flagstaff

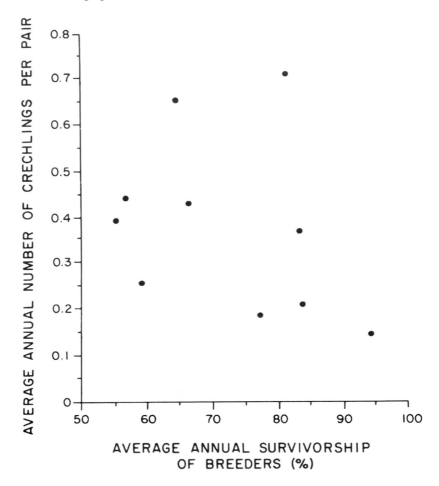

Figure 103    *Relationship between the survivorship and productivity of breeders. No significant relationship is evident indicating that years good for breeding are not necessarily good for living (r = −0.31, N = 10, P > 0.10).*

have wreaked havoc on breeding productivity but, as these predators only eat eggs and nestlings, they do not affect breeder survival.

Let's take a closer look at the strange relationship between adult survivorship and the size of the pinyon pine seed crop (Fig. 104). Survivorship is low when pine crops are very small as expected. However, it is also low when crops are very large. We can only guess that increased activity associated with harvesting and transporting seeds plus exposure to predators in unfamiliar habitat reduces adult survivorship in years with bumper crops. This is supported by the seasonality of mortality; most adults die in the autumn (Fig. 100). Concentration of activity in a localized area such as a patch of pinyon woodland bearing abundant cones, may be the key to high mortality.

**Table 42.**    *Important correlates of survivorship in various age classes of Pinyon Jays in the Town Flock. Significant correlations follow each factor in parentheses.*

| Measure of Survivorship | Number of Years Studied (N) | Significant Correlates |
|---|---|---|
| Percentage of Juveniles surviving first year | 11 | Fall pinyon crop (+0.58)[1]* |
| | | Spring temperature (+0.86)[2]** |
| | | Spring precipitation (+0.85)[3]** |
| | | Previous spring precipitation (−0.59)* |
| | | Previous summer precipitation (+0.84)** |
| Percentage of Yearlings surviving second year | 12 | Spring temperature (+0.62)* |
| | | Spring precipitation (+0.76)** |
| | | Previous spring snowfall (−0.79)** |
| | | Previous winter snowfall (−0.52)* |
| | | Previous year's pinyon crop (+0.46)[4] |
| Percentage of Adults surviving one year | 12 | Spring precipitation (+0.55)* |
| | | Spring temperature (−0.54)* |
| | | Previous spring snowfall (−0.54)* |

\* $P < 0.05$
\*\* $P < 0.01$
[1] *Partial correlation holding spring precipitation and spring temperature constant*
[2] *Partial correlation holding spring precipitation constant*
[3] *Partial correlation holding spring temperature constant*
[4] $P = 0.06$

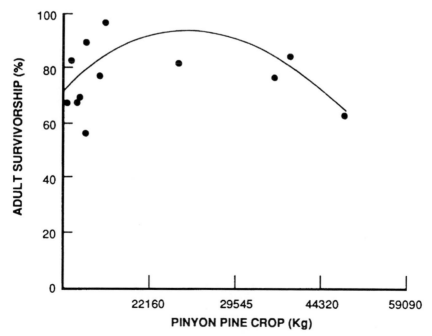

Figure 104    *Adult survivorship as a function of the size of the fall pinyon pine crop. Survival is highest when crops are of intermediate size. Solid line is the best fitting curve to these data. From Marzluff and Balda (1988a).*

Jays predictably return to these areas to harvest seeds and use the same flyways to travel between their caching grounds and the source of seed. Cooper's Hawk populations are abundant in the pinyon-juniper woodland at this time and they may concentrate in harvesting areas, caching areas, and even along travel corridors connecting them. Although few Pinyon Jays are taken by hawks, a bird with a throatful of seeds is probably more vulnerable to predators because of reduced maneuverability. Predators may also have an easier time catching jays during bumper crop years because birds mentally, as well as physically, concentrate on the harvest. Few sentries surround the harvesting flock as all birds are intent on getting their share of the harvest. In years of modest cone crops, birds may not be as susceptible to predation because they take a variety of routes back and forth from caching to harvesting areas. Fall may be a time of bounty for seed-eaters and cachers but in nature no feast is free.

Weather factors have the strongest influences on survivorship of young age classes in the Town Flock. Juveniles experience their highest survivorship when they are born after warm, wet springs. As we suggested in Chapter 5, cold and snowy springs are detrimental to juveniles. Partial correlation

*A raven soars low over the pinyon-juniper country searching for a jay nest to prey upon.*

analyses of juvenile survivorship support this claim. Spring temperature and precipitation are not significantly correlated with juvenile survivorship until we hold one constant and examine the effect of the other. Strong partial correlations indicate that springs that are either warm or wet are not beneficial to survival. Springs that are both warm and wet are the key to high juvenile survival, presumably because these conditions create a summer flush of insects that parents can stuff into growing young jays. The overriding importance of weather to the survival of juvenile Pinyon Jays in Flagstaff is also revealed by partial correlation analysis. The importance of pinyon crop is not evident until we statistically remove the masking effects of spring weather. Large pinyon crops are associated with slightly higher juvenile survivorship than are small crops (correlation between pinyon crop and juvenile survivorship: $r = 0.31$, P $= 0.17$). However, if we remove the influence of spring temperature and precipitation the benefits of large pine crops are more obvious (the correlation rises to 0.58 and is significant at $P = 0.05$, Table 42). Large pinyon crops enhance juvenile survivorship, but the full benefit is rarely realized in the cold climate of northern Arizona. We believe that these relationships are due to our flock living at the upper elevational edge of the species' range.

The survivorship of yearlings is also strongly correlated with spring conditions (Table 42). As with juveniles, yearlings survive best over warm, wet springs. Interestingly, environmental conditions when yearlings were juveniles continue to have strong correlations with survival. Juveniles reared in lean snow years survive best as yearlings. Additionally, juveniles reared after large pinyon crops continue to have marginally higher ($P = 0.06$, Table 42) survival as yearlings than juveniles reared without the benefit of abundant seeds.

As expected, adults are least influenced by the vagaries of weather (Table 42). Their high dominance status in the flock ensures that they will be the last to suffer any hard times. They survive best if springs are wet, even if they are cold, though this relationship is anything but impressive as it only accounts for one quarter of the annual variation in their survivorship. As we suggested above, the predictability of their movements in the fall may account for an additional fraction of the year-to-year variation in adult survival.

*Sex and survivorship*

Male Pinyon Jays have greater longevity than females. It is difficult for us to pinpoint exactly when males begin to outlive females because we typically must observe breeding behavior before we can ascertain a jay's sex. This usually does not occur until a bird is two years old. This biases our sex-specific estimates of survival upward for the first few years of a jay's life (most of the jays that are of known sex must have lived two years and bred). However, it does not preclude comparisons of survival between the sexes as we must wait to identify males as well as females. Survival rate of the few yearlings we sexed suggests that high mortality of young female breeders initiates sex differences in survivorship (Fig. 105). All 13 male yearling breeders lived to be two years old, but eight of the 28 female yearling breeders died after their initial year of

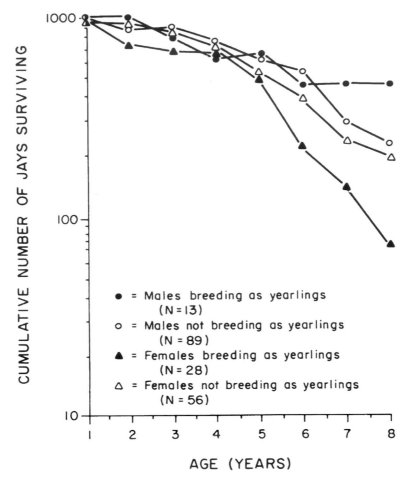

Figure 105   *Survivorship curves for yearlings of known sex and breeding experience. Curves are based on N = 13 male yearling breeders, 89 males delaying breeding beyond their yearling year, 28 female yearling breeders, and 56 females delaying breeding.*

breeding. Non-breeding male and female yearlings had survivorship on a par with breeding males. Survivorship of breeding female yearlings was significantly lower than survivorship of breeding males and non-breeders (chi-square test comparing number living *versus* number dying between first and second year: $\chi^2 = 21.4$, DF = 2, P < 0.001). This initial male advantage is maintained among adults. Males continue to outlive females until they are seven years old (Fig. 106). Eight-year-old females survive better than eight-year-old males, but this advantage is slight (and likely to be a result of small sample size) as older males have annual survival rates 25% higher than females.

Why do males live longer than females? Differential investment in repro-

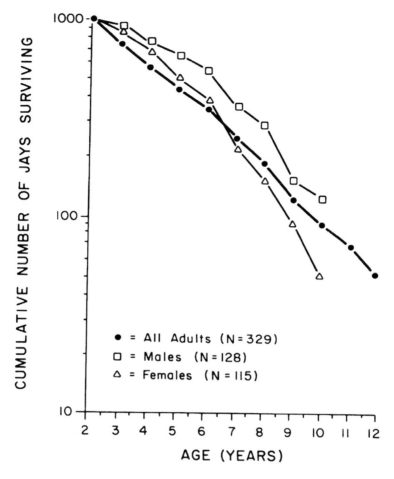

Figure 106  *Survivorship curves for adults of known and unknown sex. Curves are based on N =
329 jays of known and unknown sex (solid line), 128 males, and 115 females.*

duction and the importance of sex to a jay's status in the flock (Chapter 6) are
likely to act together to tip the scales of survivorship in favor of males. As we
argued above, the extra energetic drain on breeding females associated with
egg production is unlikely to account for their low survivorship (see also Breitwisch 1989). Probably of more importance is the fact that females are less
mobile during breeding because they must incubate their eggs and brood their
nestlings. While on the nest they are literally "sitting ducks" for predators,
especially keen-eyed Great-horned Owls and armed teenagers! Young females
would be especially vulnerable as their nests are more exposed than those of
older females (Chapter 8). Males would be less susceptible to predation
because they remain in mobile flocks while outside the nesting colony and

roost communally. Any sex-specific differences in the costs of breeding appear to be imposed most severely on yearling breeders. Adult females probably suffer fewer costs of breeding as they are no more likely to die during the breeding season than are adult males. Lower dominance rank of females (especially yearling females) relative to males may explain why females continue to have low survivorship throughout adulthood. The largest difference in adult male and female survivorship occurs during the winter, when resources are likely to be in shortest supply and any competitive advantages accrued from dominance and body size would be most influential on survival.

Regardless of its cause, lower female survival has the important consequence of introducing a male skew or bias into the flock's sex ratio. As we argued in Chapters 7 and 11, this skew and the variability imposed on it by annual fluctuations in male and female survival has important ramifications for the options that young jays pursue and it may enable females to enforce monogamy.

## THE LIFE TABLE

A life table is an age-specific listing of the survivorship probabilities (lx) and reproductive productivities (mx) of individuals in a population. We combined the information on survivorship presented in the previous section with our estimates of fecundity from Chapter 9 (Table 25) into life tables for the Town Flock (Table 43). Traditionally, life tables are constructed for females only. However, because male and female Pinyon Jays differ in their survivorship as well as their fecundity, we thought it would be best to present a separate life table for each sex. Additionally, we present a combined male and female table because we can increase our sample of birds if we include those of unknown sex. All three tables give similar conclusions and are best used to indicate a range of likely values for each parameter we calculate.

If you have ever purchased life insurance you have been influenced by a life table. Insurance agents rely upon human survivorship probabilities to calculate an individual's expectation of future life. Their premiums reflect these expectations. For example, premiums are higher for males than for females and they increase as you age because mortality is higher for males than for females and increases as senescence takes hold. We will use life tables to calculate similar statistics for Pinyon Jays. In addition, we will use life tables to estimate our population's vigor.

### An insurance agent's view of Pinyon Jays

Life insurance for Pinyon Jays would be really expensive because most jays are expected to die soon after purchasing their policy (Table 44). Rates would be particularly steep for crechlings because they can expect to live only about two and a half additional years. Beyond this age, rates would remain fairly constant. One-year-olds, two-year-olds, and older jays can expect to live about four more years. However, females would pay higher premiums than males

**Table 43.**   *Life table entries for the Town Flock. Thirteen cohorts born from 1972 to 1984 were followed for 16 years (1972–1986). Survivorship is calculated by averaging the annual survival of known-aged birds across the years of study. Only birds whose exact age was known were used to calculate survival of crechling to four-year-olds. Survival of older birds includes birds of known exact age and birds banded as adults who are known to be at least as old as the birds in the exact age sample. Average survival is used to calculate the table entry (lx) which equals the cumulative proportion of birds alive at each age. Sex-specific survivorship is not available until birds are two years old. Raw data and sample sizes for survival are presented in Figure 106. Fecundity (mx) is a measure of age-specific reproduction. Mx equals the average production of crechlings by known-aged breeders (see Table 25 for raw data) divided in half (we assume a 50:50 sex ratio of crechlings) and then multiplied by the age-specific probability of breeding (see Figure 46 for raw data). Mx can be interpreted as the number of offspring that an average jay of a given age replaces itself with in the population. Due to small samples we used the average survival of all birds nine or more years old to calculate the proportion of nine-year-old and older birds surviving each year. We also averaged the fecundity of seven-year-old and older jays as an estimate of fecundity by old jays. We terminated the life tables at the maximum observed age of females (14 years) and males (16 years).*

| Age | Females | | Males | | Females and Males | |
|---|---|---|---|---|---|---|
| | lx | mx | lx | mx | lx | mx |
| crechling | 1.000 | 0.000 | 1.000 | 0.000 | 1.000 | 0.000 |
| 1 year | 0.413 | 0.179 | 0.413 | 0.058 | 0.413 | 0.115 |
| 2 year | 0.256 | 0.485 | 0.256 | 0.421 | 0.256 | 0.541 |
| 3 year | 0.223 | 0.535 | 0.230 | 0.510 | 0.192 | 0.523 |
| 4 year | 0.176 | 0.515 | 0.198 | 0.420 | 0.145 | 0.482 |
| 5 year | 0.131 | 0.425 | 0.162 | 0.550 | 0.113 | 0.492 |
| 6 year | 0.103 | 0.395 | 0.141 | 0.305 | 0.093 | 0.342 |
| 7 year | 0.058 | 0.595 | 0.093 | 0.565 | 0.065 | 0.575 |
| 8 year | 0.040 | 0.595 | 0.074 | 0.565 | 0.049 | 0.575 |
| 9 year | 0.024 | 0.595 | 0.041 | 0.565 | 0.032 | 0.575 |
| 10 year | 0.013 | 0.595 | 0.033 | 0.565 | 0.024 | 0.575 |
| 11 year | 0.007 | 0.595 | 0.026 | 0.565 | 0.019 | 0.575 |
| 12 year | 0.004 | 0.595 | 0.021 | 0.565 | 0.013 | 0.575 |
| 13 year | 0.002 | 0.595 | 0.017 | 0.565 | 0.010 | 0.575 |
| 14 year | 0.001 | 0.595 | 0.013 | 0.565 | 0.007 | 0.575 |
| 15 year | — | — | 0.011 | 0.565 | 0.005 | 0.575 |
| 16 year | — | — | 0.009 | 0.565 | — | — |

**Table 44.**   *Town Flock vital statistics calculated from life tables (see Table 43 for raw data). Dr David McDonald of Archbold Biological Station used Leslie matrix techniques and a Macintosh Pascal program to calculate these values. The program uses equations from Caswell (1989) for Tc, $R_0$, A and expectation of further life. R is calculated by taking the natural logarithm of lambda.*

| | Lifetable calculated using: | | |
|---|---|---|---|
| Population parameter | Females | Males | Females and Males |
| Expectation of further life (years): | | | |
| of crechlings | 2.4 | 2.7 | 2.4 |
| of 1-year-olds | 3.5 | 4.2 | 3.5 |
| of 2-year-olds | 4.1 | 5.2 | 4.0 |
| Generation time (years): | | | |
| cohort (Tc) | 3.9 | 5.2 | 4.5 |
| weighted (A) | 4.7 | 6.1 | 5.7 |
| Population growth: | | | |
| net fecundity ($R_0$) | 0.59 | 0.66 | 0.57 |
| geometric (lambda) | 0.88 | 0.93 | 0.89 |
| instantaneous (r) | −0.12 | −0.07 | −0.11 |

owing to their lower annual survivorship. Each year a male can expect to live nearly a full year longer than a female.

### The generation gap

In humans it takes a mother nearly 20 years to replace herself in the population. As a result, parents and their children grow up in very different social environments and this so-called "generation gap" often hinders communication between them. Pinyon Jays are unlikely to have such problems. A Pinyon Jay generation is between four and six years long (depending on whether you weigh your estimate by the population's growth rate, Table 44). This is the time it takes the average jay to produce a surviving crechling of the same sex. The generation gap is wider between young Pinyon Jays and their fathers than between young jays and their mothers because males delay breeding longer than females (Chapter 7). Acorn Woodpeckers and Florida Scrub Jays also have generation gaps of four to six years (Woolfenden and Fitzpatrick 1984, Koenig and Mumme 1987).

### How much is a Pinyon Jay worth?

We can use our life table to determine how important the average Pinyon Jay of a given age is to the population. Expected fecundity (the product of the proportion of jays surviving to an age [lx] and their fecundity at that age [mx]) gives one measure of an individual's contribution to the flock. We can interpret this product as the reproductive output that the average jay can expect at each age of its life. Expected fecundity is highest for two- to four-year-olds because most jays survive this long and breed (Fig. 107). Yearlings have lower expected fecundity because few breed. Expected fecundity declines with age after four or five years because few jays beat the odds and live beyond this age. From the standpoint of expected fecundity we can conclude that young jays, especially those two to four years of age, are the most important because they contribute the majority of crechlings to the population each year. Young females, especially yearlings, contribute more to the population than young males because females typically reproduce at an earlier age than males (Fig. 107). The expected fecundity curves we have presented are identical in shape to those derived from another long-lived, co-operative species, the Acorn Woodpecker (Koenig and Mumme 1987). In contrast, older individuals contribute more to populations characterized by extremely delayed reproduction such as in Florida Scrub Jays and Yellow-eyed Penguins (Woolfenden and Fitzpatrick 1984, Ricklefs 1973).

A similar picture of a jay's contribution is obtained by quantifying its reproductive value (Fig. 108). Reproductive value has been used as a measure of the evolutionary importance of individuals since 1930 when Sir Ronald Fisher (the same genius we discussed in Chapter 7) devised it. Reproductive value is the sum of an individual's current fecundity (mx) and its expected future fecundity $(\sum_{t=x+1}^{t=\infty} lt/lx \times mt)$. Our favorite way to think of reproductive value is to

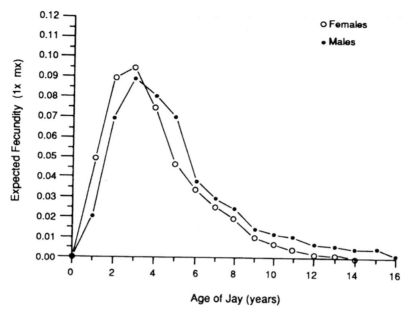

Figure 107   *Expected fecundity of males and females. High probability of survival and successful reproduction lead to high expected fecundity. Raw data are given in Table 43.*

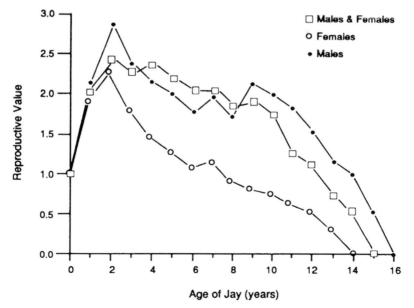

Figure 108   *Reproductive value curves for Pinyon Jays in the Town Flock. These data were calculated by Dr David McDonald of Archbold Biological Station using Fisher's (1930) formula. Curves of males, females, and the combined sexes are plotted separately.*

look at the consequences of removing an individual from a population. Reproductive value measures the relative reduction in future population growth caused by removing an individual of a given age (Ricklefs 1979). The higher an individual's reproductive value, the greater will be the impact of its removal from the population. Reproductive value peaks near the age of first reproduction because an individual's expectation of future productivity is greatest at this age (they have their entire reproductive lives ahead of them). Before this age, reproductive value is lower because death claims many individuals before they can reproduce. After reproduction begins, reproductive value declines as more and more of an individual's reproductive life is used up.

Two-year-old Pinyon Jays have the highest reproductive value (Fig. 108). This is a reflection of the usual onset of reproduction at age two and a high probability of three to five more years of productive breeding. Greater male survivorship makes males slightly more valuable than females. In general, both measures of a jay's worth give the same answer: one- to four-year-old Pinyon Jays are the most valuable to a population.

Reproductive value can also be used to suggest with whom individuals should cooperate. Rubenstein (1982) makes two important points in his summary of recent work on this topic. First, in order to assure the most genetic return for altruistic acts animals should aid relatives who have high reproductive values. Second, animals should avoid risks when their reproductive values are high by withholding potentially risky altruism until their reproductive values are low and declining.

There is evidence that some animals behave in these predicted ways. In the Rhesus macaque monkey, for example, mothers support daughters that are approaching sexual maturity (those nearing peak reproductive value) more than they support older daughters (those with declining reproductive values, Schulman and Chapais 1980). Young Florida Scrub Jays remain in the protection of their natal territories thereby avoiding risks until they reach peak reproductive value (Woolfenden and Fitzpatrick 1984).

In Pinyon Jays we predict that one- to three-year-olds should be especially conservative in their actions. These jays have very high reproductive values which should not be jeopardized by actions that might reduce their survival. Perhaps this is one reason that only the largest (and presumably the most dominant) yearlings breed; the costs of breeding are too risky for less competitive yearlings. The latter may float in the yearling flock or help their parents because these options are less risky and function to increase their chances of surviving to breed.

Let's re-examine the decision made by some birds to help, this time in light of predictions from reproductive value theory. Yearlings have higher reproductive value than either crechlings or adults aged four to six years (depending on which curve in Figure 108 (male, female, or both sexes) we follow). Clearly, if reproductive value influences behavior, then yearlings should help themselves more than they should help crechlings. Furthermore as most parents with helpers are older than four to six years, yearlings are unlikely to jeopardize their high reproductive value to aid their parents. These two points agree with our earlier conclusions (Chapter 9) that helping has evolved because it enhances the survivorship of *helpers*, not because it benefits parents or nestlings.

Our interpretation of parents facilitating their weakest young by allowing them to "help" is also supported by an analysis of reproductive value. The high reproductive value of yearlings (especially relative to crechlings) suggests that parents should aid their one-year-old offspring more than their crechlings. Therefore, the lower reproductive output that is the price parents may pay by allowing helpers at their nests (Table 34) is well worth it because older offspring with higher reproductive value (the helpers) benefit from this form of extended parental care. If possible, parents should extend their care even longer to aid their two- and three-year-old offspring as they have the highest reproductive values. We have no evidence of this and suggest that it may be unlikely because parents are unable to offer much in the way of support to older offspring. Yearlings are ripe for extended prental care because parents may still be able to supplement any of the yearlings' shortcomings and the aid given by parents will be repaid genetically because yearlings have high reproductive value. This may explain why, at our feeders, we often observed yearling females feeding next to their fathers.

We are not suggesting that parents should invest in their older offspring instead of rearing nestlings. Investing in nestlings is unlikely to lower the survivorship of a pair's older offspring with higher reproductive value. Rather, we suggest that parents should do everything in their power to continue nurturing their older offspring because such care will quickly be repaid genetically in the form of grandchildren.

All yearlings do not have equal reproductive values. Rubenstein (1982) points out that young animals have extremely variable reproductive values and our results suggest that body size and dominance may influence a yearling's reproductive value. Studs have high juvenile survival and often breed as yearlings. Duds are expected to have poor survival and rarely breed until they are two or three years old. As a result, the reproductive value curve for studs will lie above the curve for duds, at least over their first two or three years of life (Fig. 109). This may explain why parents tolerate helping by their weak sons but not by their strong sons. Parents that allow their weak sons to help are raising the reproductive values of these sons primarily by increasing their survival and perhaps also by increasing their future reproduction (note the line through triangles in Figure 109). On the other hand, if parents allowed their stud sons to help, they would lower their sons' reproductive values because helping necessitates a delay in reproduction. A flexible system where parents allow their studs to breed and shelter their duds maximizes the combined reproductive value of all a parent's sons. Parents would receive a lower genetic payback if all their sons were forced into service as helpers because the combined reproductive values of these sons would be lower than the combined values of helping duds and breeding studs.

## A population in peril?

We have used the life table as a supplementary way to interpret the behavior of individuals, but now we must turn to its more traditional use, as a model of population growth. We presented three traditional measures of population

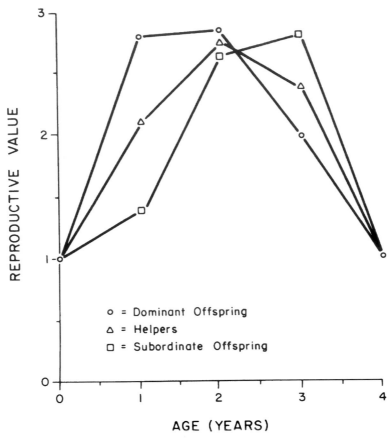

Figure 109    *Theoretical exploration of differences in reproductive value of young Pinyon Jays. Dominant yearlings have earlier and higher peak values than subordinates because dominants breed earlier and more successfully than subordinates. If parents extend their parental care by allowing yearlings to become helpers at the nest rather than breed the reproductive value of these young is intermediate. Thus, extended parental care reduces the reproductive value of dominants, but augments the reproductive value of subordinates.*

growth in Table 44. The first, and most intuitive measure of population growth is $R_0$, or the net reproductive rate. $R_0$ is the sum of the product (lx × mx) over an individual's lifetime, or the total number of offspring the average individual is expected to produce in its lifetime. We usually speak in terms of females, so using only females in the life table we calculate $R_0$ to be equal to 0.59. In other words, the average female in the Town Flock is expected to produce 0.59 females during her lifetime. That's not a good rate! If a population is to remain stable in size, then each individual must replace itself, so $R_0$ must equal 1.0. Population growth during one generation equals $R_0$ multiplied by the number of individuals alive at the start of the generation. Thus we can see that an $R_0$ of 0.59 means that the Town Flock is declining by 41% $(1 - 0.59)$ each generation.

Another measure of population growth is lambda $(\lambda)$. This is the geometric rate of population change each year and is simply the population size one year divided by the population size the previous year. If lambda equals 1, then the population is stable. If it is between 0 and 1 the population is declining. If it is greater than 1, then the population is increasing. Our calculated value of lambda is between 0.88 and 0.93 meaning that the Town Flock is declining by 7–12% each year. This rate of decline is greater than the current estimate for the threatened Northern Spotted Owl (Lande 1988).

A final measure of population growth is r, the intrinsic rate of natural increase. This gives a measure of population growth at each instant in time which is directly comparable between populations. The intrinsic rate of increase and $R_0$ are linked by the equation: r = (natural log of $R_0$) / generation time. Therefore, an r value of 0 corresponds to an $R_0$ of 1 and means the population is stable. Negative r values indicate declining populations. Positive r values indicate increasing populations. This measure also suggests that the Town Flock is in trouble as r is negative.

Our life table clearly shows that breeders in the Town Flock are not replacing themselves one for one, therefore the population will decline. But, in Chapter 5 we did not see the dramatic drop in flock size predicted by the life table statistics. What is the cause of this apparent discrepancy? The life table deals only with birth and death, yet over one third of the yearlings in the Town Flock were not born there. They immigrated into the flock (Fig. 33).

Are these immigrants able to sustain our population? If the population drops by 12% each year and we start with an average winter flock size of 140 jays, then we can expect to loose 17 jays per year. We expect seven yearlings to immigrate each year, so we have a net loss of 10 jays per year. That gives a new estimate of lambda equal to 0.93 (130 / 140) and $R_0$ = 0.67. That is closer to a stable population, but still indicates an annual population decline of 7%. If we use the larger value of $R_0$ from the male life table, then immigration balances the loss of all but three jays per year.

Could we have missed something in our calculations and overestimated the Town Flock's decline? A likely source of error is our estimate of crechling survival. We know some crechlings emigrate rather than die, so our estimate of 41.3% survival is surely low. To stabilize our population, (using the female life table and including immigration) juvenile survivorship needs to be 62% (0.413 / $R_0$, Ricklefs 1973). We cannot imagine that juvenile survivorship is that high (remember that even yearling survival is only 62%). However, it may approach 50%. It is therefore unlikely that immigration into the Town Flock is sufficient to offset completely the low productivity of jays in the flock. The fact that immigration is largely driven by mating opportunity rather than population size also suggests that it will not act to stabilize population declines (Chapter 11).

Our recent censuses indicate that the Town Flock is beginning to show a slight, but steady, decline (Fig. 110 top). Furthermore, the age composition of the flock is reflecting a drop in productivity; adults are comprising a larger and larger share of the population (Fig. 110 bottom).

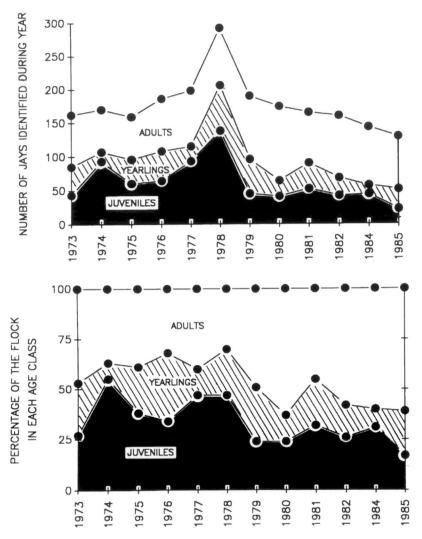

Figure 110    *Trends in the size (upper graph) and age composition (lower graph) of the Town Flock over the course of 14 years. Population sizes overestimate true flock size in any given season (see Chapter 5 for estimates within seasons) because here we plot the total number of individual birds observed in the flock anytime during the year. This enables us to supplement earlier censuses with data from recent years. No censuses were possible in 1983 because the flock rarely visited feeding stations.*

The most likely culprit responsible for the Town Flock's decline is the continual growth of Flagstaff. Massive habitat changes have eliminated some breeding grounds that the birds regularly used in the early 1970s. Nesting areas presently used may not provide all the benefits of previously used areas. People are also indirectly at the root of the Town Flock's decline. The rubbish we generate translates into crows and ravens and these birds are serious nest predators whose increasing numbers make successful reproduction by Pinyon Jays more and more difficult (see also Chapter 9).

The Town Flock may be making one last desperate attempt to survive. Starting in 1986, the flock shifted its foraging and nesting areas to the very edge of Flagstaff where human interference is minimal. Maybe this move will eventually result in a population increase. However, it will not take long for crows and ravens to discover these new nesting areas. In the spring of 1990 almost no crechlings were present in the Town Flock. Like many animals, Pinyon Jays may be unable to co-exist with humans.

*American Crows and Common Ravens forage on rubbish at a dump. Large numbers of these scavengers survive times of food shortage by relying on humanity's refuse. (J. Marzluff).*

## POPULATION REGULATION

In the previous sections we have documented the survivorship of individuals, identified when mortality is most likely, suggested which environmental attributes are most influential on mortality, and examined the life table as a means of predicting individual behavior as well as understanding population dynamics. We will conclude this chapter by investigating how mortality may act as a regulatory process in the Town Flock. Here we are not just interested in whether our population is growing, stable, or declining. Instead, we want to determine the stage of life where mortality is most likely to cause annual fluctuations in population size. To this end we employ two techniques known as key-factor analysis (Begon *et al.* 1986) and elasticity analysis (Caswell 1989).

Key-factor analysis quantifies the strength of mortality at any number of stages in a species' life cycle. For our purposes we have quantified mortality at several stages from eggs to six-year-olds (Appendix 4). A k-value (k here refers to killing power at a given stage) is calculated by simple subtraction: $k_n$ = (log number of individuals at start of stage n) − (log number of individuals at end of stage n). Big differences, and therefore large k-values, result when mortality during a stage is great. The average k-values we present in Appendix 4 allow us to identify the stages where mortality is greatest. These values only reinforce what we already knew, namely, that mortality is greatest when fledglings first leave the nest and continues to be high in juveniles during their first year. Additionally, we are reminded that during nesting predation on eggs and nestlings is the major source of mortality, but this pales in comparison to the loss of fledglings during their first months of free flight. One last suspicion is confirmed. Many jays die as six-year-olds, making this age a major contributor to total mortality.

Elasticity analysis also suggests that the survivorship of young jays is the primary factor influencing the dynamics of the Town Flock. Elasticity measures the influence of age-specific survivorship and fecundity on lambda, the geometric rate of population growth (Caswell 1989, p. 134). Elasticities thus enable us to quantify the relative importance of mortality and fecundity of each age class on population growth.

We plot the elasticities derived from the combined male and female life table in Figure 111. Three results are particularly striking in this figure. First,

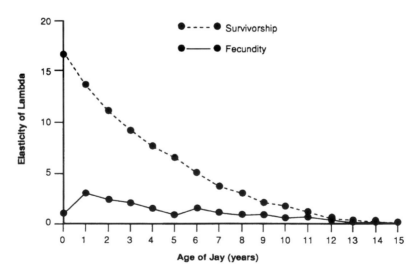

Figure 111    *The elasticities of lambda with respect to changes in age-specific fecundity (solid line) and survivorship (dashed line). Elasticities are reported as percentages and can be interpreted as the percentage of lambda accounted for by each measure of fecundity or survival. Survival of crechlings, for example, accounts for 16.7% of lambda and it therefore has the single greatest contribution to population growth in the Town Flock.*

survivorship of crechlings, one-year-olds, and two-year-olds has the greatest influence on population growth. Second, fecundity has a small influence on population growth relative to the large influence of survivorship. Third, the fecundity of young jays is more influential on population growth than the fecundity of older jays. It should be noted that survivorship elasticities usually decline with age. Therefore, if our life table started with eggs rather than with crechlings then egg survivorship would be the primary force in population growth. The important point brought out by these analyses is that young jays are the key to population growth. Their survival and reproduction are paramount to population productivity.

We begin to learn new things with key-factor analysis when we search for key factors. These are the mortality sources that correlate most closely with annual fluctuations in total mortality. They are the forces behind annual changes in mortality, causing year-to-year changes in population size.

Several key factors are evident in Appendix 4. Our sample of nests during 1981–1985 identifies three: death of crechlings before color banding, loss of eggs to snow, and loss of nestlings when nests fall apart. Death of crechlings accounts for most of the mortality each year and, as shown in Fig. 112, fluctuates in parallel with total mortality. Loss of eggs to snowfall is a surprising key factor. It contributes little to total mortality each year, but its variable importance closely tracks changes in overall mortality. Loss of nestlings when nests deteriorate is probably too insignificant to influence changes in overall mortality even though it fluctuates in concert with overall mortality. Predation was not a key factor from 1981 to 1985 because it was consistently high in these years. Our larger sample of mortality later in life again indicates that crechling mortality is a major driver of population size (Fig. 113). Mortality of six-year-

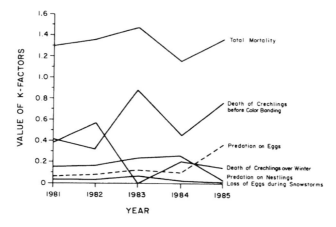

Figure 112   *Annual variation in the importance of various sources of mortality during a jay's first year of life. Death of crechlings before color banding stands out as an important key factor because it accounts for much of the total mortality each year and fluctuations in crechling mortality closely match fluctuations in total mortality.*

olds also stands out as a key factor which was especially important in 1974.

In summary, annual variation in mortality of crechlings is responsible for most of the year-to-year change in mortality of the Town Flock. This is a major force influencing population size. The importance of juvenile abundance to total flock size was evident when we plotted annual fluctuations in the size of the Town Flock (e.g. Fig. 26). Juvenile abundance is variable and this variation drives the year-to-year variation in flock size. Two factors of lesser overall importance, death of six-year-olds and loss of eggs during cold and snowy conditions, may play secondary roles in setting population size.

Once we have identified factors that play key roles in changing population size we can ask whether or not they are capable of regulating populations

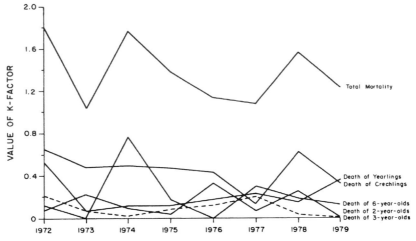

Figure 113    *Annual variation in importance of various sources of mortality during a jay's later years. As in the previous figure the importance of crechling mortality stands out as a key factor.*

around some stable level. Stabilizing regulation, better known as density-dependent regulation, occurs when the strength of a key factor is low when populations are small and increases when they are large. These factors regulate populations like thermostats regulate heaters; they take a large toll as populations grow but lessen their effect as populations dwindle. As a result they dampen population fluctuations. Strong, positive correlations between the number of individuals in a stage and the resultant mortality due to a key factor are evidence of density-dependence.

The survival of crechlings before color banding is not density-dependent, but the survival of crechlings over winter is density-dependent (Appendix 4, Fig. 114A, B). Crechlings survive well in years when few are produced, but they have poor survival in years when many are produced. Therefore, fluctuations in the number of juveniles recruited from within the flock are reduced. Given this effect why is there still year-to-year variation in the number of juveniles? First, although mortality increases with density, its ameliorating

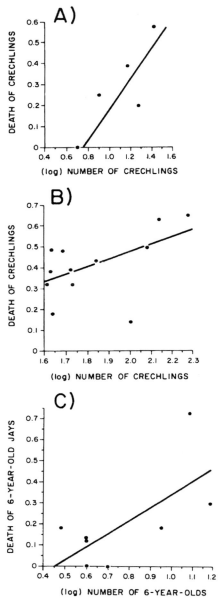

Figure 114   *Density dependence of three key factors. Strong positive correlations suggest that as density increases the mortality due to these factors increases. (A) Crechling mortality in 1981–1985 as a function of the number of crechlings (r = 0.87, N = 5, P = 0.002). Solid line based on regression of y = 0.68x − 0.48. (B) Crechling mortality in 1972–1984 as a function of the number of crechlings (r = 0.51, N = 12, P < 0.001). Solid line based on regression of y = 0.34x − 0.207. (C) Mortality of six-year-old jays as a function of the number of six-year-olds (r = 0.65, N = 8, P < 0.001). Solid line based on regression of y = 0.61x − 0.27.*

influence does not keep pace with rising density (note that the slopes in Figure 114 are less than unity). As a result of this "undercompensation" annual fluctuations in juveniles recruited from the Town Flock still occur, albeit to a lesser extent than would occur without any density-dependent control. Second, all juveniles do not originate from within the flock, each year roughly half are recruited through immigration. Immigration is closely associated with sex ratio, but it is not density dependent (Chapter 11). Variability in sex ratio provides an impetus for fluctuation in immigration which results in variable abundances of juveniles in the flock. Mortality of six-year-olds, another key factor, is also density dependent (Fig. 114C), which may help reduce annual fluctuations in flock size.

Not all key factors are density-dependent. Unpredictable storms usually kill all vulnerable organisms regardless of how many are in a population. Their action is density-independent. Spring snow storms act on Pinyon Jay eggs in such a manner. Regardless of how many eggs are exposed to the cold, the vast majority are killed. Accordingly, annual loss of eggs to snowfall is not strongly correlated with the number of eggs exposed to this danger (Appendix 4). Snow storm-induced mortality may have a slight effect on overall population size but it cannot act to stabilize populations.

CHAPTER 13

# Conclusions: sociality in a variable environment

*The Pinyon Jay "is also said to be restless and impetuous, as if of an
unbalanced mind."* (Baird *et al.* 1874)

Most long-term studies of animals raise many more questions than they
answer. Our study of Pinyon Jays is obviously no exception. We know much
more about the social life of these animals than we did 15 years ago, but we
remain in the dark about many aspects of the environment that have shaped
the evolution of this fascinating species. As Scott noted over 30 years ago
(1956), understanding the social organization of a species is a process that may
*never* be entirely realized.

In this final chapter we will review many of our findings as we address three
major questions: How are Pinyon Jay societies organized? What is the structure of Pinyon Jay populations? What are the major selective forces influencing the evolution of Pinyon Jays?

## HOW ARE PINYON JAY SOCIETIES ORGANIZED?

Pinyon jay societies represent a pinnacle in the evolution of avian sociality.
Many species which we characterize as "social" live in groups for only part of

the year or for part of their lives. Many of the species of birds that are social throughout the year live only in small groups comprised of extended families. However, Pinyon Jays live in large flocks throughout the year and even nest colonially. Socialization begins at an early age as nestlings preen each other and fledglings join together to form creches after they have been out of the nest for only a few days. Creches form due to the social interactions of these, barely three-week-old, birds. Once an individual settles in a flock (through immigration or remaining in its natal flock) it remains there for life. Even when we transport birds out of the flock they quickly return. This prolonged association between flock members allows jays to develop lasting relationships, such as the lifelong pair-bond we have documented that forms between mates (Chapter 7).

We can begin to understand this social system by identifying social relationships between flock members. These relationships indicate how a society is structured from our human point of view (Scott 1956). In order to determine social structure from a Pinyon Jay's own point of view we must also study the discriminatory abilities of individuals (Cheney and Seyfarth 1982), and that is a big order.

Several types of social relationships are evident to the human observer. First, all flock members are organized by dominance-subordination relationships (Chapter 6). Adult males dominate adult females, which dominate yearling males, which dominate yearling females. Juveniles share a special status in the flock through their first fall. They are deferred to by most older jays. Among adult males, one individual is most dominant, and 10–15 males are subordinate to him but dominant over all other males. Social status may be established very early in life. Second, members of age cohorts interact more with each other than with differently-aged jays. This is especially true of immatures integrating into the flock (Chapter 6). It may also extend through adulthood as mates are typically of the same age (Chapter 7). Third, members of extended families are tied together by pair-bonds and kinship bonds (Chapter 5). Parent-young bonds probably persist at least for one year because yearling sons occasionally help their parents at the nest (Chapter 9). Pair-bonds typically endure for life and appear to be the strongest bond between flock members (Chapter 7). Fourth, social bonds link flock members together and act to co-ordinate flock movements. Birds within a flock maintain constant contact and move as a unit. The flock flies, roosts, forages, and breeds together. Most flock members even die together.

Our studies of vocal recognition (Chapter 4) suggest that many of these social relationships are also evident to Pinyon Jays. Members of a pair recognize each other's Near vocalizations throughout the year and use this call to co-ordinate their activities during courtship, nest-building and breeding. Parents and young recognize each other at least through the creching stage. Flock members are readily distinguished from members of other flocks which may enable flocks to remain together even when they mix with others during the fall seed harvest.

Despite this well developed ability to recognize individuals in the flock and the occurrence of related jays within a flock, there appears to be no special (or at least no obvious) association, and only rudimentary co-operation, between relatives (except for parents and their dependent young). An exciting exception concerns nest placement within a colony. Relatives rarely nest near each other and therefore rarely attract predators to their relatives' nests (Chapter 8). For this to occur, birds must be aware of their family ties. The general lack of apparent co-operation between relatives and the fact that most jays in a flock are not related suggests that kin selection currently plays at best a minor role in the maintenance of sociality (Chapter 5).

Pinyon Jay flocks are conspicuously organized into age and sex classes. The first three age classes are easily distinguished by humans and jays as plumage differences are great (Chapter 1). Slight plumage and size differences also indicate a jay's sex. These factors influence social relationships by determining social status (Chapter 6) and influencing pair-bonding (Chapter 7). Moreover, from a jay's perspective, its social environment also changes during its lifetime. Young jays are in the company of more mature relatives than old jays (Chapter 5). Old jays should support young jays more than *vice versa* (even if they are not related) because reproductive value peaks early in life and then declines with age (Chapter 12). We should therefore expect young jays to be nurtured to a greater extent than old jays, which may explain why juveniles have such an elevated status within the flock (Chapter 6).

In summary, Pinyon Jay flocks are organized into social strata defined by a jay's sex and age. Age may further structure the flock as jays of the same age interact frequently and mates are usually selected from one's age cohort. Kinship organizes young jays into family groups, but this structure quickly fades as many families from a breeding colony mix in the creche. Although we can delineate the flock into clans of nuclear families (Chapter 5), the jays do not seem to observe this delineation in any obvious way. They rarely co-operate preferentially with their relatives beyond typical parental behavior. But, there is probably much more going on within a flock than we have been able to identify and study.

## PINYON JAY POPULATION STRUCTURE

Out of convenience, we usually define a population as a group of organisms of a single species that live in our study area. In this way, the Town Flock is a population. However, evolutionary geneticists usually employ a more rigorous definition of a population. Namely, the collection of all individuals of a species which regularly interbreed. This definition usually transcends a study area's boundaries. It is an important definition because individuals that regularly interbreed share a common gene pool, and it is the entire gene pool that evolves in response to natural selection. It is important to know where your study "population" lies relative to its genetic population because natural selection acting upon individuals influences the composition of genetic populations.

Is the Town Flock a genetic population? We must take another look at our information on dispersal (Chaper 11) in order to answer this question. Dispersing individuals move genes between social flocks. Therefore, in order to calculate the genetic population size we must know how far and how often individuals move before they breed. We first need to determine where 85% of the offspring produced in a flock end up breeding. An estimate of the genetic (or "effective") population size is simply the number of breeders located within this area (Chepko-Sade and Shields *et al.* 1987). Most Pinyon Jays born in the Town Flock do not disperse before breeding (N = 122). We observed 22 immigrate to the neighboring CCC flock and it is likely that an equal number also immigrated to the other neighboring flock (which we did not regularly census). If no more than 27 others immigrated to more distant flocks we can conclude that 85% of offspring born in the Town Flock bred there or in neighboring flocks. This seems reasonable given the rarity of long-distance dispersal in this species (Chapter 11, and Whitney 1982).

Our first estimate of the effective population size of the Town Flock is the number of breeding individuals in the flock plus the number of breeders in neighboring flocks. For the Town Flock, this equals approximately 300 birds. For flocks in richer pinyon-juniper woodland, effective population size is probably closer to 500 jays. Variation in lifetime reproductive success reduces the effective population size because all individuals do not contribute equally to the gene pool. When we modified our estimate of effective population size by taking this into account we came up with a value of 222 for the Town Flock and 370 for a flock in pinyon-juniper woodland (Marzluff and Balda 1989). Regardless of which estimate we use the conclusion is the same: Pinyon Jay genetic populations comprise breeders in a flock plus those in neighboring flocks. As a result, a group of flocks, not an individual flock, evolves in response to selective pressures.

We provide a model of genetic and social population structure in Figure 115. Genetic populations are subdivided into social groups with slightly overlapping home ranges. Genetic populations may be partially isolated from each other by small patches of unsuitable habitat (such as high altitude habitat to the north of the Town Flock). Depending on the sex ratio of young birds in the fall and in the breeding season, social groups either act as sources of emigrants or as sinks for immigrants. For example, the flock in the center of Figure 115 has a slightly female-biased sex ratio and thus acts as a source, while two adjoining flocks are strongly male-biased and thus act as sinks. As we argued in Chapter 11, whether a flock acts as a sink or as a source varies from year to year because sex ratios vary (Chapter 5). This model of Pinyon Jay population structure is very similar to a model of newt populations where ponds containing permanent groups of adults act to subdivide the genetic population, and immatures disperse between ponds (Gill 1978). Ponds vary in productivity between years which results in the same social group acting as a sink for immigrants in some years and as a source for emigrants in other years.

Genetic population size in Pinyon Jays is intermediate in comparison to the few data on birds (Barrowclough 1980, Woolfenden and Fitzpatrick 1984, Koenig and Mumme 1987) and mammals (Chepko-Sade and Shields *et al.* 1987).

Why are we so interested in the effective size of our population? The relative influence of random genetic drift (random changes in gene frequencies in a population due to breeding by a small sample of individuals) and natural selection differ depending on effective population size. Species fragmented into

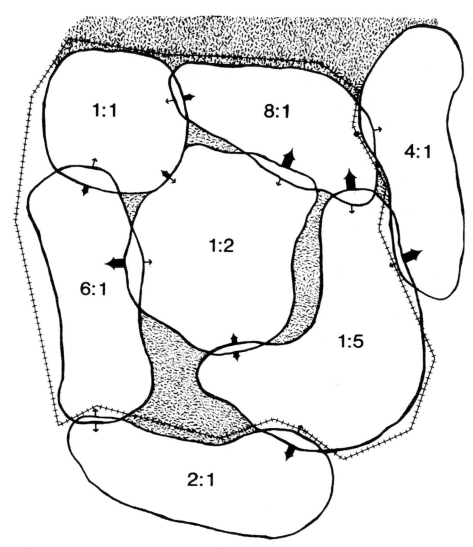

Figure 115   *Model of Pinyon Jay population structure. Home ranges of social populations (flocks) are outlined with solid lines. One genetic population is outlined with the hatched line. Shaded areas represent patches of unsuitable habitat (upper elevation alpine areas or low elevation deserts). Arrows indicate the direction of female transfer between neighboring flocks. The size of the arrow is proportional to the amount of transfer as determined by the flocks' sex ratios (numbers in each flock are ratios of males to females). After Marzluff and Balda (1989).*

genetic populations between 50 and 500 individuals, such as Pinyon Jays, are not expected to show interpopulational differences due to random drift, if natural selection is acting to homogenize differences (Wright 1931, 1932). This theoretical prediction appears valid in Pinyon Jays because the species has not been partitioned into any morphological races or subspecies (Phillips 1989). This is consistent with the effects of drift being minimized by the occurrence of similar selective pressures on morphology throughout the species' range. Selective pressures should be similar because all populations harvest, eat, and cache pine seeds. This specialization has favored striking morphological divergence from other jays and striking morphological convergence with other seed-caching specialists (Chapter 3).

Natural selection may minimize some differences between populations as suggested above, but it may also exaggerate other differences. Subdivided populations, such as Pinyon Jays, are able to quickly change in response to strong selection (Wright 1931, 1932). Selective pressures on breeding behavior (especially on the timing of breeding) may be quite different between nearby populations (Chapter 8). An ability to evolve rapidly in response to these differences may explain why some populations breed in the fall following bumper pine crops, but other nearby flocks do not (Chapter 8).

In summary, the "population" of Pinyon Jays that is most obvious to the observer is the social population or the flock. However, flocks are not populations in the genetic sense, because dispersal joins a flock's gene pool with its neighboring flocks' gene pools. Neighboring flocks may be quite similar genetically because of this dispersal and because successful dispersers also typically have many siblings in their natal flock. Even though the breeders in several flocks together form a genetic population, effective population size is still rather small (usually less than 500). Accordingly, natural selection can cause rapid shifts in gene frequencies and/or negate random shifts that may result from genetic drift.

## RESPONSE TO SELECTIVE PRESSURES

We have identified many features of the Pinyon Jay's environment that have molded its lifestyle. Four features seem to stand above the rest in importance – the fall pine seed crop, predation, the flock's sex ratio, and permanent membership in a large flock.

Pinyon Jays are aptly named because of their mutualistic relationship with the pinyon pine tree. These two species have probably co-evolved in a type of arms race that modified both of them (Chapter 3). Natural selection molded the Pinyon Jay's morphology to enable efficient utilization of a well protected, nutritious pine seed. A long bill devoid of feathering and an expandable esophagus are testaments to the power of natural selection. Morphological convergence with other seed-eating specialists, most notably the Clark's Nutcracker, is striking and caused much initial confusion over the Pinyon Jay's place within the corvid family (Chapter 2).

The variable nature of pine seed production poses a special problem for seed specialists. This powerful selective force is likely to be the reason that Pinyon Jays cache seeds and are not territorial (Chapter 1). Moreover, the reproductive cycle of Pinyon Jays has evolved to take advantage of the occasional bumper seed crops. Pinyon Jays gauge their timing of breeding and their investment in egg production by the size of the seed crop (Chapters 8, 9).

Other aspects of the environment also influence reproduction. However, the pine crop appears to be fundamental as our study flock continues to adjust its behavior to pine seed abundance despite the frequent and conflicting influences of a cold and snowy climate (Chapters 8, 9).

Predation may currently be the strongest selective force shaping Pinyon Jays in the Town Flock. Nearly half of all Pinyon Jay eggs end up as fuel for an exploding population of nest predators, most notably Common Ravens and American Crows. This is likely to be a common scenario in many western towns where human refuse has allowed ravens and crows to build large populations. The low productivity of jays in the Town Flock is the single most important factor responsible for this population's current decline (Chapter 12). Pinyon Jays have thwarted predation in the past and they probably will adjust to current predators as well. In the past, predation on adults may have been a primary factor favoring the formation of large flocks (Chapter 1). At present, predation on eggs and nestlings is favoring concealed nest placement and an ability to adjust nest placement to prevailing conditions (Chapter 8).

As we have seen, many factors external to the flock have molded the physique and behavior of Pinyon Jays. However, once these external factors favored group life this aspect of the environment – predictable association with group members – became a strong selective agent. As a manifestation of mate availability, the flock's sex ratio is a primary force shaping the options pursued by young jays (Chapter 11). Life in a closed society may also have constrained the mate choice options available to Pinyon Jays. We have argued that many factors affect the choice of a mate. However, the constraining influence of prior reproductive success may supersede the response to these factors (Chapter 7). The use of prior success is only possible in species like Pinyon Jays that select subsequent mates from a pool of *familiar* individuals.

Group life has also placed a premium on co-ordinating the activities of individuals in the group. Many nuances of social life appear to be mediated primarily through vocalization. Pinyon Jays have an extremely variable and graded repertoire of calls (Chapter 4) which they use to keep in nearly continuous vocal contact with their flock members. They frequently utter calm contact calls as they fly and forage. As soon as one jay detects danger, it gives an alert call and the entire flock seeks shelter in dense foliage or congregates to mob the source of the alarm. Our results in Chapter 4 indicate that calls convey the identity of the caller in addition to a variety of messages about the environment. An ability to discriminate among the myriad of calls given by a flock of jays appears to be integral to maintaining organization in such a large society. This complex vocal system may give each jay a detailed auditory sense of its flock. A jay should be able to determine the location, identity, and motivation of all its flock members within hearing distance. A jay could then respond to these flock members in ways that enhance its indirect and direct fitness.

Co-ordination of activities may be costly to some group members. Waiting is one cost of co-ordination that appears often in the lives of Pinyon Jays but has, to our knowledge, not been stressed in other species. Co-ordination necessitates waiting by individuals that are ready to perform some behavior before other members of the group. For example, individuals that locate abundant food sources do not feed from them until other flock members have been alerted. Once feeding begins, dominant males feed first, but then wait for less dominant flock members to feed so that the flock can leave as a group. An extreme form of waiting may occur in early spring when jays are entering reproductive condition. Pairs ready to nest before the rest of the flock may wait to facilitate synchronized breeding. All of these cases of waiting are presumably advantageous for those waiting because lone jays may be more vulnerable to predators and may be less efficient at tracking variable resources than groups of jays. We find it interesting that in order for seemingly efficient members of a society to garner the benefits of social life they may often have to wait for their less efficient (i.e. slower, subordinate) group members.

As a summary of Pinyon Jay social life, we have illustrated the most obvious characteristics of Pinyon Jay societies that appear to facilitate the maintenance of sociality in ecological and evolutionary time (Fig. 116). A peaceful disposition, habitual sentinel duty, waiting by efficient flock members, and individual recognition are especially important in co-ordinating and synchronizing individuals without overt aggression. Synchronization and co-ordination, made possible by extensive communication, maintain sociality on a daily basis. Individual recognition, memory of past events, and long life span enable learned behaviors to be culturally transmitted which may be especially important in maintaining a stable social system over a time span of several hundreds or thousands of years (ecological time). These same hallmarks of

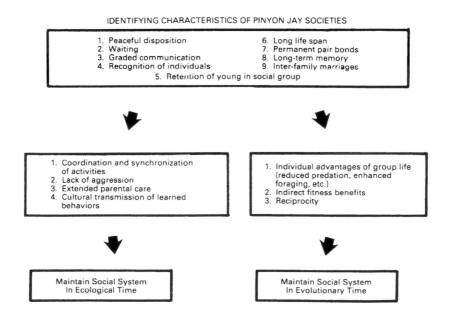

Figure 116    *Characteristics of Pinyon Jays that promote advantages of group life and therefore favor the maintenance of sociality.*

Pinyon Jay sociality also may enable individuals living in groups to obtain higher fitness than isolated individuals. Higher fitness may be obtained through individual, indirect, or reciprocated benefits of group life. Regardless of the mechanism, higher fitness associated with group life maintains sociality throughout evolutionary time.

### THE TAKE-HOME MESSAGE: ENVIRONMENTAL VARIABILITY IS A SELECTIVE FORCE

We have discussed many patterns and provided many tidy summaries of the average behavior of Pinyon Jays. However, average behavior is never observed. The most important point we can make in summarizing our findings is that nearly every factor we investigated is better characterized by its variable nature rather than by its average value. Variability is truly the spice of Pinyon Jay life.

We documented one prevalent form of variability: that occurring from year to year. We have emphasized the variable nature of the environment in the southwestern United States and suggest that this variability is the force behind variability within a flock. We have focused on three major sources of annual variability: the size of the pinyon pine crop, the amount of nest predation, and the harshness of the physical environment, especially the amount of snow falling during the nesting season. Variability in these factors is likely to be the major reason that we observed great variation in: flock movement, the age and sex composition of the flock, mortality, reproductive success, nest placement, and the influence of age and experience on breeding performance. Variation in the age and sex composition of the flock acts further to force variation in the options taken by young jays.

A variable physical and social environment, such as we observed, is an important force on the life history options pursued by members of a population. A common denominator of the variability we documented is that it is more influential on young individuals than on older individuals. This is especially important when considering demography. Young jays have more variable survivorship than old jays and as such old jays are expected to behave more conservatively than young jays (Goodman 1979). Delayed reproduction, small clutch size, and extended parental care are some examples of the conservative nature of older Pinyon Jays. They do not put all their eggs in one basket, but rather they spread out their efforts over many seasons. When variation is unpredictable, animals cannot afford to compromise their chances of surviving by going for big fitness pay-offs in one year. The importance of lifespan to lifetime fitness in our population underscores this point. However, when good years are predictable they should be capitalized upon. Remember, as we argued with respect to the timing of nesting, a big pay-off in a few years can compensate for many poor years. Indeed, Pinyon Jays are quick to invest heavily in a good opportunity: they breed early and lay large clutches of eggs after bumper pine crops. Pinyon Jays, like successful politicians, take the least risky road when conditions are uncertain, but quickly jump on the bandwagon when success is a sure bet!

# Number and spacing of nests with eggs within colonies of the Town Flock

Minimum distance between neighboring nests in 1981 represents the only time two nests were built in the same tree. They were 9.5 m apart (vertically) in the tree. Multiple colonies within a year are listed in chronological order. Data for 1973–1977 are from Gabaldon (1978).

| Year | Number of Nests in Colony | Distance (m) Between Neighboring Nests | | | |
|---|---|---|---|---|---|
| | | Average | SD | Minimum | Maximum |
| 1973 | 18 | — | — | — | — |
| 1974 | 28 | 63.7 | 21.0 | 18.3 | 94.6 |
| 1975 | 23 | 73.4 | 50.7 | 11.6 | 183.0 |
| 1976 | 32 | 75.2 | 50.2 | 20.7 | 144.0 |
| 1977 | 20 | 66.5 | 40.7 | 15.6 | 164.7 |
| 1981 | 4 | 212.5 | 168.0 | 80.0 | 430.0 |
| 1981 | 26 | 45.8 | 33.8 | 0.0 | 150.0 |
| 1981 | 8 | 92.3 | 53.7 | 29.0 | 160.0 |
| 1982 | 2 | 130.0 | 0.0 | 130.0 | 130.0 |
| 1982 | 6 | 61.5 | 48.5 | 32.0 | 159.0 |
| 1982 | 11 | 55.2 | 36.7 | 22.0 | 151.0 |
| 1982 | 8 | 108.5 | 162.5 | 16.0 | 500.0 |
| 1982 | 2 | 97.0 | 0.0 | 97.0 | 97.0 |
| 1982 | 11 | 110.2 | 108.6 | 28.0 | 400.0 |
| 1982 | 4 | 93.0 | 28.1 | 75.0 | 135.0 |
| 1982 | 8 | 423.3 | 245.6 | 118.0 | 600.0 |
| 1982 | 2 | 204.0 | 0.0 | 204.0 | 204.0 |
| 1983 | 10 | 58.4 | 45.6 | 18.0 | 170.0 |
| 1983 | 12 | 38.3 | 27.8 | 20.0 | 110.0 |
| 1983 | 2 | 135.0 | 0.0 | 135.0 | 135.0 |
| 1983 | 3 | 76.7 | 63.5 | 40.0 | 150.0 |
| 1983 | 3 | 51.7 | 11.6 | 45.0 | 65.0 |
| 1983 | 5 | 218.0 | 93.1 | 150.0 | 320.0 |
| 1983 | 2 | 180.0 | 0.0 | 180.0 | 180.0 |
| 1983 | 2 | 130.0 | 0.0 | 130.0 | 130.0 |
| 1984 | 29 | 36.7 | 15.6 | 10.0 | 75.0 |
| 1984 | 2 | 430.0 | 0.0 | 430.0 | 430.0 |
| 1984 | 8 | 267.5 | 224.7 | 20.0 | 500.0 |
| 1984 | 3 | 196.7 | 262.7 | 45.0 | 500.0 |
| 1985 | 21 | 37.9 | 16.3 | 20.0 | 100.0 |
| 1985 | 16 | 50.3 | 20.1 | 20.0 | 82.2 |
| 1986 | 15 | 33.7 | 10.1 | 20.0 | 60.0 |
| 1986 | 11 | 62.7 | 69.1 | 10.0 | 200.0 |
| 1986 | 20 | 36.0 | 25.7 | 10.0 | 100.0 |
| 1986 | 2 | 100.0 | 0.0 | 100.0 | 100.0 |
| 1986 | 8 | 36.9 | 22.2 | 10.0 | 60.0 |

| Year | Number of Nests in Colony | Distance (m) Between Neighboring Nests | | | |
|------|------|---------|------|---------|---------|
| | | Average | SD | Minimum | Maximum |
| 1986 | 5 | 53.0 | 27.1 | 35.0 | 100.0 |
| 1986 | 11 | 51.8 | 38.8 | 25.0 | 150.0 |
| 1987 | 31 | 29.8 | 19.2 | 10.0 | 100.0 |
| 1987 | 3 | 133.3 | 57.7 | 100.0 | 200.0 |
| 1987 | 12 | 40.4 | 37.0 | 15.0 | 150.0 |

# APPENDIX 2

# Average productivity of Pinyon Jays in the Town Flock

Mean, sample size and standard deviation are listed for each year.

| | Clutch Size | Number of Nestlings | Number of Fledglings | Number of Crechlings | Number of Yearlings | Percent of Eggs Hatching | Percent of Nestlings Fledging | Percent of Fledglings in Creche | Percent of Crechlings overwintering |
|---|---|---|---|---|---|---|---|---|---|
| 1971 | 3.88 | | | | | | | | |
| | 17 | | | | | | | | |
| | — | | | | | | | | |
| 1973 | 3.50 | — | — | 1.71 | 0.67 | — | — | — | 0.40 |
| | 20 | | | 21 | 21 | | | | 13 |
| | 0.95 | | | 1.62 | 0.91 | | | | 0.34 |
| 1974 | 4.32 | | | 1.51 | 0.43 | | | 0.51 | 0.29 |
| | 25 | | | 35 | 35 | | | 17 | 22 |
| | 0.75 | | | 1.46 | 0.74 | | | 0.29 | 0.36 |
| 1975 | 3.75 | | | 0.63 | 0.37 | | | 0.29 | 0.56 |
| | 16 | | | 27 | 27 | | | 11 | 13 |
| | 0.86 | | | 0.79 | 0.63 | | | 0.30 | 0.50 |
| 1976 | 3.00 | | | 0.87 | 0.44 | | | 0.25 | 0.51 |
| | 15 | | | 45 | 45 | | | 8 | 17 |
| | 1.20 | | | 1.22 | 0.81 | | | 0.39 | 0.41 |
| 1977 | | | | 1.35 | 0.71 | | | | 0.44 |
| | | | | 31 | 31 | | | | 19 |
| | | | | 1.40 | 1.19 | | | | 0.43 |
| 1978 | | | | | | | | | 0.33 |
| | | | | | | | | | 11 |
| | | | | | | | | | 0.33 |
| 1979 | | | | 1.19 | 0.65 | | | | 0.62 |
| | | | | 21 | 20 | | | | 13 |
| | | | | 1.08 | 0.75 | | | | 0.42 |
| 1981 | 3.65 | 2.79 | 1.68 | 1.14 | 0.39 | 0.79 | 0.57 | 0.39 | 0.31 |
| | 26 | 19 | 19 | 28 | 28 | 19 | 14 | 13 | 16 |
| | 0.75 | 1.44 | 1.77 | 1.33 | 0.79 | 0.37 | 0.51 | 0.43 | 0.40 |
| 1982 | 3.71 | 2.43 | 1.44 | 0.84 | 0.21 | 0.64 | 0.58 | 0.49 | 0.26 |
| | 35 | 42 | 41 | 44 | 42 | 42 | 27 | 16 | 16 |
| | 0.71 | 1.84 | 1.87 | 1.26 | 0.47 | 0.46 | 0.50 | 0.32 | 0.31 |
| 1983 | 3.77 | 1.87 | 1.05 | 0.15 | 0.15 | 0.48 | 0.53 | 0.13 | 1.00 |
| | 31 | 39 | 39 | 39 | 39 | 39 | 20 | 12 | 5 |
| | 0.72 | 1.96 | 1.69 | 0.43 | 0.43 | 0.48 | 0.47 | 0.17 | 0.00 |
| 1984 | 3.97 | 2.41 | 1.12 | 0.40 | 0.26 | 0.60 | 0.45 | 0.36 | 0.65 |
| | 35 | 41 | 43 | 43 | 43 | 41 | 28 | 13 | 8 |
| | 0.62 | 1.94 | 1.79 | 0.95 | 0.69 | 0.47 | 0.50 | 0.36 | 0.44 |
| 1985 | 3.88 | 1.78 | 1.45 | 0.26 | 0.19 | 0.46 | 0.86 | 0.17 | 0.70 |
| | 16 | 32 | 31 | 31 | 31 | 32 | 14 | 12 | 5 |
| | 0.81 | 2.00 | 1.93 | 0.68 | 0.60 | 0.50 | 0.36 | 0.25 | 0.45 |
| 1986 | 3.08 | 1.23 | 0.63 | — | — | 0.39 | 0.46 | — | — |
| | 38 | 62 | 64 | | | 62 | 26 | | |
| | 1.12 | 1.66 | 1.36 | | | 0.49 | 0.51 | | |

|  | Clutch Size | Number of Nestlings | Number of Fledglings | Number of Crechlings | Number of Yearlings | Percent of Eggs Hatching | Percent of Nestlings Fledging | Percent of Fledglings in Creche | Percent of Crechlings overwintering |
|---|---|---|---|---|---|---|---|---|---|
| 1987 | 3.64 | 1.93 | 1.10 | — | — | 0.47 | 0.50 | — | — |
|  | 33 | 40 | 42 |  |  | 40 | 22 |  |  |
|  | 1.08 | 2.08 | 1.90 |  |  | 0.48 | 0.51 |  |  |
| Average | 3.70 | 2.06 | 1.21 | 0.91 | 0.41 | 0.55 | 0.56 | 0.32 | 0.51 |
|  | 11 | 7 | 7 | 11 | 11 | 7 | 7 | 8 | 12 |
|  | 0.38 | 0.52 | 0.34 | 0.52 | 0.20 | 0.14 | 0.14 | 0.14 | 0.22 |

# APPENDIX 3

# Annual variation in the survival of Pinyon Jays in the Town Flock

The figures for juveniles and yearlings include those dying and dispersing under the category "number dying". Immigrants are not included in juvenile counts and are included in yearling and adult counts only after they remained in the flock for at least one year. Adults of unknown age were banded in 1972 as birds that were at least two years old. All other birds were banded as juveniles or yearlings and followed until their death or disappearance. Juvenile survivorship was unavailable in 1984 and 1986 because censuses were incomplete. CV indicates the coefficient of variation for year-to-year survivorship. $\chi^2$ gives the value of a chi-square test for the significance of annual variation in survival within an age class. Degrees of freedom for each test equals the number of years minus 1. Tests on juveniles, yearlings and adults of known age are significant at $P < 0.001$. The test performed on adults of unknown age only uses data from 1974 to 1980 because of small samples in later years. This test is not significant ($P = 0.20$).

| | Juveniles | | | Yearlings | | |
|---|---|---|---|---|---|---|
| | Number | | Percentage | Number | | Percentage |
| Year | Surviving | Dying | Surviving | Surviving | Dying | Surviving |
| 1974 | 15 | 28 | 34.9 | 12 | 30 | 28.6 |
| 1975 | 36 | 67 | 35.0 | 13 | 3 | 81.3 |
| 1976 | 16 | 32 | 33.3 | 29 | 18 | 61.7 |
| 1977 | 20 | 43 | 31.7 | 20 | 24 | 45.5 |
| 1978 | 58 | 31 | 65.2 | 17 | 8 | 68.0 |
| 1979 | 32 | 106 | 23.2 | 41 | 31 | 56.9 |
| 1980 | 17 | 28 | 37.8 | 36 | 15 | 70.6 |
| 1981 | 19 | 22 | 46.3 | 11 | 14 | 44.0 |
| 1982 | 21 | 31 | 40.4 | 30 | 8 | 78.9 |
| 1983 | 17 | 25 | 40.5 | 24 | 2 | 92.3 |
| 1984 | — | — | — | 7 | 10 | 41.2 |
| 1985 | 29 | 15 | 65.9 | 5 | 1 | 83.3 |
| 1986 | — | — | — | 16 | 13 | 55.2 |
| average | | | 41.3 | | | 62.1 |
| CV | | | 32% | | | 31% |
| $\chi^2$ | | 57.2 | | | 51.2 | |

| Year | Adults of Known Age | | | Adults of Unknown Age | | |
|---|---|---|---|---|---|---|
| | Number | | Percentage | Number | | Percentage |
| | Surviving | Dying | Surviving | Surviving | Dying | Surviving |
| 1974 | — | — | — | 56 | 29 | 65.9 |
| 1975 | 10 | 2 | 83.3 | 45 | 11 | 80.4 |
| 1976 | 13 | 10 | 56.5 | 39 | 6 | 86.7 |
| 1977 | 34 | 8 | 81.0 | 27 | 12 | 69.2 |
| 1978 | 50 | 4 | 92.6 | 19 | 8 | 70.4 |
| 1979 | 43 | 24 | 64.2 | 13 | 6 | 68.4 |
| 1980 | 66 | 18 | 78.6 | 10 | 3 | 76.9 |
| 1981 | 56 | 46 | 54.9 | 8 | 2 | 80.0 |
| 1982 | 56 | 11 | 83.6 | 7 | 1 | 87.5 |
| 1983 | 81 | 5 | 94.2 | 5 | 2 | 71.4 |
| 1984 | 62 | 43 | 59.0 | 3 | 2 | 60.0 |
| 1985 | 64 | 19 | 77.1 | 3 | 0 | 100.0 |
| 1986 | 49 | 20 | 71.0 | 1 | 2 | 33.3 |
| average | | | 74.7 | | | 73.1 |
| CV | | | 18% | | | 22% |
| $\chi^2$ | 72.0 | | | 8.7 | | |

# Key-factor analysis of mortality in the Town Flock

Average k-values indicate the importance of each stage of mortality. Correlations between annual k-values for each stage and the total mortality (the sum of all k-values each year) indicate key factors in population regulation. Correlations between k-values and density indicate the density-dependency of each mortality stage. Losses at all early stages were known only from 1981 to 1985, therefore we present these data separately in the first portion of the table. Data from 1972 to 1985 were used in the latter part of the table, but truncation of these data in 1985 means that 1979 was the last year k-factors for death of six-year-olds could be computed. As a result, calculations of importance for late sources of mortality are based on fewer years than early sources and correlations are based only on data from 1972 to 1979.

| Stage of Mortality | Importance of Mortality (average K-value) | Key Factors (correlation of K with K-total) | Density Dependence (correlation of K with preceding abundance) |
|---|---|---|---|
| 1981–1985 | | | |
| Infertile eggs | 0.012 | −0.69 | 0.14 |
| Eggs lost to snow/cold | 0.031 | 0.57 | −0.18 |
| Eggs lost to wind | 0.015 | 0.06 | 0.46 |
| Eggs abandoned | 0.030 | −0.02 | −0.78 |
| Eggs preyed upon | 0.142 | 0.20 | −0.07 |
| Nestlings preyed upon | 0.174 | −0.33 | 0.94 |
| Nestlings abandoned | 0.048 | −0.07 | 0.07 |
| Nestlings lost to wind | 0.028 | −0.47 | −0.39 |
| Nestlings lost when nest deteriorated | 0.019 | 0.47 | 0.36 |
| Nestlings lost for unknown reasons | 0.014 | 0.14 | 0.37 |
| Crechlings dying before color-banding | 0.566 | 0.62 | −0.37 |
| Crechlings dying over winter | 0.259 | −0.24 | 0.87 |
| 1972–1985 | | | |
| Crechlings dying over winter | 0.456 | 0.67 | 0.51 |
| Death of yearlings | 0.221 | 0.34 | 0.15 |
| Death of 2-year-olds | 0.140 | −0.23 | 0.14 |
| Death of 3-year-olds | 0.098 | −0.03 | 0.29 |
| Death of 4-year-olds | 0.118 | −0.05 | 0.15 |
| Death of 5-year-olds | 0.138 | 0.30 | 0.08 |
| Death of 6-year-olds | 0.210 | 0.55 | 0.65 |

# Family trees of Pinyon Jays

Family trees of Pinyon Jays in the Town Flock presented as a genealogical graph (the graph has been split into two halves for reproduction so the S and T regions have been repeated in the second half of the graph which, in the original form, would run continuously from A through Z to LL).

All extended families (clans) are linked by pair-bonds between some of their members. Two nodes are shown: O represents an individual, and ● represents a pair-bond between two individuals. Marriage arcs →— join mated individuals and reproductive arcs →→— join parents and their offspring. Adjacent to each individual we list its color band combinations, its sex (⊖ = female or ⊕ = male, or O = unknown), whether it helped its parents breed (H = yes), and whether it emigrated (E = yes). A reproductive arc with a question mark preceding it (?→→) indicates the parents of this individual were not known. This notation follows Cannings and Thompson (1981). From Marzluff and Balda (1990).

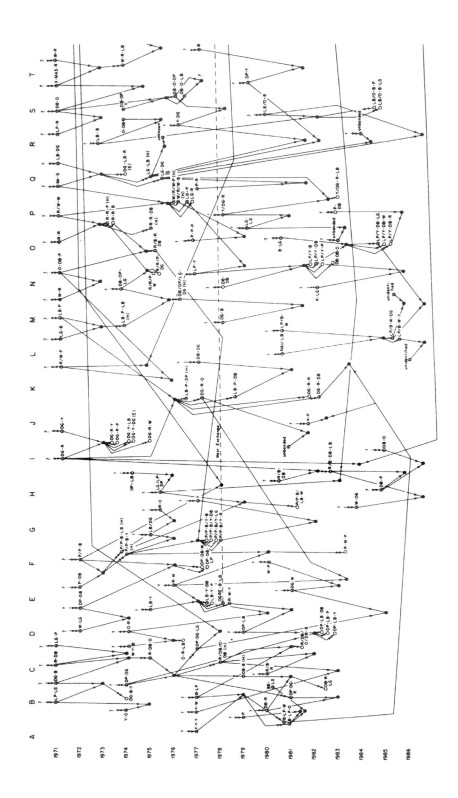

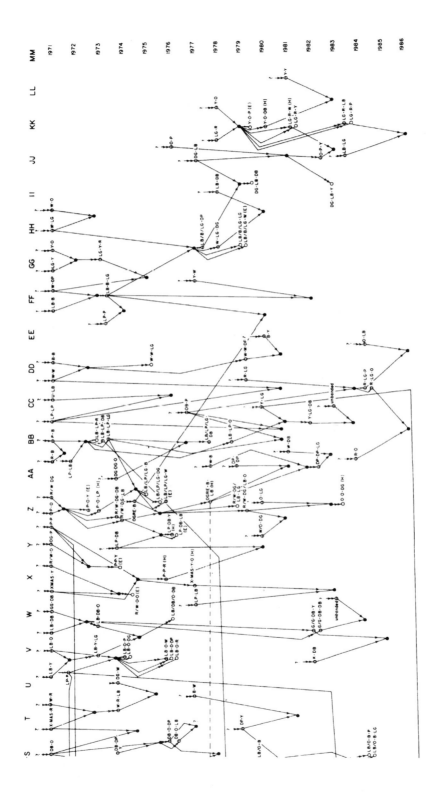

# Common and latin names of species
# mentioned in the text

Common Name

Latin Name

BIRDS

| Common Name | Latin Name |
|---|---|
| Cattle Egret | *Bubulcus ibis* |
| Bewick's Swan | *Cygnus columbianus* |
| Lesser Snow Goose | *Chen caerulescens* |
| Mallard | *Anas platyrhynchos* |
| Ruff | *Philomachus pugnax* |
| Ring-billed Gull | *Larus delawarensis* |
| Black-legged Kittiwake | *Rissa tridactyla* |
| Common Tern | *Sterna hirundo* |
| Northern Fulmar | *Fulmarus glacialis* |
| Yellow-eyed Penguin | *Megadyptes antipodes* |
| Adelie Penguin | *Pygoscelis adeliae* |
| Cooper's Hawk | *Accipiter cooperii* |
| Northern Goshawk | *Accipiter gentilis* |
| European Sparrowhawk | *Accipiter nisus* |
| Osprey | *Pandion haliaetus* |
| Prairie Falcon | *Falco mexicanus* |
| Japanese Quail | *Coturnix japonica* |
| Rock Dove | *Columba livia* |
| Wood Pigeon | *Columba palumbus* |
| Mourning Dove | *Zenaida macroura* |
| Great-horned Owl | *Bubo virginianus* |
| Northern Spotted Owl | *Strix occidentalis* |
| Green Woodhoopoe | *Phoeniculus purpureus* |
| Northern Flicker | *Colaptes auratus* |
| Acorn Woodpecker | *Melanerpes formicivorus* |
| House Martin | *Delichon urbica* |
| Cliff Swallow | *Hirundo pyrrhonota* |
| Barn Swallow | *Hirundo rustica* |
| Scrub Jay | *Aphelocoma coerulescens* |
| Mexican (Gray-breasted) Jay | *Aphelocoma ultramarina* |
| Pinyon Jay | *Gymnorhinus cyanocephalus* |
| Blue Jay | *Cyanocitta cristata* |
| Steller's Jay | *Cyanocitta stelleri* |
| European Jay | *Garrulus glandarius* |
| Clark's Nutcracker | *Nucifraga columbiana* |
| European Nutcracker | *Nucifraga caryocatactes* |
| Black-billed Magpie | *Pica pica* |
| Rook | *Corvus frugilegus* |
| American Crow | *Corvus brachyrhynchos* |
| Common Raven | *Corvus corax* |
| Great Tit | *Parus major* |
| House Wren | *Troglodytes aedon* |

| Common Name | Latin Name |
|---|---|
| Grey-crowned Babbler | *Pomatostomus temporalis* |
| Western Bluebird | *Sialia mexicana* |
| American Robin | *Turdus migratorius* |
| European Starling | *Sturnus vulgaris* |
| Song Sparrow | *Melospiza melodia* |
| Bobolink | *Dolichonyx oryzivorus* |
| Red-winged Blackbird | *Agelaius phoeniceus* |
| Zebra Finch | *Poephila guttata* |

OTHER ANIMALS

| | |
|---|---|
| Abert's Squirrel | *Sciurus aberti* |
| Bullsnake | *Pituophis melanoleucus* |
| Chimpanzee | *Pan troglodytes* |
| Coyote | *Canis latrans* |
| Elephant Seal | *Mirounga leonina* |
| Gorilla | *Gorilla gorilla* |
| Grey Fox | *Urocyon cinereoargenteus* |
| Pinyon Mouse | *Peromyscus trueii* |
| Rhesus Macaque | *Macaca mulatta* |
| Wild Dog | *Lycaon pictus* |

PLANTS

| | |
|---|---|
| Colorado Pinyon Pine | *Pinus edulis* |
| Juniper | *Juniperus sp.* |
| Pinyon Pine | *Pinus ponderosa* |
| Single-needled Pinyon Pine | *Pinus monophylla* |

# Bibliography

Amadon, D. (1944) The genera of Corvidae and their relationships. *American Museum Novitates* **1251**:1–21.

Appleby, M. C. (1983) The probability of linearity in hierarchies. *Animal Behaviour* **31**:600–608.

Arnold, S. J. and Wade, M. J. (1984a) On the measurement of natural and sexual selection: Theory. *Evolution* **38**:709–719.

Arnold, S. J. and Wade, M. J. (1984b). On the measurement of natural and sexual selection: Applications. *Evolution* **38**: 720–734.

Ashley, J. F. (1941) A study of the structure of the humerus in the Corvidae. *Condor* **43**:184–194.

Aubin, T. (1989) The role of frequency modulation in the process of distress calls recognition by the Starling (*Sturnus vulgaris*). *Behaviour* **101**:57–72.

Baily, F. M. (1928) Birds of New Mexico. New Mexico Department of Game and Fish, Washington, DC.

Baird, S. F., Brewer, T. M. and Ridgway, R. (1874) *A history of North American birds*. Little, Brown & Co., Boston.

Baker, J. R. (1938) The evolution of breeding seasons. In Beer, C. R. (ed) *Evolution: Essays on aspects of evolutionary biology*, pp. 161–77. Oxford University Press, London.

Baker, R. R. (1978) *The evolutionary ecology of animal migration*. Hodder and Stoughton, London, England.

Balda, R. P. and Balda J. (1978) The care of young piñon jays and their integration into the flock. *Journal für Ornithologie* **119**:146–171.

Balda, R. P. and Bateman, G. C. (1971) Flocking and annual cycle of the Piñon Jay, *Gymnorhinus cyanocephalus*. *Condor* **73**: 287–302.

Balda, R. P. and Bateman, G. C. (1972) The breeding biology of the Piñon Jay. *Living Bird* **11**: 5–42.

Balda, R. P. and Bateman, G. C. (1973) Unusual mobbing behavior by incubating Piñon Jays. *Condor* **75**:251–252.

Balda, R. P. and Bateman, G. C. (1976) Cannibalism in the Piñon Jay. *Condor* **78**:562–564.

Balda, R. P., Bateman, G. C. and Foster, G. F. (1972) Flocking associates of the Piñon Jay. *Wilson Bulletin* **84**:60–76.

Balda, R. P. and Kamil, A. C. (1989) A comparative study of cache recovery by three corvid species. *Animal Behaviour* **38**:486–495.

Balda, R. P., Morrison, M. L. and Bement, T. R. (1977) Roosting behavior of the Piñon Jay in autumn and winter. *Auk* **94**:494–504.

Barrowclough, G. F. (1980) Gene flow, effective population sizes, and genetic variance components in birds. *Evolution* **34**:789–798.

Bateman, G. C. and Balda, R. P. (1973) Growth, development, and food habits of young Piñon Jays. *Auk* **90**:39–61.

Bateson, P. (1978) Sexual imprinting and optimal outbreeding. *Nature* **273**:659–660.

Bateson, P. (ed) (1983) *Mate choice*. Cambridge University Press, Cambridge, United Kingdom.

Begon, M., Harper, J. L. and Townsend, C. R. (1986) *Ecology*. Sinauer Associates, Inc., Sunderland, Massachusetts.

Bendire, C. (1895) *Life history of North American birds*. Government Printing Office, Washington, D.C.

Bengtsson, B. O. 1978. Avoiding inbreeding: at what cost? *Journal of Theoretical Biology* **73**:439–444.

Bent, A. C. (1946) Life histories of North American jays, crows, and titmice. *U.S. National Museum Bulletin* **191**, Part II.

Berger, L. R. and Ligon, J. D. (1977) Vocal communication and individual recognition in the Piñon Jay, *Gymnorhinus cyanocephalus. Animal Behaviour* **25**:567–584.

Boag, P. T. and Grant, P. R. (1978) Heritability of external morphology in Darwin's finches. *Nature* **274**:793–794.

Bock, W. J., Balda, R. P. and Vander Wall, S. B. (1973) Morphology of the sublingual pouch and tongue musculature in Clark's Nutcracker. *Auk* **90**:491–519.

Bollinger, P. B., Bollinger, E. K. and Malecki, R. A. (1990) Tests of three hypotheses of hatching asynchrony in the Common Tern. *Auk* **107**:696–706.

Botkin, C. W. and Shires, L. B. (1948) The composition and value of piñon nuts. *New Mexico Experimental Station Bulletin* **344**:3–14.

Bray, E. O., Kennelly, J. J. and Guarino, J. L. (1975) Fertility of eggs produced on territories of vasectomized Red-winged Blackbirds. *Wilson Bulletin* **87**:187–195.

Braly, J. C. (1931) Nesting of the piñon jay in Oregon. *Condor* **33**:39.

Breitswisch, R. (1989) Mortality patterns, sex ratios, and parental investment in monogamous birds. In Power, D. (ed) *Current Ornithology Volume 8*, pp. 1–50. Plenum Press, New York.

Brown, C. R. (1986) Cliff swallow colonies as information centers. *Science* **220**:212–214.

Brown, J. L. (1963) Aggressiveness, dominance and social organization in the Steller's Jay. *Condor* **65**:460–484.

Brown, J. L. (1969) Territorial behavior and population regulation in birds. *Wilson Bulletin* **81**:293–329.

Brown, J. L. (1975) *The evolution of behavior*. Norton, New York.

Brown, J. L. (1986) Cooperative breeding and the regulation of numbers. *Proceedings of the International Ornithological Congress* **18**:774–782.

Brown, J. L. (1987) *Helping and communal breeding in birds*. Princeton University Press, Princeton, New Jersey.

Brown, J. L. and Brown, E. R. (1981) Extended family system in a communal bird. *Science* **211**:959–960.

Brown, J. L. and Brown, E. R. (1984) Parental facilitation: parent-offspring relations in communally breeding birds. *Behavioural Ecology and Sociobiology* **14**:203–209.

Bruggers, D. J. (1988) *The behavior and ecology of the Common Raven in northeastern Minnesota*. PhD thesis, University of Minnesota, Minneapolis, Minnesota.

Burger, J. (1987) Selection for equitability in some aspects of reproductive investment in Herring Gulls *Larus argentatus. Ornis Scandinavica* **18**:17–23.

Burley, N. (1977) Parental investment, mate choice and mate quality. *Proceedings of the National Academy of Sciences (USA)* **74**:3476–3479.

Burley, N. (1981) Mate choice by multiple criteria in a monogamous species. *American Naturalist* **117**:515–528.

Burley, N. (1985) Leg-band color and mortality patterns in captive breeding populations of zebra finches. *Auk* **102**:647–651.

Burley, N. (1986) Sexual selection for aesthetic traits in species with biparental care. *American Naturalist* **127**:415–445.

Cannings, C. and Thompson, E. A. (1981) *Genealogical and genetic structure*. Cambridge University Press, Cambridge, United Kingdom.

Cannon, F. D., Jr (1973) *Nesting energetics of the Piñon Jay*. MSc Thesis, Northern Arizona University, Flagstaff, Arizona.

Cary, M. (1901) Birds of the Black Hills. *Auk* **18**:231–238.

Caswell, H. (1989) *Matrix population models: construction, analysis, and interpretation*. Sinauer Associates, Inc. Sunderland, Massachusetts.

Cheney, D. L. and Seyfarth, R. M. (1982) Recognition of individuals within and between groups of free-ranging vervet monkeys. *American Zoologist* **22**:519–529.

Chepko-Sade, B. D. and Shields, W. M. with Berger, J., Halpin, Z., Jones, W. T., Rogers, L. L., Rood, J. and Smith, A. (1987) The effects of dispersal and social structure on effective population size. In Chepko-Sade, B. D. and Halpin, Z. T. (eds) *Mammalian dispersal patterns: the effects of social structure on population genetics*, pp. 287–321. University of Chicago Press, Chicago, Illinois, USA.

Clark, L. and Gabaldon, D. J. (1979) Nest desertion by the Piñon Jay. *Auk* **96**:796–798.

Clark, L. and Balda, R. P. (1981) The development of effective endothermy and homeothermy by nestling Piñon Jays. *Auk* **98**:615–619.

Clutton-Brock, T. H. (ed) (1988) *Reproductive success: studies of individual variations in contrasting breeding systems*. The University of Chicago Press, Chicago, Illinois.

Colgan, P. (1983) *Comparative social recognition.* John Wiley and Sons, Inc., New York.

Collias, N. E. (1943) Statistical analysis of factors which make for success in initial encounters between hens. *American Naturalist* **77**:519–538.

Cooke, F., Bousfield, M. A. and Sadura, A. (1981) Mate changes and reproductive success in the Lesser Snow Goose. *Condor* **83**:322–327.

Coulson, J. C. (1966) The influence of the pair-bond and age on the breeding biology of the Kittiwake Gull (*Rissa tridactyla*). *Journal of Animal Ecology* **35**:269–279.

Coulson, J. C. and White, E. (1958) The effect of age on the breeding biology of the Kittiwake *Rissa tridactyla*. *Ibis* **100**:40–51.

Cox, C. R. and LeBoeuf, B. J. (1977) Female incitation of male competition: a mechanism in sexual selection. *American Naturalist* **111**:317–335.

Cully, J. F. and Ligon, J. D. (1986) Seasonality of mobbing intensity in the Pinyon Jay. *Ethology* **71**:333–339.

Darling, F. F. (1938) *Bird flocks and the breeding cycle. A contribution to the study of avian sociality.* Cambridge University Press, Cambridge, United Kingdom.

Darwin, C. (1871) *The descent of man and selection in relation to sex.* John Murray, London.

Dhindsa, M. S., Komers, P. E. and Boag, D. A. (1989). Nest height of Black-billed Magpies: is it determined by human disturbance or habitat type? *Canadian Journal of Zoology* **67**:228–232.

Dunn, P. O. and Hannon, S. J. (1989) Evidence for obligate male parental care in Black-billed Magpies. *Auk* **106**:635–644.

Emlen, S. T. and Oring, L. W. (1977) Ecology, sexual selection, and the evolution of mating systems. *Science* **197**:215–223.

Fisher, R. A. (1930) *The genetical theory of natural selection.* Clarendon Press, Oxford, United Kingdom.

Ford, N. L. (1983) Variations in mate fidelity in monogamous birds. In Johnston, R. F. (ed.) *Current Ornithology Volume 1*, pp. 329–356. Plenum Press, New York.

Frame, L. H. and Frame, G. W. (1976) Female African wild dogs emigrate. *Nature* **263**:227–229.

Frame, L. H., Malcolm, J. R., Frame, G. W., and van Lawick, H. (1980) Social organization of African wild dogs (*Lycaon pictus*) on the Serengeti plains, Tanzania 1967–1978. *Zeitschrift für Tierpsychologie* **50**:225–249.

Freed, L. A. (1987) The long-term pair-bond of tropical House Wrens: advantage or contraint? *American Naturalist* **130**:507–525.

Gabaldon, D. J. (1978) *Factors involved in nest site selection by Piñon Jays.* PhD thesis. Northern Arizona University, Flagstaff, Arizona.

Gill, D. E. (1978) Effective population size and interdemic migration rates in a metapopulation of the red-spotted newt, *Nototphthalmus viridescens* (Rafinesque). *Evolution* **32**:839–849.

Goodall, J. (1986) *The chimpanzees of Gombe.* Belknap Press, Cambridge, Massachusetts.

Goodman, D. (1979) Regulating reproductive effort in a changing environment. *American Naturalist* **113**:735–748.

Goodwin, D. (1951) Some aspects of the behaviour of the jay *Garrulus glandarius*. *Ibis* **93**:414–442, 602–625.

Goodwin, D. (1986) *Crows of the world*. 2nd Edition. British Museum of Natural History Publications, London.

Gould, S. J. (1980) *The panda's thumb*. W. W. Norton, New York.

Gould, S. J. and Lewontin R. C. (1979) Spandrels of San Marco and the Panglossian paradigm – a critique of the adaptationist program. *Proceedings of the Royal Society*, **B**, **205**:581–598.

Grafen, A. (1988). On the uses of data on lifetime reproductive success. In Clutton-Brock, T. H. (ed) *Reproductive success: studies of individual variations in contrasting breeding systems*, pp. 454–471. The University of Chicago Press, Chicago, Illinois.

Greene, E. (1987) Individuals in an osprey colony discriminate between high and low quality information. *Nature* **329**:239–241.

Greenwood, P. J. (1980) Mating systems, philopatry and dispersal in birds and mammals. *Animal Behaviour* **28**:1140–1162.

Gwinner, E. (1964) Untersuchungen über das Ausdrucks- und Sozialverhalten des Kolkraben (*Corvus corax* L.). *Zeitschrift für Tierpsychologie* **21**:657–748.

Halliday, T. R. (1983) The study of mate choice. In Bateson P. (ed) *Mate choice*, pp. 3–32. Cambridge University Press, Cambridge, United Kingdom.

Hamilton, W. D. (1964) The genetical evolution of social behavior. I, II. *Journal of Theoretical Biology* **7**:1–52.

Harcourt, A. H., Stewart, K. S., and Fossey, D. (1976) Male emigration and female transfer in wild mountain gorillas. *Nature* **263**:226–227.

Hardy, J. W. (1961) Studies in behavior and phylogeny of certain New World Jays (Garrulinae). *University of Kansas Science Bulletin* **42**, Lawrence, Kansas.

Hardy, J. W. (1969) A taxonomic revision of the New World jays. *Condor* **71**: 360–375.

Harrington, F. H. (1978) Ravens attracted to wolf howling. *Condor* **80**:236–237.

Hediger, H. (1950) *Wild animals in captivity*. Butterworth and Company, London.

Heinrich, B. (1988) *Ravens in winter*. Summit Books, New York.

Hope, S. (1989) Phylogeny of the avian family Corvidae. Ph.D. thesis, City University of New York.

Horn, H. S. (1968) The adaptive significance of colonial nesting in the Brewer's Blackbird (*Euphagus cyanocephalus*). *Ecology* **48**:682–694.

Hunt, G. L., Jr. (1981) Mate selection and mating systems in seabirds. In Burger J. and Olla, B. L. (eds) *Behavior of marine animals Volume 4. Behavior of marine birds*, pp. 113–151. Plenum Press, New York.

Hussell, D. T. (1988) Supply and demand in Tree Swallow broods: a model of parent-offspring food-provisioning interactions in birds. *The American Naturalist* **131**:175–202.

Johnson, K. (1988a) Sexual selection in pinyon jays I: female choice and male–male competition. *Animal Behavior* **36**:1038–1047.

Johnson, K. (1988b) Sexual selection in pinyon jays II: male choice and female–female competition. *Animal Behaviour* **36**:1048–1053.

Johnson, M. S. and Brown, J. L. (1980) Genetic variation among trait groups and apparent absence of close inbreeding in Grey-crowned Babblers. *Behavioral Ecology and Sociobiology* **7**:93–98.

Klump, G. M. and Shalter, M. D. (1984) Acoustic behavior of birds and mammals in the predator context. I. Factors affecting the structure of alarm signals. II. The functional significance and evolution of alarm signals. *Zeitschrift für Tierpsychologie* **66**:189–226.

Koenig, W. D. and Mumme, R. L. (1987) *Population ecology of the cooperatively breeding Acorn Woodpecker*. Princeton University Press, Princeton, New Jersey.

Kovacs, K. M. and Ryder, J. P. (1981) Nest-site fidelity in female–female pairs of Ring-billed Gulls. *Auk* **98**:625–627.

Krebs, J. R. and Dawkins, R. (1978) Animal signals: mind-reading and manipulation. In Krebs, J. R. and Davies, N. B. (eds) *Behavioural ecology*, pp. 380–402. Sinauer Associates, Sunderland, Massachusetts.

Lack, D. (1968) *Ecological adaptations for breeding in birds*. Methuen, London.

Lande, R. (1988) Demographic models of the Northern Spotted Owl. *Oecologia* **75**:601–607.

Lanner, R. M. (1981) Avian seed dispersal as a factor in the ecology and evolution of limber and whitebark pines. Sixth North American Forest Biology Workshop, Edmonton, Alberta.

Lanner, R. M. (1981) *The pinyon pine: a natural and cultural history*. University of Nevada Press, Reno, Nevada.

LeBoeuf, B. J. and Peterson, R. S. (1969) Social status and mating activity in elephant seals. *Science* **163**:91–93.

Ligon, J. D. (1971) Late summer–autumnal breeding of the piñon jay in New Mexico. *Condor* **73**:147–153.

Ligon, J. D. (1974a) Green cones of the piñon pine stimulate late summer breeding in the piñon jay. *Nature* **250**:80–82.

Ligon, J. D. (1974b) Comments of the systematic relationships of the piñon jay (*Gymnorhinus cyanocephalus*). *Condor* **76**:468–470.

Ligon, J. D. (1978) Reproductive interdependence of piñon jays and piñon pines. *Ecological Monographs* **48**:111–126.

Ligon, J. D. and Ligon, S. H. (1978) Communal breeding in green woodhoopoes as a case for reciprocity. *Nature* (London) **276**:496–498.

Ligon, J. D. and Martin, D. J. (1974) Piñon seed assessment by the piñon jay. *Animal Behaviour* **22**:421–429.

Ligon, J. D. and White, J. L. (1974) Molt and its timing in the piñon jay, *Gymnorhinus cyanocephalus*. *Condor* **76**:274–287.

Lima, S. L. (1986) Predation risk and unpredictable feeding conditions: determinants of body mass in birds. *Ecology* **67**:377–385.

Linden, M. and Møller, A. P. (1989) Cost of reproduction and covariation of life history traits in birds. *Trends in Ecology and Evolution* **4**:367–371.

Little, E. L., Jr. (1950) Southwestern trees, a guide to the native species of New Mexico and Arizona. *United States Department of Agriculture Handbook* **9**.

Lorenz, K. (1966) *On aggression*. Harcourt, Brace & World, Inc., New York.

Lorenz, K. (1981) *The foundations of ethology*. Springer-Verlag, New York.

MacArthur, R. H. (1972) *Geographical ecology*. Harper and Row, New York.

MacRoberts, B. R. and MacRoberts, M. H. (1972) Social stimulation of reproduction in Herring and Lesser Black-backed Gulls. *Ibis* **114**:495–506.

Maholt, R. K. and Trost, C. (1989) Self-advertisement: relations to dominance in black-billed magpies. *Animal Behaviour* **38**:1079–1081.

Marler, P. (1955) Characteristics of some animal calls. *Nature* **176**:6–8.

Marler, P. (1969) Tonal quality of bird sounds. In Hinde, R. A. (ed) *Bird vocalizations*, pp. 5–18. Cambridge University Press, Cambridge, United Kingdom.

Marshal, A. J. and Coombs, C. J. F. (1957) The interaction of environmental, internal and behavioural factors in the Rook, *Corvus f. frugilegus* L. *Proceedings of the Zoological Society of London* **128**:545–589.

Marzluff, J. M. (1983) *Factors influencing reproductive success and behavior at the nest in Pinyon Jays (Gymnorhinus cyanocephalus)*. MSc thesis, Northern Arizona University, Flagstaff, Arizona.

Marzluff, J. M. (1985) Behavior at a Pinyon Jay nest in response to predation. *Condor* **87**:559–561.

Marzluff, J. M. (1987) Individual recognition and the organization of Pinyon Jay (*Gymnorhinus cyanocephalus*) societies. PhD thesis. Northern Arizona University. Flagstaff, Arizona.

Marzluff, J. M. (1988a) Do pinyon jays use prior experience in their choice of a nest site? *Animal Behaviour* **36**:1–10.

Marzluff, J. M. (1988b) Vocal recognition of mates by breeding pinyon jays. *Animal Behaviour* **36**:296–298.

Marzluff, J. M. and Balda, R. P. (1988a) Resource and climatic variability: influences on sociality of two southwestern corvids. In Slobodchikoff, C. N. (ed) *The ecolology of social behavior*, pp. 225–283. Academic Press, New York.

Marzluff, J. M. and Balda, R. P. (1988b) Pairing patterns and fitness in a free-ranging population of pinyon jays: what do they reveal about mate choice? *Condor* **90**:201–213.

Marzluff, J. M. and Balda, R. P. (1988c) The advantages of, and constraints forcing, mate fidelity in pinyon jays. *Auk* **105**:286–295.

Marzluff, J. M. and Balda, R. P. (1989) Causes and consequences of female-biased dispersal in a flock-living bird, the pinyon jay. *Ecology* **70**:316–328.

Marzluff, J. M. and Balda, R. P. (1990) Pinyon Jays: making the best of a bad situation by helping. In Stacey, P. B. and Koenig, W. D. (eds) *Cooperative breeding in birds*, pp. 198–237. Cambridge University Press, Cambridge, United Kingdom.

Maynard Smith, J. (1977) Parental investment: a prospective analysis. *Animal Behaviour* **25**:1–9.

McArthur, P. D. (1979) *Parent-young recognition in the piñon jay: mechanisms, ontogeny, and survival value.* PhD thesis, Northern Arizona University. Flagstaff, Arizona.

McArthur, P. D. (1982) Mechanisms and development of parent-young vocal recognition in the piñon jay (*Gymnorhinus cyanocephalus*). *Animal Behaviour* **30**: 62–74.

Mock, D. W. (1985) Avian monogamy: the neglected mating system. *Ornithological Monographs* **37**:6–15.

Mock, D. W. and Fujioka, M. (1990) Monogamy and long-term pair bonding in vertebrates. *Trends in Ecology and Evolution* **5**:39–42.

Morton, E. S. (1977) On the occurrence and significance of motivation-structural rules in some bird and mammal sounds. *American Naturalist* **111**:855–869.

Murton, R. K., Isaacson, A. J. and Westwood, N. J. (1971) The significance of gregarious feeding behaviour and adrenal stress in a population of wood-pigeons *Columba palumbus*. *Journal of Zoology, London* **165**:53–84.

Nisbet, I. C. T. (1973) Courtship-feeding, egg size and breeding success in Common Terns. *Nature* **241**:141–142.

Nishida, T. (1979) The social structure of chimpanzees of the Mahale mountains. In Hamburg, D. A. and McGown, E. R. (eds) *The great apes*, pp. 73–121. Benjamin/Cummings Publishing, Menlo Park, California, USA.

Phillips, A. (1989) *A checklist of the birds of Arizona.* University of Arizona Press, Tuscon, Arizona.

Pitelka, F. A. (1951) Speciation and ecologic distribution in American jays of the genus *Aphelocoma*. *University of California Publications in Zoology* **50**.

Pusey, A. (1979) Intercommunity transfer of chimpanzees in Gombe national park. In Hamburg, D. A. and McGown, E. R. (eds) *The great apes*, pp. 465–479. Benjamin/Cummings Publishing, Menlo Park, California, USA.

Redondo, T. and Arias de Reyna, L. (1988) Locatability of begging calls in nestling altricial birds. *Animal Behaviour* **36**:653–661.

Ricklefs, R. E. (1973) Fecundity, mortality, and avian demography. In D. S. Farner (ed) *Breeding biology of birds*, pp. 366–435. National Academy of Sciences, Washington, D.C.

Ricklefs, R. E. (1979) *Ecology*. Chiron Press, Portland, Oregon.

Rockwell, R. F., Findlay, C. S. and Cooke, F. (1985) Life history studies of the Lesser Snow Goose V.: temporal effects on age-specific fecundity. *Condor* **87**:142–143.

Rogers, C. M. (1987) Predation risk and fasting capacity: do wintering birds maintain optimal body mass? *Ecology* **68**:1051–1061.

Røskaft, E. and Espmark, Y. (1982) Vocal communication by the rook *Corvus frugilegus* during the breeding season. *Ornis Scandinavica* **13**:38–46.

Rubenstein, D. I. (1982) Reproductive value and behavioral strategies: coming of age in monkeys and horses. In Bateson, P. P. G. and Klopfer, P. H. (eds) *Perspectives in ethology* Vol 5, pp. 469–487. Plenum Press, New York.

Ryder, J. P. (1981) The influence of age on the breeding biology of colonial nesting seabirds. In Burger, J. and Olla, B. L. (eds) *Behavior of marine animals Volume 4. Behavior of marine birds*, pp. 153–168. Plenum Press, New York.

Schulman, S. R. and Chapais, B. (1980) Reproductive value and rank relations among macaque sisters. *American Naturalist* **115**:580–593.

Scott, J. P. (1956) The analysis of social organization in animals. *Ecology* **37**:213–221.

Searcy, W. A. (1982) The evolutionary effects of mate selection. *Annual Review of Ecology and Systematics* **13**:57–85.

Seyfarth, R. M., Cheney, D. L., and Marler, P. (1980) Monkey responses to three different alarm calls: evidence of predator classification and semantic communication. *Science* **210**:801–803.

Sherry, D. F. (1985) Food storage by birds and mammals. *Adv. Study Behavior*. **15**: 153–188.

Shields, E. M. (1982) *Philopatry, inbreeding, and the evolution of sex*. State University of New York Press, Albany, New York, USA.

Shields, W. M., Crook, J. R., Hebblethwaite, M. L. and Wiles-Ehmann, S. S. (1988) Ideal free coloniality in the swallows. In Slobodchikoff, C. N. (ed) *The ecology of social behavior*, pp. 189–228. Academic Press, New York.

Short, L. L. and Mayr, E. (1970) Species and taxa of North American birds: a contribution to comparative systematics. *Nutall Ornithological Club Publication* **9**. (R. A. Paynter, Jr., ed), Cambridge, Massachusetts.

Skutch, A. F. (1976) *Parent birds and their young*. University of Texas Press, Austin, Texas.

Smith, S. M. (1980) Demand behavior: a new interpretation of courtship feeding. *Condor* **82**:291–295.

Smith, C. C. and Balda, R. P. (1980) Competition among insects, birds and mammals for conifer seeds. *American Zoologist* **19**:1065–1083.

Stotz, N. (1991) Cache-recovery behavior of Pinyon Jays in a social context. Paper presented at the 109th American Ornithologists' Union Annual Meeting, Montreal, Canada.

Tenaza, R. (1971) Behavior and nesting success relative to nest location in Adelie Penguins. *Condor* **73**:81–92.

Tinbergen, L. (1960) The natural control of insects in pinewoods. 1. Factors influencing the intensity of predation by songbirds. *Archives Neerlandaises de Zoologie* **13**:265–343.

Tomback, D. F. and Linhart, Y. B. (1990) The evolution of bird-dispersed pines. *Evolutionary Ecology* **4**:185–219.

Trivers, R. L. (1972) Parental investment and sexual selection. In Campbell, B. (ed) *Sexual selection and the descent of man*, pp. 136–179. Heinemann, London.

Trost, C. H. and Webb, C. L. (1986) Egg moving by two species of corvid. *Animal Behaviour* **34**:294–295.

Vander Wall, S. B. and Balda, R. P. (1977) Co-adaptations of the Clark's nutcracker and the piñon pine for efficient seed harvest and dispersal. *Ecological Monographs* **47**:89–111.

Vander Wall, S. B. and Balda, R. P. (1981) Ecology and evolution of food-storage behavior in conifer-seed-caching corvids. *Zeitschrift für Tierpsychologie* **56:** 217–242.

Vleck, C. M. (1981) Energetic cost of incubation in the Zebra Finch. *Condor* **83**:229–237.

Wade, M. J. (1980) An experimental study of kin selection. *Evolution* **34**:844–855.

Ward, P. and Zahavi, A. (1973) The importance of certain assemblages of birds as information centers for food finding. *Ibis* **115**:517–534.

Westneat, D. F., Sherman, P. W. and Morton, M. L. (1990). The ecology and evolution of extra-pair copulations in birds. In Power, D. M. (ed) *Current ornithology Vol. 7*, pp. 331–370. Plenum Press, New York.

Whitney, N. R., Jr (1982) Twenty-five years of Pinyon Jay studies. *South Dakota Bird Notes* **34**:29–30.

Wilson, D. S. (1975) A theory of group selection. *Proceedings of the National Academy of Sciences, USA* **72**:143–146.

Wilson, D. S. (1977) Structured demes and the evolution of group-advantageous traits. *American Naturalist* **111**:157–185.

Wittenberger, J. F. (1979) The evolution of mating systems in birds and mammals. In Marler, P. and Vandenbergh, J. (eds) *Handbook of behavioral neurobiology, Volume 3. Social behavior and communication*, pp. 271–349. Plenum Press, New York.

Wittenberger, J. F. (1981) *Animal social behavior.* Wadsworth, Inc., Belmont, California.

Woolfenden, G. E. and Fitzpatrick, J. W. (1984) *The Florida Scrub Jay.* Princeton University Press, Princeton, New Jersey.

Wright, S. (1931) Evolution in mendelian populations. *Genetics* **16**:97–159.

Wright, S. (1932) The roles of mutation, inbreeding, crossbreeding, and selection in evolution. *Proceedings of the VI International Congress of Genetics* **1**:356–366.

Zahavi, A. (1982) The pattern of vocal signals and the information they convey. *Behaviour* **80**:1–8.

Zusi, R. L. (1987) A feeding adaptation of the jaw articulation in New World jays (Corvidae). *Auk* **104**:665–680.

# Index